FUELS FROM
BIOMASS AND WASTES

Copyright © 1981 by Ann Arbor Science Publishers, Inc.
230 Collingwood, P.O. Box 1425, Ann Arbor, Michigan 48106

Library of Congress Card Catalog Number 81-68245
ISBN 0-250-40418-4

Butterworths, Ltd., Borough Green, Sevenoaks, Kent TN15 8PH, England

FUELS FROM BIOMASS AND WASTES

Edited by
Donald L. Klass
George H. Emert

ANN ARBOR SCIENCE
PUBLISHERS INC / THE BUTTERWORTH GROUP

PREFACE

As we approach the end of the twentieth century, the availability of fuel supplies continues to be a major determinant of economic growth and development. Increasing fossil energy costs and shortages are as inevitable as death and taxes. Some believe we are approaching the twilight of the fossil fuel era and that we are now in the process of returning to the use of nonfossil, renewable energy sources derived directly or indirectly from the sun. Those energy scientists and engineers that have a clean crystal ball (or, possibly, a dirty one, since the complete story has not yet been told), who adopted this position in the early 1970s, decided to resurrect an old but nevertheless feasible technology for harvesting chemically fixed solar energy for the production of organic fuels and chemicals. The intermediate raw material is biomass — land- and water-based vegetation in the form of trees, grasses, plants and algae — which can be converted to a broad range of useful energy forms. This idea makes sense because as a first approximation, the net annual biomass energy stored worldwide is about ten times the total energy used by man each year.

This book evolved from a symposium on this subject — fuels from biomass — sponsored by the American Chemical Society, Division of Fuel Chemistry Inc. and Division of Industrial and Engineering Chemistry, and the American Institute of Chemical Engineers at the Second Chemical Congress of the North American Continent in Las Vegas, Nevada, August 24–29, 1980. This meeting was a one-of-a-kind event because just two weeks before the conference began, it was moved from San Francisco to Las Vegas, a move which was beyond our control. Travel schedules and hotel reservations were rearranged at the last minute. Thus, a good percentage of the symposium audience was lost, but most of the speakers were on hand to make their presentation. We thank all the speakers and we are extremely grateful for their participation in the symposium.

All 18 papers on the symposium program, suitably revised, and 13 invited papers are included in this book. They are organized into chapter groupings: Introduction, Biomass Procurement and Production, Biological and Thermal Gasification, Hydrolysis and Extraction, Fermentation Ethanol, Natural and Thermal Liquefaction, Environmental Effects, and Systems and Case Studies. Many of the chapters have extensive reference lists which serve as an entree to the literature on the subject addressed. It is apparent that the technology of renewable energy from biomass and wastes is advancing on all fronts. We expect this source of primary energy will meet more and more of our energy and chemical needs as time passes. In our opinion, this course of events is also inevitable.

Donald L. Klass
George H. Emert

Donald L. Klass is Vice President of the Institute of Gas Technology (IGT), which has conducted energy education and research programs since 1941. Dr. Klass administers basic research and grants conducted by staff and graduate students, IGT's academic and industrial education programs, and contract education services. His research and development experience has been concentrated on petrochemical and refinery processes, fuels and lubricants, conversion of biomass and wastes to synthetic fuels, and gas processing. Dr. Klass' publications number over 200 papers and patents in these fields. He received his BS in chemistry at the University of Illinois and his AM and PhD degrees in organic chemistry at Harvard University.

George H. Emert is Director of the Biomass Research Center at the University of Arkansas at Fayetteville. The Center has responsibilities for teaching, research and extension service in biomass utilization throughout the state. The research objective of the Center is to develop technologies for utilization of cellulosic materials as feedstocks for chemicals production, e.g., ethanol from forest, agricultural or municipal wastes.

Dr. Emert received his bachelor's degree from the University of Colorado, his Master of Science from Colorado State University, and his doctorate in biochemistry and nutrition at Virginia Polytechnic Institute and State University. Dr. Emert's current efforts include scaling-up cellulose-to-ethanol technology to demonstration plant operating level.

CONTENTS

ix

Section 3
Hydrolysis and Extraction

Section 4
Fermentation Ethanol

Section 5
Natural and Thermal Liquefaction

Section 6
Environmental Effects

Section 7
Systems and Case Studies

CHAPTER 1

FUELS FROM BIOMASS AND WASTES
– AN INTRODUCTION

Donald L. Klass

Institute of Gas Technology
Chicago, Illinois

THE CONCEPT

The concept of producing synfuels and energy products from biomass and wastes is a simple one, as shown in Figure 1. The capture of solar energy as fixed carbon in organic plant material via photosynthesis is the key initial step in this scheme and is represented by the equation:

$$CO_2 + H_2O + \text{light} \xrightarrow{\text{chlorophyll}} (CH_2O) + O_2$$

Carbohydrate is the primary product. For each gram atom of carbon fixed, about 470 kJ (112 kcal) are absorbed. Oxygen liberated in the process comes exclusively from the water, according to radioactive tracer studies. Although there are still many unanswered questions regarding the detailed molecular mechanisms of photosynthesis, the prerequisites for biomass production are well established; carbon dioxide, water, light in the visible region of the electromagnetic spectrum, a sensitizing catalyst and a living plant are essential. The upper limit of the capture efficiency of the incident solar radiation in biomass has been variously estimated to range from about 8% to as high as 15%, but in most real situations, it is generally in the 1% or less range.

The idea of using biomass as a raw material for conversion to synfuels and energy products is certainly not a new one. Wood was a major

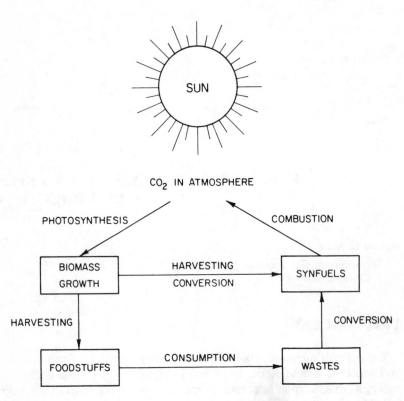

Figure 1.   Schematic representation of synfuel production from biomass and wastes.

source of primary energy for the United States only a relatively few years ago (Figure 2). The main difference between this "old technology" and the concept as it is being developed today is that we are applying more advanced methodologies and processes to produce organic fuels and chemicals in more desirable forms. To be sure, however, combustion of certain biomass and wastes for heat, steam, and electric power production also an important application of nonfossil renewable carbon and will continue to grow.

The reasons for utilizing biomass and wastes as energy resources are numerous. Some of the more obvious ones are:

- Nonfossil forms of fixed carbon are not depletable, in contrast to fossil fuels such as oil, natural gas and coal.
- Biomass is available in large quantities, and provides a raw material for conversion to major supplies of synfuels.

- Combining waste disposal and energy recovery processes offers recycling opportunities as well as improved disposal technology, often at lower cost.
- A synfuels industry based on biomass and wastes is independent of foreign price controls and regulations.
- Technological breakthroughs are not required to develop commercial systems and processes.

These are some of the more important reasons why momentum to develop nonfossil carbon-to-energy technology is increasing. The purpose of this chapter is to introduce the subject and to summarize the many technologies now under development and in commercial use.

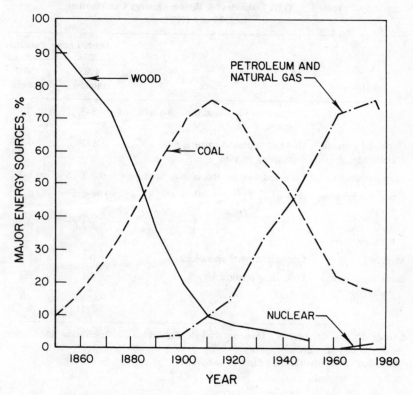

Figure 2. Historical U.S. energy consumption pattern.

ENERGY IMPACT

As might be expected, recent U.S. projections of the potential of energy from biomass and wastes span the entire range from small to large impact. One of the forecasts made by the Office of Technology Assessment (OTA) indicated that in the year 2000, as few as 4–6 quads (1 quad = 10^{15} Btu = 1.05×10^{18} J), or as many as 12–17 quads could be derived from biomass and selected wastes depending on cropland availability, crop yield improvements, the development of efficient conversion processes and the level of policy support [1]. The high estimate was made up of 10 quads from wood, 0–5 quads from grasses and legume herbage (depending on cropland needs for foodstuffs), 1 quad from crop residues, 0.3 quad from manure (as methane) and 0.2 quad from grains (as ethanol); more details are shown in Table I for the years 1985 and 2000.

Table I. OTA Estimates for Biomass Energy Contribution
in 1985 and 2000[a]

Feedstock	Source	Gross Energy Potential 1985 (quads)	Gross Energy Potential 2000 (quads)
Wood	Commercial forestland including mill wastes	3–5	10
Grass and Legume Herbage	Hayland, cropland pasture, non-cropland pasture	1–3[b]	0–5[b]
Crop Residues	Cropland used for intensive agriculture	0.7–1	0.8–1.2
Grain and Sugar Crops	(For ethanol production)	0.08–0.2[b]	0–1[b]
Grass and Short-Rotation Trees		0.3–1.6[b]	0–5[b]
Manure	Confined animal operations	0.1	0.1–0.3
Other Agricultural Wastes	Processing operations	0.1	0.1
Total		5.3–11	6–17

[a] Adapted from Reference 1. Does not include deductions for cultivation and harvest energy, losses or end-use efficiency.
[b] Not additive because they use some of same land.

OTA's higher estimate of the contribution of biomass energy in the year 2000 is four to six times that of a recent estimate of the U.S. Department of Energy (DOE) [2], but it should be noted that DOE's rather pessimistic estimate (3 quad/yr) is less than twice the current commercial contribution of wood energy alone (1.8 quad/yr; 850,000 barrels oil equivalent per day) [3]. DOE's forecast is also quite inconsistent when viewed in the context of the large federal investment to develop and commercialize biomass energy. Indeed, DOE's Office of Alcohol Fuels has a production goal of 3.0 quads of fuel ethanol alone by the year 2000 (Table II).

Despite the uncertainties encountered in conducting energy forecasts, they are still necessary to determine which of the various renewable organic raw materials have the greatest potential and the magnitude of the contribution that energy from biomass and wastes can make to meet future U.S. demands. Most energy forecasts are subject to continuous revision, particularly when there is a major technological improvement that decreases costs or when there is substantial price increase in competitive fuels.

Table II. Fuel Ethanol Production Goals of DOE Office of Alcohol Fuels

Year	Production Goals 10^9 liter/yr	bbl/day[a]	Percent of 1978 Gasoline Consumption[b]	Capital Investment Required ($ billion)	Quads[c]
1980	1.21	20,900	0.3	0.64	0.02
1981	1.89	32,600	0.5	1.1	0.04
1982	3.48	60,000	0.8	2.2	0.07
1985	7.57	130,000	1.8	5.7	0.15
1990	37.85	652,000	9.1	28.5	0.76
2000	151.4	2,610,000	36	100	3.0

[a]One bbl = 42 gal = 159 liters.
[b]U.S. gasoline consumption in 1978 was 110 billion gal (7.2 million bbl/day). Percent of gasoline consumption calculated on basis of replacement of equivalent volume of gasoline by ethanol.
[c]Calculated on basis of 21.08 MJ/l (75,000 Btu/gal) ethanol heating value.

On a worldwide basis, traditional fuel forms of biomass and wastes such as firewood, wood-derived charcoal, crop residues and dried animal dung now supply large portions of the total energy used—50–65% in Asia and 70–90% in Africa—and overall, they account for about 8.5 million barrels of oil-equivalent per day, or approximately 20–25% of energy consumption in the developing world [4]. Biomass and wastes have also been targeted to supply large portions of the energy demand in Europe. For example, a national program has been launched in France to obtain about 8–10% of its energy needs from biomass by 1990 [5]. Other European countries are developing similar plans.

BIOMASS PRODUCTION

Silviculture

Activities on tree species selection (Table III), from accumulated data, and management techniques for wood fuel farms are in progress, and emphasis is being given to the costs of different production strategies, improvement of existing forest stands, development of specialized harvesting equipment and the assessment of environmental effects [6]. The maximum delivered wood cost in the DOE program is targeted at $25.00/green metric ton ($22.68/green ton) [6]. This corresponds to a cost of about $2.87/GJ ($3.02/10^6 Btu) for green wood at 50 wt % moisture content and heating value of 17.43 GJ/metric ton (15 × 10^6 Btu/dry short ton).

Table III. Candidate Woody Biomass Species for Growth in the United States as Energy Crops [6]

Northeast and North Central
  Hybrid Poplar   American Sycamore
  Black Locust   Boxelder
  Black Alder   Sugar Maple
  Willow Species   Ash Species
  Eastern Cottonwood   Birch Species

South
  Hybrid Poplar   American Sycamore
  Black Locust   Slash Pine
  Tree of Heaven   Sand Pine, Loblolly Pine
  Sweet Gum   Australian Pine
  Eucalyptus   Chinese Tallow Tree

Great Plains
  Black Locust   Eastern Cottonwood
  Ash Species   Silver Maple
  Elm Species   Catalpa

Northwest
  Black Cottonwood   Red Alder

Southwest
  Athel Tree   Rabbitbush
  Salt Cedar   Greasewood
  Mesquite   Big Sagebrush
  Palo Verde   Creosote Bush
  Jojoba   Lucaena
  Saltbush   Ironwood

Hawaii
  Eucalyptus   Mimosa

For energy applications, it has generally been felt that short-rotation coppice growth would provide the maximum amount of biomass carbon in the shortest time. Data are now being published that will permit real comparisons to be made of short-rotation growth with other silviculture methods. The results of an exemplary study are shown in Table IV [7]. Harvesting of American sycamore in this field test gave average above-ground yields of 7.1 and 9.7 oven-dry metric ton/ha-yr (2.9 and 4.0 short ton/ac-yr) at one- and two-year rotations, while the average yield was 10.9 oven-dry metric ton/ha-yr (4.5 short ton/ac-yr) for only one harvest in four years; spacing had little effect. Additional growth data are being collected, but these results suggest that the optimum rotation has not yet been selected for this particular field test. The incremental yield improvement at the optimum rotation must of course justify the additional cost and energy expenditures of more frequent harvesting before short-rotation coppice growth methods are used on a large scale for wood fuel farms.

Table IV.   Aboveground Biomass (dry metric ton/ha-yr)
(without Foliage) of American Sycamore
at Different Rotations in the South [7]

| Spacing and Rotation | Growing Season after Initial Coppice | | | | Four-Year Total |
	First	Second	Third	Fourth	
0.3 × 1.2 m					
1 yr	4.4	7.1	7.8	7.4	26.7
2 yr		14.0		19.5	33.5
4 yr				49.5	49.5
0.6 × 1.2 m					
1 yr	4.4	9.2	8.0	6.5	28.1
2 yr		15.1		24.3	39.4
4 yr				28.8	28.8
1.2 × 1.2 m					
1 yr	4.0	9.0	9.1	8.2	30.3
2 yr		16.5		27.1	43.6
4 yr				52.4	52.4

Nonwoody Herbaceous Plants

Some of the nonwoody herbaceous plants considered as potential energy crops from the production standpoint are the sorghums [8]; sugarcane [9]; sugar beets [10]; the arid and semi arid land plants guayule, jojoba and buffalo gourd [11]; Jerusalem artichoke [12]; *Euphorbia lathyris* [13]; and the tropical grasses and legumes [14]. Hybrids of sweet sorghums and grain sorghums—the sweet-stemmed sorghums—have large grain heads and sugar in their stalks, and are believed to be good candidates as energy crops because the fermentable sugars require no cooking for fuel ethanol production; the grain is easily stored for extended processing; and the energy-rich stalks can be used for plant fuel [8]. Sugarcane is believed to be one of the most promising energy crops for subtropical climates because of the potential of increasing the yield by at least 50% through breeding and narrow-row culture [9]. In addition, the energy cane concept, which refers to the growth of sugarcane to maximize biomass content rather than sugar alone, offers the possibility of producing a multiproduct plant (sugar, molasses, fiber) at total yields about three times those of conventional sugarcane [15]. Fodder beets and particularly fodder beet–sugar beet hybrids have been proposed as fuel crops for the northern states because of their growth characteristics and high fermentables content [10]. Jerusalem artichoke has also shown excellent potential as a carbohydrate-rich energy crop for the northern states [12]. Several continuing projects indicate that it may be possible to grow certain arid and semiarid plants in the southwestern states for their organic liquids. Efforts are in progress to improve the productivities of guayule, which contains extractable amounts of polyisoprene identical with natural rubber [11], and *Euphorbia lathyris,* which contains extractable triterpenoid hydrocarbons [13]. Large numbers of hydrocarbon-containing plants exist in the United States, and some of them may have potential as energy crops [16]. However, the key to commercial success with one or more of these crops is to increase the hydrocarbon yield to the point where it is economically feasible to grow and harvest the plant and to extract the hydrocarbons. For example, under the best agronomic conditions, the yield of rubber from guayule is 560 kg/ha-yr (500 lb/ac-yr), a level that is not competitive with *Hevea* rubber [11]. The basic ways of improving yields are to develop new cultivars with higher hydrocarbon content, to grow existing cultivars at higher yields, and to optimize hydrocarbon yields per unit growth area, which may not necessarily correspond to maximum biomass yields per unit growth area. Which of these techniques or combinations thereof are the most useful is expected to evolve from the work now in progress.

Probably one of the best biomass categories to consider as an energy crop is the grasses, because of their abundance and growth characteristics [16]. Relatively little work is in progress on the production of the common warm- and cool-season grasses as energy crops, although grass as a family includes sugarcane and sorghum, which are already attracting a great deal of attention as alluded to above. Continuing screening programs on tropical grasses have resulted in the selection of several species as potential energy crops [14], and opened the possibility of coproduction with tropical legumes to reduce fertilization costs [14]. In temperate zones, one of the best nitrogen-fixing grasses has been found to be salt-marsh cordgrass, *Spartina alterniflora,* which can fix up to about 70 kg of nitrogen/ha-yr [17.] Nonnitrogen-fixing grasses coproduced with *Spartina alterniflora* have been reported to develop small amounts of nitrogen-fixing activity, and it also appears that certain forage grasses that have heretofore been classified as nonnitrogen fixers can fix limited amounts of nitrogen in the presence of the proper bacteria [17]. Further development of these discoveries could ultimately have a significant beneficial effect on the costs of both energy and food crops. The use of nitrogenous fertilizers to promote crop growth is one of the major cost factors in biomass production; it is also one of the largest nonsolar energy inputs, since most nitrogenous fertilizers are made from fossil fuels. Other techniques for reducing external nitrogen requirements, such as no-tillage agriculture [18], are being developed, but the use of inherently nitrogen-fixing plants is the most energy- and cost-efficient approach. In addition to coproduction of nitrogen- and nonnitrogen-fixing plants, recombinant DNA techniques that transfer the nitrogen-fixing characteristic from certain bacteria to plants may also be feasible [17].

Terrestrial plant production for energy applications cannot be evaluated properly without assessing the potential conflicts with foodstuffs production. Many studies and detailed analyses have been conducted to examine the question over the last several years. There are many situations where competition could exist such as might occur for land, equipment and labor. Many of the analyses that have been carried out compare the land area requirements for biomass energy production with the data on the land areas needed for commercial crop production, the 1979 statistics of which are presented in Table V [19]. It is evident that corn is the largest food crop in the United States, but it is also currently used for commercial production of fuel ethanol. Conflicts could arise because of competitive energy and food uses for commercial crops, such as corn, in the near future, so these and related problems should be assessed early in any commercialization program for biomass energy. Several studies of small and large areas have been performed which indicate that biomass energy, food and feed ventures can coexist with benefits to society. For example, a recent study indicates that the grain crop residues generated in Indiana, a state which has about 52% of its land area under cultivation, could supply a substantial portion of its energy needs [20]. In this case, commercial crops could simultaneously serve as

Table V. U.S. Production of Selected Commercial
Crops in 1979[a]

	Area Planted			Production[c]			
	10^6 ha	10^6 ac	% U.S. Land Area[b]	(bu/ac)	Total (10^3 bu/ac)	$/bu	Value (10^9 $)
Corn	32.379	80.011	3.535	109.4[d]	7,763,771[d]	2.44[d]	18.9[d]
Soybeans	28.970	71.586	3.163	32.2[e]	2,267,647[e]	6.12[e]	13.9[e]
Wheat	28.959	71.558	3.161	34.2	2,141,732	3.77	8.1
Oats	5.725	14.146	0.625	54.4	534,386	1.31	0.7
Barley	3.262	8.060	0.356	50.6	378,067	2.28	0.9
Rye	1.245	3.077	0.136	25.9	24,549	2.10	0.053
Sorghum	6.232	15.399	0.680	62.9[f]	814,308[f]	2.31[f]	1.9[f]
Hay	24.751	61.162[c]	2.702	2.39[g]	145,878[g]	59.20[g]	7.3
Total	131.523	324.999	14.358				

[a] Adapted from Reference 19.
[b] Land area of the United States in 1970 was 9,160,454 km² (3,536,855 mi²).
[c] Acreage harvested.
[d] Includes corn for grain, but excludes corn for silage.
[e] Soybeans for beans.
[f] Includes sorghum for grain but excludes sorghum for silage.
[g] Tons, rather than bushels, for hay.

food and energy resources. It is beyond the scope of this chapter to treat this subject in any detail, but there still appear to be no insurmountable barriers that preclude commercial use of terrestrial biomass energy resources. Note that of the 850,000 bbl/day of oil-equivalent currently consumed as biomass, most of it is wood [3]. With the exception of the fuel alochol plants that use corn, there is only minimal commercial interest now in using herbaceous plants as energy resources. Without heavy subsidies, there will have to be significant improvement in the costs of producing and harvesting herbaceous biomass (or large increases in fossil fuel costs) before it can seriously be considered as a strong candidate for energy purposes. This seems distant at the present time when one considers that even hay was valued at $65.26/metric ton ($59.20/ton) by the U. S. Department of Agriculture (USDA) in 1979 (Table V); this corresponds to a raw energy cost of about $4.74/GJ ($5.00/10^6 Btu) at 15 wt % moisture content.

Aquatic Biomass

The attractive features of aquatic biomass as energy resources are that they grow rapidly; most species have little or no commercial market and hence few competitive uses; and land availability is not a limiting factor for marine biomass. The types of aquatic biomass currently under evaluation are the microalgae, marsh plants, submerged and floating macrophytes, and marine biomass.

Research and pilot projects on the growth of microalgae such as *Chlorella* and *Scenedesmus* have been performed for many years, and, with the exception of filamentous organisms such as *Spirulina,* the basic problem continues to be separation difficulties [21]. Costs of $551/ metric ton ($500/ton) are currently believed to be minimal without wastewater treatment credits [21, 22]. Consequently, recent efforts have been directed to development of microalgae-based processes in which the chemical-producing properties of microalgae are utilized for synthesis of high-value chemicals such as lipids, polyols and glycerol. It has also been reported that certain microalgae have the potential of manufacturing hydrocarbons in controlled environments on a continuous basis [23-25]. Conceptually, at least, hydrocarbons might be manufactured in this manner from only carbon dioxide, light and minerals. It is probable, however, that what is reported as hydrocarbons is actually lipid which could not be refined in the same way as hydrocarbons.

Recent experimental data indicate that the maximum practical storage of solar energy under controlled conditions by algal biomass in vitro is 18% [26]. This is about twice the maximum previously predicted for conventional agriculture and suggested to the investigators that a precision-engineered algal culture might be operated on any open space over land or sea at high efficiencies (yield). The technical feasibility of using microalgae for combined wastewater treatment and methane production is well established, and demonstration is expected to provide considerable improvement over conventional processes [22].

Aquatic macrophytes such as the submerged species *Hydrilla* and the floating varieties water hyacinth and duckweed are often widespread on lakes and ponds. Harvesting of these materials presents a much less difficult problem than microalgae collection, and numerous projects have been conducted to evaluate their food, feed and energy potential. An interesting characteristic of certain macrophytes is the property of luxuriant uptake of heavy metals as well as nutrients from aqueous media, and it is this property that is being utilized to develop sewage treatment systems. Water hyacinth has shown considerable promise in this application and the combined processes of wastewater treatment and methane production [27]. Currently, water hyacinth wastewater treatment plants are under construction in several locations; the first municipal plant is reported to be operational in Hercules, CA [28]. The water hyacinth is sold as a soil conditioner, and the system is claimed to use only one-third as much power as a conventional sewage treatment plant.

Wetland or marsh plants such as cattail and reed grass are also being studied as energy crops. Cattail is one of the most promising species because natural stands often exceed 40 metric ton/ha-yr (17.8 ton/ac-yr); it is easily propagated using seeds or rhizomes, grows in monoculture and has few insect pests [29]. Work is in progress in the northern states to develop more information on harvesting methods, land use considerations, possible environmental constraints and costs [29].

Marine algae, such as the floating brown seaweed *Sargasum* and giant brown kelp, also have potential as energy crops. Such biomass would not be subject to terrestrial limitations and would not compete with foodstuffs for agricultural land. Work is continuing to develop growth data in small- and moderate-scale systems. A vegetative clone of the red seaweed *Gracilaria tikvahiae* has now been grown continuously for over two years in 2600-liter tanks with yields averaging 12 g ash-free dry weight/ m^2-day (19.5 ton/ac-yr) with four exchanges of water per day [30]. With an improved media, the same yield can be achieved with only one volume change per day [30]. This observation is believed to be a major accomplishment. Work is now underway to further reduce the flowrate consistent with high yields. The ocean farm concept with giant brown kelp, a concept that involves total integration of kelp growth in upwelled deep ocean water, harvesting, and conversion to methane and other products, is progressing in several areas [31]. As reported in 1979, juvenile plants began to develop on the solid structures of a test farm off the Southern California coast from spores liberated by adult transplants [3]. Subsequent monitoring of the growth of these plants has shown that adult plants are formed and can grow in the open ocean when fertilized with upwelled water [32]. This is considered to be a major milestone, and it is expected that within three years, the feasibility of the marine farm concept can be determined [32].

CONVERSION

Combustion

Direct combustion for heat, steam and electric power generation is the most obvious route to energy from biomass and wastes. It is therefore no surprise to find that direct combustion is used throughout the world with these materials at various levels of technological development and sophistication. Wood and wood wastes comprise the bulk of current U.S. primary energy production from biomass as already mentioned. Three of the largest wood-powered electric plants scheduled for startup in the early 1980s are listed in Table VI. The plant to be operated by Dow Corning

Table VI.   Wood-Fueled Electric Plants Scheduled for Startup in 1980s[a]

Location	Operator	Fuel	Feed Rate (metric ton/day)	Capital Cost (10⁶ $)	Electric Output (MW)	Scheduled Startup
Midland, MI	Dow Corning Corp.	Wood	~450	30	22.4[b]	1982
Madera, CA	California Power & Light Corp.	Pellets[c]	635	70	40–50	1981
Burlington, VT	Burlington Electric Dept.	Green wood	~1240	80	50	1983

[a] Adapted from References 33–36.
[b] Will cogenerate with 124,700 kg/hr (275,000 lb/hr) of 8620-kPa (1250-psi) steam.
[c] Woodex.

Corp. is expected to save the company 30–50% on fuel as compared to the cost of coal [33–36]. The economics of wood vs fossil fuels in recent estimates seem to be quite favorable for wood at this time, as illustrated in Table VII. The key factor in any analysis such as that depicted in Table VII is fuel cost, so local wood markets and hauling distances can have substantial negative or positive impact on the economic feasibility of any given project.

An important concern of increased usage of wood combustion systems is environmental impact. Particulate emissions have been a special concern because they contain relatively large amounts of polycyclic organic matter (POM), some of which has been identified as carcinogenic hydrocarbons. Even though wood fuel is still a very small fraction of the total fuels burned in the United States, total POM emissions from wood are much larger than from most other sources [37]. Current total POM emission estimates relative to other combustion emissions in the United States are shown in Table VIII. It is apparent that wood contributes more than any other source with the possible exception of residential coal-fired systems.

Municipal solid wastes (MSW) represent another renewable energy source currently undergoing a resurgence in interest for energy and resource recovery projects [38, 39]. Although there are only 23 operating plants in the United States [3], more than 100 cities in 37 states are planning waste-to-energy systems which could yield 143,000 bbl/day of oil-equivalent if all were placed on-line [38, 39]. Unfortunately, many of the plants that were designed and built over the last ten years have been plagued with operating difficulties caused by excessive wear, poor design, corrosion and failure of some units, particularly front-end equipment for resource recovery. For example, the 1000-ton/day Chicago, IL, plant for recovery of ferrous metal and refuse-derived fuel (RDF) has been inactive for more than a year. Problems such as plugging of pneumatic lines for conveying RDF to the power plant for cocombustion with coal, inoperability of the fine shredder, short-term fan blade loss because of the abrasive nature of RDF and difficult retrieval of RDF from the storage bins are just some of the reasons why the plant was shut down. The Chicago system is only one of the several first-of-a-kind resource recovery plants that are having operating difficulties.

The current trend of MSW energy recovery systems seems to be in the direction of direct combustion of raw MSW with minimal or no front-end processing for resource recovery and recycling. The apparent reason for this is the success of the Wheelabrator Frye plant in Saugus, MA, which continues to exhibit the best overall performance of all the energy-from-MSW plants in the United States. The plant uses direct combustion of raw unsorted MSW for steam production, and since going on-line in

Table VII.   Cost Comparison of Fuels[a]

Fuel	Heating Value	Boiler Efficiency (%)	Cost ($) Current	Cost ($) Per Energy Unit	Cost ($) Adjusted Per Energy Unit[c]
Wood (13% moisture)	17.41 MJ/kg	78.0	16.53/metric ton	0.95/GJ	1.22/GJ
	7,490 Btu/lb	78.0	15.00/ton	1.00/MBtu	1.28/MBtu
Wood (50% moisture)	10.00 MJ/kg	66.7	13.23/metric ton	1.32/GJ	1.98/GJ
	4,300 Btu/lb	66.7	12.00/ton	1.40/MBtu	2.09/MBtu
Natural Gas	39.25 MJ/normal m³	82.5	2.24/GJ	2.24/GJ	2.72/GJ
	1,000 Btu/SCF	82.5	2.36/10⁶ Btu	2.36/MBtu	2.86/MBtu
No.2 Fuel Oil	38.71 MJ/l	82.5	0.177/l[b]	4.57/GJ	5.54/GJ
	139,000 Btu/gal	82.5	0.67/gal[b]	4.87/MBtu	5.84/MBtu
			(0.264/l)	(6.82/GJ)	(8.27/GJ)
			(1.00/gal)	(7.19/MBtu)	(8.72/MBtu)

[a] Adapted from References 33–36; original source, Georgia Institute of Technology.
[b] Considerably higher now; values in parenthesis estimated at $0.264/l or $1.00/gal.
[c] Adjusted for boiler efficiency.

Table VIII.   Total Relative Polycyclic Organics Emissions
in the United States[a]

Wood	100
Coal-Fired Utilities	1–11
Industrial Coal	0.4–3
Oil-Fired Boilers	4–33
Oil-Fired Furnaces	2–14
Residential Coal	33–1000

[a]Adapted from Budiansky [37].

1975, has processed over 1.5 million tons of MSW and generated over 8 billion lb of turbine-quality steam [40]. In addition, the plant emissions are about one-third the maximum permitted by EPA. The same technology has been used worldwide in more than 160 facilities since 1954 when the first plant was placed in operation in Bern, Switzerland.

More development work is apparently necessary to perfect front-end designs for large-scale resource recovery. The few small- to moderate-scale plant designs that have shown satisfactory performance cannot, evidently, be easily extrapolated to large-scale plants. Based on the reported performance of successful large-scale systems, energy recovery can best be accomplished with MSW by direct combustion until reliable processes are available for large-scale RDF preparation.

One of the possible solutions to the reliability problem is to densify the combustible fraction of MSW. This approach is under intensive development with RDF, wood, wood wastes and other residues (Woodex, Frajon, Eco-Fuel II, K-Fuel, ROEMMC). However, a dry, powdered refuse (Eco-Fuel II) produced in a 1800-ton/day plant in Bridgeport, CT, was reported to present combustion difficulties when cocombusted with oil in utility boilers [38, 39], but the plant is now operating satisfactorily.

Gasification

Anaerobic Digestion Research

Research on the development of anaerobic digestion systems for biomass and wastes has continued, with emphasis on the evaluation of different feedstocks and development of advanced digester and system designs [41]. Terrestrial and aquatic biomass and various industrial, municipal and agricultural wastes are being evaluated alone [41] and in mixtures [42]. The energy, moisture, volatile solids and ash values are often quite different between the various feedstocks, but when balanced digestion is attained, the gas production parameters, volatile solids reductions, and energy recovery efficiencies appear to span a relatively narrow range when the digestion conditions are the same. This is illustrated by the data shown in Table IX. However, there are often differences between the gasification characteristics of the same biomass species when grown under varied conditions, as, for example, was found with water hyacinth [43]. This plant exhibited higher methane yields and energy recovery efficiencies under conventional high-rate digestion conditions when grown in sewage-fed lagoons as compared to the corresponding values of material harvested from a freshwater pond.

Stirred-tank reactors are still the mainstay of most digestion systems, but significant advances are being made to develop other designs for methane production. Examples are plug-flow, two-phase, and fixed-film digesters [41]. Recently, new research results have been published on fixed-film digesters [44-47]. The laboratory data indicate that in general, fixed-film digesters permit digestion to occur at shorter hydraulic residence times and higher gas production rates with soluble and several insoluble feedstocks than conventional high-rate systems. The composition of the support and its size as well as direction of liquid flow affect digestion performance. Small-size solid supports that can be partially expanded or "fluidized" by the flowing feed and product gas seem to exhibit better performance than the packed-bed systems containing large stationary supports.

Other new developments at the research stage that continue to show promise are combined anerobic digestion and thermal gasification systems such as the IGT BIOTHERMGAS™ process [41, 48]. In this process, anerobic digestion, dewatering, steam-oxygen gasification and catalytic methanation or biomethanation of the hydrogen-rich gas from the thermal gasifier are combined in that order to produce methane. The refractory organics from the digester are thus completely gasified, and waste heat, ammonia and inorganic nutrients can be recycled from the thermal gasifier to the digesters. Overall thermal efficiency is high because heat for digestion is supplied by the thermal gasifiers. Another combination process in which anerobic digestion and pyrolysis are used together is at the design stage [49]. A 200-ton/day demonstration plant is being designed in which refuse, dairy manure, agricultural residues and food-processing wastes are proposed for conversion to gaseous fuel that will be converted to 3-5 MW of electric power.

Anaerobic Digestion Commercialization

Commercial methane recovery and use via anaerobic digestion is growing at a modest rate with sewage, landfilled refuse, manure and industrial

Table IX.   Comparison of Digester Performance under High-Rate Mesophilic Conditions[a]

Feed	Gas Production Rate (vol/vol-day[b])	Methane in Gas (mol %)	Methane Yield (normal m^3/kg VS added)	Volatile Solids Reduction (%)	Feed Energy Recovery as CH_4 (%)
Primary Sewage Sludge	0.78	68.5	0.314	41.5	46.2
Primary-Activated Sewage Sludge[c]	0.89	65.5	0.328	49.0	54.4
RDF[d]-Sewage Sludge[e]	0.62	60.0	0.210	36.7	39.7
Biomass-Waste[d,f]	0.55	62.0	0.201	33.3	38.3
Coastal Bermuda Grass[d,g]	0.59	55.9	0.208	37.5	41.2
Kentucky Blue Grass[h]	0.55	60.4	0.150	25.1	27.6
Giant Brown Kelp[d]	0.66	58.4	0.229	43.7	49.1
Water Hyacinth[d,i]	0.50	62.8	0.185	29.8	35.7

[a] Adapted from Klass [41]. Conditions were 35°C, daily feeding, continuous mixing, pH 6.7-7.2, 12-day detention time, 1.6 kg VS/m^3-day (0.10 lb VS/ft^3-day) loading rate except for kelp which was 2.08 kg VS/m^3-day (0.13 lb VS/ft^3-day).

[b] Standard volume of gas produced per culture volume per day.

[c] Blend of 50 wt % primary-activated sludge on a total solids basis.

[d] Ground with an Urschel Laboratory Grinder (Comitrol 3600) to give particle sizes 1 mm or less.

[e] Blend of 80 wt % RDF and 20 wt % 50:50 primary-activated sludge on a total solids basis.

[f] Blend of 32.3 wt % water hyacinth, 32.3 wt % Coastal Bermuda grass, 32.3 wt % RDF, and 3.1 wt % primary-activated sewage sludge (30 wt % and 70 wt % on a total solids basis) on a volatile solids basis.

[g] Feed slurry supplemented with NH_4Cl to overcome nitrogen deficiency.

[h] Passed through a 2-mm sieve.

[i] Harvested from a sewage-fed lagoon.

wastes; little has been done with biomass. Many wastewater treatment plants throughout the United States are now making maximum use of digester gas for process heat, steam production or electric power generation [41]. One of the largest plants, the South Shore Waste Water Treatment plant in Milwaukee, WI [50], anticipates a yearly savings of over $700,000 in purchased natural gas and electric power. This 454-million-liter/day plant operates three Allis Chalmers 350/437.5 kW generators with three digester gas-fueled turbocharged dual gas engines each rated at 635 hp. Digester gas consumption at full load is 193 m^3/hr. Digester gas is also used to supply 4000 m^3/min of air at 59 kPa for the activated sludge system. Gas engine waste heat is recovered for digester temperature maintenance at 35°C.

No anaerobic digestion plants supplied with MSW or MSW-sludge blends have been commercialized yet. The largest demonstration plant is the proof-of-concept, 100-ton/day plant sponsored by DOE in Pompano Beach, FL. In contrast with liquid-slurry digestion of MSW, eight commercial methane-from-landfill recovery plants are now in operation that produce medium-Btu gas and substitute natural gas (SNG), and others are in the testing stage (Table X). The technology was developed with private capital and appears to be well established [41]. All of the systems listed in Table X would provide about 3500 bbl/day of oil equivalent, but the potential of landfill gas has not been fully developed yet. Many landfill sites remain to be evaluated, and many gas companies are now participating in several landfill gas programs which in the past, have usually been carried out by nonutility companies. The theoretical availability of landfill gas has been estimated to be about 1 quad, but the consensus is that much less — between 0.05 and 0.2 quads — can be economically developed [51].

Although not developed in the United States, an interesting in situ process similar to landfill gas recovery, except that it uses peat instead of municipal refuse, is under test in Sweden [53]. A pilot plant has been in operation since 1978 in which natural anaerobic digestion is promoted in a peat bog to generate methane. Water containing dissolved methane is pumped to a degassing unit, the methane is separated, and the water is recycled to the bog. The temperature of the recirculated groundwater is almost always between 6 and 11°C (43 and 52°F) at the pilot site, and the dissolved methane concentration is about 20 mg/1 (0.17 scf/bbl of water). One of the module sizes evaluated required a peat bog of 2–4 ha (5–10 ac) and a water circulation rate of 100 liter/sec. The net energy production rate of methane was calculated to be 41 metric ton/yr (2.14 × 10^6 scf/yr), and the capital and operating costs were estimated to be $41,000/yr. This is believed to be a technically and financially feasible

Table X. Landfills Used for Commercial Methane Recovery[a]

Location	Well Type[b]	Product Gas[c]	System Cost (10⁶ $)	Gas Production (10³ m³/day)	Startup Time	Operator	Gas Use
Azuza, CA[d]	D	MBG	1.2	20	1978	Azuza Land Reclamation Co.	Sold to Reichold Chemical Co.
Cinnaminson, NJ[d]		MBG		8	1979	Public Service Electric & Gas Co.	Industrial sales
Industry, CA[d]	S	MBG	0.4	19	1978	SCS Engineers	Boiler fuel
Los Angeles, CA[d] (Sheldon Arleta)	D	MBG	1.75	70	1979	City	City electric power
Monterey Park, CA[d]	D	SNG		110	1979	Getty Synthetic Fuels, Inc.	Sold to Southern California Gas Co.
Mountain View, CA[d]	S	MBG	0.84	19	1978	Pacific Gas & Electric Co.	Blended into natural gas
Palos Verdes, CA[d]	D	SNG		20	1975	Getty Synthetic Fuels, Inc.	Sold to Southern California Gas Co.
Wilmington, CA[d] (Ascon)	S	MBG		40	1978	Watson Energy Systems	Sold to Shell Oil refinery
Carson, CA[e] (South Bay 6)	S	MBG		16	Late 1980	Syuly Enterprises	Engine fuel for electric power
Chicago, IL[e]	S	SNG		54–80	Late 1980	Getty Synthetic Fuels, Inc.	Sold to Natural Gas Pipeline Co. of America
Glendale, CA[e] (Scholl Canyon)	D	MBG		32	June 1981	City	Engine fuel for electric power

Location	Depth		Gas	Date	Operator	Use
Los Angeles, CA[e] Puente Hills	D		MBG	Early 1981	Los Angeles County Sanitary District	Boiler fuel for heating and cooling
Bradley East	D	54	MBG	Late 1980	Conrock Co.	Sold to city for electric power
Lopez Canyon	D	3	MBG	Early 1981	City	Engine fuel for electric power
Martinez, CA[e] (Acme)			MBG	Late 1980	Getty Synthetic Fuels, Inc.	Boiler fuel for electric power
Orange County, CA[e] (Coyote Canyon)				Late 1980 (Testing)	Irvine Co.	
San Leandro, CA[e] (Davis Street)			MBG	Late 1980	Getty Synthetic Fuels, Inc.	Boiler fuel for electric power
Staten Island, NY[e] (Freshkills)	S		SNG	March 1982	Getty Synthetic Fuels, Inc.	Sold to Brooklyn Union Gas Co.
Staten Island, NY[e]	S	2	MBG	July 1980	Brooklyn Union Gas Co.	Engine fuel for electric power

[a] Source: U.S. Environmental Protection Agency, Getty Synthetic Fuels, Inc., and Reference 52.
[b] S, shallow, less than 30.5 m (100 ft); D, deep, more than 30.5 m (100 ft).
[c] MBG, medium-Btu gas; SNG, high-Btu gas.
[d] Operational.
[e] Planned.

project under Swedish conditions, but it is doubtful that it would be feasible in the United States given the same operating conditions and results.

Commercial manure gasification projects utilizing anaerobic digestion are being built at both the small and moderate scale. Examples are the farm-scale plug flow system in Gettysburg, PA [54], and the Lamar, CO plant that will be sized to digest 308 metric ton/day (340 ton/day) of manure from 50,000 head of cattle and produce about 26,864 m³/day (10⁶ ft³/day) of methane [55]. Unfortunately, the country's first large-scale manure gasification plant that was designed to supply 16 million m³/yr of SNG in Guymon, OK, to Natural Gas Pipeline Co. will not manufacture SNG in its future operations [56]. Only cattle feed and liquid fertilizer will be sold because Calorific Recovery Anaerobic Processes, Inc. has not been able to produce SNG at a competitive price [56]. This is somewhat expected because various economic analyses of cattle manure gasification by anaerobic digestion indicate the digested solids (cattle feed) are a very important product; in many cases, methane might even be considered to be a by-product. An example of one economic analysis is shown in Table XI. In this table, the preferred thermophilic digestion conditions and the estimated capital and operating costs for a 3220-normal-m³/day (120,000-scf/day) SNG plant were estimated. Without any credit for the residual solids as an animal feed, the methane cost, not including feed or feed transportation costs, is projected to be $8.85/GJ ($9.33/10⁶ Btu). If credit is taken for the residuals, methane cost can be substantially reduced. The development of suitable technology for recovering the digested solids in manure-to-methane plants and determination of their actual feed value thus seem mandatory if large-scale commercial plants are to compete. This work is in progress at several locations.

Table XII summarizes the basic characteristics of the Gettysburg, PA, and the Lamar, CO, plants. Innovative features of the plug-flow system include a plastic cover which serves both to maintain anaerobic conditions and collect the gas, and the use of a special diesel engine–generator set. The engine uses a small amount of diesel fuel (15%) when operating mainly on gas to ignite the gas and to lubricate the fuel injectors. The 1082-ha (2675-ac) farm produces all of its own power with this system as well as a surplus which will be sold to the local power company. The Lamar, Colorado plant is scheduled for startup in 1982 [41].

An anaerobic digestion process that utilizes a sludge blanket for industrial waste stabilization and methane recovery was recently introduced into the United States [58-60]. Called the upflow anaerobic sludge blanket process (UASB), it has been used commercially for several years in the Netherlands with soluble beet sugar processing wastes. The

Table XI.   Economic Estimate of Methane Cost from Cattle
Manure Digestion (10,000 head of cattle)[a]

Installed Equipment Cost ($)	
Without Centrifugation of Effluent	901,000
With Centrifugation of Effluent	1,357,000
Annual Operating Cost ($)	
Without Centrifugation of Effluent	291,900
With Centrifugation of Effluent	390,400
Salable Products	
SNG (normal m^3/day)	3,216
Residual Solids in Effluent (dry metric ton equiv/day)	22.7
Revenue Requirements[b]	
For SNG Only, No Centrifugation ($/GJ)	8.85
($/yr)	407,500
For SNG, No Centrifugation, with Effluent	
Credit of $0.022/kg TS for Animal Feed	
($/GJ)	4.89
($/yr)	225,000
Feed Credit ($/yr)	182,500
For SNG, with Centrifugation, with Dry Cake Credit	
of $0.055/kg TS for Animal Feed, 50% TS Recovery	
($/GJ)	7.31
($/yr)	336,400
Feed Credit ($/yr)	228,100

[a]Developed from Hashimoto et al. [57]. Operating assumptions: 2.98 kg dry equiv VS/head-day; digester volume, 1.86×10^6 m^3; digestion temperature, 55°C; detention time, 5 days; loading rate, 16 kg VS/m^3-day; gas production rate, 6.6 vol gas/culture vol-day; 50 mol % methane in gas; methane yield, 0.195 normal m^3/kg VS added-day; 5818 m^3 gross methane/day; 45% of methane for plant heating; feed solids are 85 wt % VS; volatile solids destruction 55 wt %.
[b]Feed cost and collection not included; calculated for 20-year life, straight-line depreciation, 15% DCF return on equity and 9% return on debt (65% debt 35% equity), 52% income tax.

digesters employ a specially designed baffle at the top of each unit for separation of the gas, liquid and solid phases. The sludge particles settle back into the bed and are retained for long periods. This permits long effective solids retention times (SRT) at hydraulic retention times (HRT) of only a few hours. Typical results from a 30-m³ pilot plant and a 200-m³ industrial plant are shown in Table XIII. It is apparent that the units function at reasonably high efficiencies and short HRT. Although commercial use of this process has been limited to beet sugar processing wastes, further work is in progress to apply it to potato, corn wet milling, brewing and other industrial wastes [60]. Note that the UASB process is similar in many respects to the anaerobic attached film expanded bed

Table XII.   Cattle Manure Gasification Plants [54,55]

	Plug-Flow Digester	Standard High-Rate
Location	Gettysburg, PA	Lamar, CO
Cattle	1,700 dairy animals	50,000 steer
Feed Rate	49,200 liter/day	310 metric ton/day
Digestion Temperature (°C)	38	35–38
Gas Output (m^3/day)	806	27,900
Gas Use	Detroit Diesel 6-71 Dual Fuel Engine, 125 kW Delco Generator	City boiler for electric power
By-Products	Fertilizer	Refeed cake
Capital Cost		$14.2 million (1977)
Gas Value		$2.85/GJ $3.00/$10^6$ Btu (feed at $1.00/ton)
By-Product Value		$49.60/metric ton $45.00/ton

reactor (AAFEB) [61], and the anaerobic fixed-film bioreactor (ANFLOW) [62]. Each is an upflow unit. The AAFEB contains a lower, slightly expanded solid phase of small particles (0.5–1.5 mm) such as plastic particles which serve as a support for organisms. Total volatile solids destruction efficiencies of 40% were achieved with diluted animal manure at an HRT of 1.2 days while the complete-mix control reactor required an HRT of 10 days to obtain the same performance. ANFLOW uses a stationary packed bed of Raschig rings and has shown promise for efficient waste treatment and methane production.

Thermochemical Gasification Research

Research on thermochemical gasification of biomass and wastes is in progress at the laboratory, PDU and pilot scales using partial oxidation, steam reforming, steam-oxygen reforming, hydrogasification and pure pyrolysis conditions with and without catalysts. Reactor configurations under investigation include fixed beds, moving beds, fluidized beds, entrained-feed solids reactors, stationary vertical-shaft reactors, inclined rotating kilns, horizontal shaft kilns, high-temperature electrically heated units with gas-blanketed walls, single- and multihearth-reactors, and other designs. It is beyond the scope of this chapter to cover the

Table XIII.   Typical Results of Upflow Anaerobic Sludge Blanket Process with Sugar Beet Wastes [59,60]

	30-m³ Pilot Plant		200-m³ Plant
	Waste Sugar Stream	Wash-Transport Water	Wash-Transport Water
COD (mg/l)	3500	2000	4200
Digestion Conditions			
pH	7.1	7.0	7.1
Temperature (°C)	33	30	30
Hydraulic Retention Time (hr)	9.0	4.4	7.1
COD Loading (kg/m³-day)	11	11	14
(lb/ft³-day)	0.69	0.69	0.87
Results			
COD Reduction (%)	93	63	90
Gas Production (vol/vol-day)	4.5	3.0	4.7
Methane (vol %)	72	90	83
Methane Yield (m³/kg COD added)[a]	0.295	0.245	0.279
(ft³/lb COD added)	4.72	3.92	4.47

[a]Calculated from reported data.

highlights of all of this work. However, much of the current research emphasizes gasification in fluidized bed systems, and recently, new reports appeared which indicate that several of the ongoing studies on biomass and waste conversion are producing information that could lead to superior processes.

Catalytic steam gasification in fluidized beds at atmospheric pressure has been reported to be effective for producing gases useful for either SNG or methanol manufacture at high efficiencies [63]. Typical laboratory data are shown in Table XIV. Methanol synthesis gas was produced

Table XIV.   Catalytic Steam Gasification of Wood [63]

	Catalyst	
	Ni:SiAl	Ni + SiAl
Ni:SiAl Ratio	1:1	3:1
Temperature (°C)	850	550
Wood/Catalyst Weight Ratio	100	
Steam/Wood Weight Ratio	1.25	0.33
Carbon Conversion (%)		
To Gas	99.6	70
To Liquid	0	0
To Char	0.4	
Gas Composition (%)		
H_2	56.7	32.3
CO	27.9	13.4
CH_4	0.5	20.7
CO_2	14.9	33.6
Feed Energy In Gas (%)	115	68

in a single-stage reactor at 750–850°C by steam gasification of wood in the presence of nickel and silica-alumina catalysts, while synthesis gas suitable for methane synthesis was formed at 500-550°C. The advantages of catalytic steam gasification are that no oxygen or air is required, little or no tar is formed, no shift reactor is required for methanol synthesis, methanation requirements are low, resulting in high conversion efficiency, and yields and efficiencies are greater than those obtained by conventional gasification [63]. Capital costs for a commercial plant are therefore expected to be significantly reduced.

Noncatalytic fluidized bed gasification is also being developed at temperatures near 800°C but at pressures of about 2100 kPa [48]. The residence time in the reactor is estimated to be 3–5 min to achieve 95% carbon conversion. For biomass containing 30% moisture, calculations indicate that external steam additions may not be necessary and that about one-half of the total methane is formed directly in the gasifier.

Another report on fluidized bed gasification of wood was performed in a countercurrent reactor in which oak sawdust was fed to the top of the reactor and was fluidized by a steam-air mixture fed to the bottom [64]. The gas yields at 600–800°C were about 1.1–1.4 liter/g dry ash-free feed. The gas had a gross heating value in all cases exceeding 11.2 MJ/m^3 (301 Btu/ft^3) and contained about 4% C_2H_4 and 11% CH_4. Previous results obtained with other biomass were compared with those obtained with oak sawdust and were accounted for based on the relative amounts of cellulose, hemicellulose and lignin in the feedstock.

Another gasification study in a fluidized bed reactor was conducted with manure particles of different sizes [65]. This work showed that superficial gas velocity did not have a significant effect on the composition and heating value of the product gas. Particle size, however, had a significant effect. Gas yield increased and heating value decreased with decreasing size fractions.

An innovative approach to biomass and coal gasification reported in 1979 [3] has been proposed for large-scale demonstration [66]. The Simplex gasification process uses briquettes of coal, MSW and dewatered sludge as a combined feed which permits the use of caking coals without pretreatment. The use of briquettes facilitates use of high gas rates and large-diameter gasifiers. The briquettes as they descend through the reactor first undergo pyrolysis and caking at 300–870°C. The wastes are converted to char and the coal is changed to coke. Then as the reactants descend further, the inorganics in the wastes are subjected to temperatures greater than 1100°C where they form a vitreous slag with ash from the coal, and then finally the coke and char react with steam and oxygen at about 1650°C. Economic analyses indicate that a 13.7-MJ/normal m^3 (350-Btu/scf) gas can be manufactured at about $2.37/GJ ($2.50/10^6 Btu) with coal at $27.56/metric ton ($25/ton) and a tipping fee of $11.02/metric ton ($10/ton) for MSW.

Thermochemical Gasification Commercialization

Commercial utilization of wastes as feedstocks in thermal gasification processes is growing slowly in the United States. The largest commercial pyrolysis plant in the country which converts about 500 ton/day of MSW to steam via low-Btu gas in a rotary kiln will probably be converted to direct combustion in the near future [3]. Cost and reliability of operation have been cited as the reasons for the change. Two 100-ton/day PUROX plants [3], which use oxygen for partial oxidation of biomass and wastes to medium-Btu gas in a three-zoned shaft furnace, have been built in Japan, but none has yet been constructed in the United States. Economic factors are the main reason for lack of commercial activity. To improve process costs, PUROX II is being developed. The modified process incorporates a 345- to 414-kPa (50- to 60-psi) pressurized reactor, a simplified front end, and a wastewater incinerator for cleanup. The use of moderate reaction pressures will permit higher throughput capacities and lower gas processing costs. Incineration of the wastewater, which usually has a BOD in the 60,000 mg/l range, will simplify wastewater treatment and provide steam. Similarly, the slagging pyrolysis process, Andco-Torrax, developed by Andco, Inc. for unsorted MSW, has been built at four locations in Europe and one in Japan; none is operating yet in the United States (Table XV) [67]. However, a 100-ton/day plant is under construction in Disney World and is scheduled for startup in late 1981. Hot water is the final product. The largest Andco-Torrax plants are in Europe (Luxembourg and Creteil, France), have design capacities of 200 ton/day, and produce electric power. Each of these plants supplies a low-Btu gas which is burned to completion in a secondary combustion chamber.

Table XV. Commercial Plant Data for
Andco-Torrax Process [67]

Location	Startup	Number of Units	Unit Design Capacity (metric ton/hr)	Energy Product
Luxembourg	Sep 1976	1	2.7–7.5	Electricity
Grasse, France	Oct 1977	1	3.4–6.4	Process Steam
Frankfort, W. Germany	Jul 1978	1	3.6–7.3	Electricity
Creteil, France	Nov 1979	2	4.9–7.5	Electricity
Hamamatsu, Japan	Oct 1980	1	2.4–3.3	Not Recovered
Lake Buena Vista, FL	Nov 1981	1	3.0–4.0	Hot Water

Biomass gasification, especially in small-scale producer gasifiers for wood, is commercially used in the United States and elsewhere and has been for many years. Units retrofit to gas-fired boilers have attractive features in areas where wood, wood residues or other suitable fuels are readily available. One of the newest systems to be built in 1980 was installed at a 650-bed hospital and is claimed to be one of the largest gasification systems of its kind in the United States [68]. The system characteristics shown in Table XVI include the innovative feature of using green wood chips obtained from whole trees as fuel. The low-Btu gas is burned in an existing furnace equipped with a specially designed burner and is expected to save the hospital about $250,000/yr in fuel costs. Another biomass producer gasifier application that has been developed and applied to commercial seed corn drying uses an air-blown,

Table XVI.   Characteristics of Whole Tree Gasification
System in Rome, GA [68]

Fuel Properties	Mixed hardwood, softwood, needles, leaves, twigs and bark (whole tree) chips
	Current price, $1.66–1.90/GJ ($1.75–2.00/10^6 Btu) as chips
	At 50 wt % moisture, 9.3–11.6 MJ/kg (4000–5000 Btu/lb)
Fuel Feed Rate	2.8 metric ton/hr
Gasifier	APCO, updraft, air-blown, 7.6 m high × 3 m diam, 26 GJ/hr output
Fuel Gas	7.26 MJ/normal m³ (dry, HHV); N_2, 39%; CO_2, 7%; CO, 25%; H_2, 26%; CH_4, 2%; misc., 1%
Final Product	8600 kg steam/hr (1000 kPa)
Project Cost	<$400,000

up-draft, negative-pressured gasifier supplied with corn cobs [69]. The product gas has a heating value of 4.79–5.61 MJ/normal m³ (122–143 Btu/scf) and is used as fuel for a seed dryer. Economic analyses indicate that total capital and operating costs are about $0.286/bu of dried corn compared to costs of $0.566/bu for propane drying. Producer gasifiers of many sizes and designs for biomass and wastes are available from several manufacturers in the United States, and it is expected that the market for this technology will grow at an increasing rate as fossil fuel costs increase. One untapped market is the production of electric power on farms. A recent estimate, based on the use of small-scale biomass producer gasifiers and gas engine-generator sets for farm residues, indicates a market of $10 billion for hardware to make 1.3 million farms energy self-sufficient in electric power [70].

Liquefaction

Much of the nonethanol fermentation and biological conversion work in progress is directed to utilization of the noncellulosic fractions of biomass to produce fuels and chemicals. Research currently emphasizes utilization of hemicelluloses from biomass via the C_5 sugars (pentoses) obtained by selective hydrolysis of the hemicelluloses in the presence of cellulose. Methods are being developed for producing butanol and butanediol, carboxylic acids and acetone, although some of the current work uses glucose for similar fermentations [71]. Of particular interest is the observation that *Clostridium* species have been found to form high yields of butanol (up to 46%, 75% of total product) as well as butyric acid, acetone and acetic acid when supplied with xylose as the sole carbon source [71]. This largely unexplored area of fermentation is expected to led to feasible processes because pentoses should be available in large quantities when some of the second-generation alcohol fermentation processes that use cellulosic feeds are commercialized. Another process concept uses nonsterile anaerobic fermentation of biomass in the presence of methane-formation suppressors, such as 2-bromoethanesulfonic acid and carbon monoxide, to form higher aliphatic carboxylic acids, separation of the acids by membranes or solvent extraction, and Kolbe electrolysis of the acids to yield hydrocarbon liquids [71]. Preliminary anaerobic experiments have been carried out with *Hydrilla* and corn meal, and the electrolytic oxidation of the organic acids has been performed with a favorable 6:1 energy balance based on the applied potential. An economic analysis of a 1000-ton/day plant using utility financing indicates a product cost of $5.20/GJ ($5.48/10^6 Btu). Delignification of biomass can be carried out by a number of techniques such as solvent extraction [72]. But biological depolymerization to generate liquid fuels also appears feasible [71]. Preliminary experiments indicate at least 30% degradation of sodium lignosulfonates by the yeast *Trichosporum fermentans*. However, it should be pointed out that once a lignin concentrate is available, depolymerization by hydrocracking may be the preferred route. Hydrocarbon Research Inc. has announced a two-step process (Lignol) in which lignin is depolymerized by hydrocracking in a fluidized bed to yield mono-cyclic aromatics which are then dealkylated to phenol and benzene [73]. Fuel oil and gas are the other products. Yields (by weight) from kraft lignin are estimated to be 20% phenol, 14% benzene, 13% fuel oil and 29% fuel gas.

Direct thermochemical liquefaction of biomass and wastes has generally been achieved by short-residence time pyrolysis (flash pyrolysis) to maximize liquid yields. A surprising number of researchers have referred to these liquids as "oils," but they cannot be upgraded to refined products by conventional refinery practice. If the term oil is used, the preferred nomenclature is "pyrolytic oil" or "product oil." Some attempts have been made to improve fuel volatility, oxygen content, corrosivity and gumming tendencies of these products by hydrotreating and cracking [74]; the results are not known. Biomass-derived pyrolytic oils have been proposed for use as substitutes for No. 6 fuel oil and diesel fuel. Several conceptual processes based on established procedures and equipment have been suggested to manufacture pyrolytic oils from wood [75]. Pyrolysis at 180–260°C with up to 20:1 recycle ratios of pyrolytic oil has been tested at the pilot scale. Some characteristics of the conceptual process are shown in Table XVII. Interestingly, the 200-ton/day flash pyrolysis plant for fine-shredded RDF in El Cajon, CA, the largest pyrolysis

Table XVII.   Pyrolytic Oil from Wood Pyrolysis:
A Conceptual Process [75]

Green Wood Feed	1814 metric ton/day (2000 ton/day)
Pyrolysis Temperature	180–260°C
Thermal Efficiency	45% to liquid, 29% to char
Capital Cost	$22.2 million
Operating Cost	$2.85/GJ ($3.00/10⁶ Btu) of liquid & char
Product Prices	
Utility Financing	
Liquid	$5.60/GJ ($29.00/bbl)
Char	$29.76/metric ton
Nonregulated Financing	
Liquid	$7.12/GJ ($38.00/bbl)
Char	$29.76/metric ton

plant of its kind in the country, has been down for about 2.5 years because of operating problems in the pyrolysis reactor and the product separation units [3,76]. The main product from this plant was also a pyrolytic oil.

Another direct liquefaction process for biomass is under pilot test at the DOE facility in Albany, OR. Basically, the process consists of subjecting a slurry of wood chips to hydrogen and carbon monoxide in the presence of sodium carbonate catalyst at 300–370°C and 13.8–27.6 MPa for 20–90 min. A viscous liquid product is formed. The two versions of the process are the original Bureau of Mines design (BOM) and the Lawrence Berkeley Laboratory design (LBL). BOM uses a heavy recycle product oil medium for the feed slurry and LBL uses water as the medium. Recent studies of the BOM product oil indicate that it lies somewhere between No. 6 fuel oil and the pyrolytic oil from the El Cajon plant from the standpoint of elemental composition and heating value (Table XVIII) [77].

Direct hydrogenation is another route to liquid fuels. Work just getting underway based on the IGT HYFLEX™ process utilizes a hydrogen atmosphere at moderate to high pressures (1380–2760 kPa) and temperatures in the 700°C range for only a few seconds in an entrained-flow reactor to convert biomass to liquids [48]. Short residence times promote maximum yields. The product produced by these hydropyrolysis conditions should have a higher intrinsic value than the heavy flash pyrolysis products described above. The initial work will evaluate the possibility of making liquid fuels from biomass such as eucalyptus trees and wastes from sugarcane and pineapple processing plants.

Indirect liquefaction of biomass proceeds through intermediates such as ethylene for the production of polymer gasoline, synthesis gas for the

Table XVIII.   Liquid Product Characterization and Comparison with No. 6 Fuel Oil [77]

Product	C (wt %)	H (wt %)	N (wt %)	O (wt %)	S (wt %)	Ash (wt %)	H_2O (wt %)	Heating Value (MJ/kg)	Density (g/ml)
LBL Liquid	72.3	8.6	0.2	17.6	0.006	0.078	8.5	33.7	1.19
Pyrolytic Oil	57.0	7.7	1.1	33.2	0.2	0.5	14	24.7	1.39
No. 6 Fuel Oil	85.7	10.5	2.0	0–3.5	0–3.5	0.05	0.20	42.3	1.02

production of methanol or hydrocarbons, or methanol for the production of hydrocarbons (Mobil Process). The closest technology to commercialization is Mobil's methanol-to-gasoline process [78]. Although the process will be commercialized in New Zealand using methanol from natural gas instead of biomass, it is worth mentioning because the process economists who evaluated the technology selected it over the established Fischer-Tropsch route. A comparative economic analysis of most of the liquefaction processes under development is summarized in Table XIX [79]. It is apparent that there is a broad range of product costs as well as intrinsic product values. The higher value products have higher costs.

Table XIX.   Estimated Costs of Liquid Fuels
from Green Wood[a] [79]

Process	BOM	LBL	Pyrolysis	Methanol	Mobil[b]	Ethylene
Product	Heavy liquid	Heavy oil	Pyrolytic oil	Methanol	Gasoline	Polymer gasoline
Capital Cost (10^6 $)	56.1	39.5	25.8	109.8	128.6	60.7
Product Price						
Industrial, ($/GJ)[c]	8.12	7.57	8.41	15.37	18.93	24.82
Utility, $/GJ[d]	6.47	6.25	6.73[e]	11.78	14.33	19.56

[a] Basis: 1800 metric ton/day feed at $12.13/metric ton ($1.19/GJ); mid-1979 dollars.
[b] Royalty for Mobil catalyst not included.
[c] 15% DCF ROI, 100% equity.
[d] 65/35 debt/equity with return at 9%/15%.
[e] Includes by-product char credit at $1.42/GJ.

Examples of natural processes for liquid fuels have already been alluded to in the sections of this chapter on nonwoody herbaceous plants (polyisoprenes from guayule, triterpenoids from *Euphorbia lathyris*) and aquatic biomass (lipids from microalgae). There are a large number of biomass species that have the potential of commercial use for liquid fuels production [3,16], and considerable work is in progress to develop optimum biomass growth methods and liquid extraction or production techniques (for those cases where plant destruction is not necessary). Utilization of the natural liquids can take several forms such as direct use, blending with other fuels, or conversion to more conventional fuels. The latter application was recently reported in which Mobil Oil's ZSM-5 zeolite catalyst was used to promote conversion of natural liquids to premium hydrocarbon fuels [80]. Liquids from jojoba, corn and castor bean were completely converted to hydrocarbons, water and carbon oxides, while extracts from *Euphorbia lathyris* and *Grindelia squarrosa* were more difficult to convert. Modest amounts of aromatics and olefins and coke were identified. These were the results of preliminary experiments, and it is expected that other applications of the shape-sensitive zeolite catalysts will be found.

Ethanol and Methanol Fuels

It is clear that the proponents of gasohol, a blend of 90 vol % unleaded gasoline and 10 vol % ethanol, have successfully promoted ethanol as a gasoline extender. Gasohol is available in almost all parts of the country, despite some of the controversy that abounds on such matters as net energy production, subsidy, and the real impact of ethanol fuel on oil imports [81]. Federal funding levels have increased substantially on ethanol fuels, and projects are underway in all parts of the country.

Much of the research in progress to develop better ethanol processes is aimed at improving pretreatment processes for low-value cellulosics to convert them to fermentable materials that can be use as alcohol feedstocks [81]. The starchy products in corn have good biodegradabilities and are easily converted to ethanol, but the cellulose polymers in wood, for example, must first be converted to lower-molecular-weight fragments such as glucose before fermentation. Significant experimental data have been reported on new cellulose solubilization techniques that yield noncrystalline cellulose which is more readily hydrolyzed to glucose, improved enzyme-catalyzed hydrolysis of low-grade cellulosics to glucose and continuous hydrolysis of cellulosics to glucose via flash hydrolysis using dilute sulfuric acid. For example, hydrolysis of water slurries of newsprint with 1% sulfuric acid in the 235–240°C range at a residence time of 0.22 min afforded 50–55% of the theoretical glucose yields in a plug-flow reactor. These results are believed to be of commercial interest. Another short–residence-time sulfuric acid hydrolysis process feeds a hydropulped slurry of newsprint or sawdust into a twin-screw extruder device which expresses water from the slurry. The resulting high-solids cellulose plug is then hydrolyzed with acid which is injected into the feeder. The residence time/temperature/glucose yield relationships are about the same as those of the plug flow reactor experiments.

Research on the fermentation process itself has been directed to increasing ethanol yields and reducing fermentation times [81]. Recent approaches include the use of reduced pressure to remove the ethanol as fast as it is formed in the fermentation broth; bacteria instead of yeasts to shorten fermentation times; continuous fermentation techniques to shorten fermentation times; simultaneous saccharification and fermentation of low-grade cellulosics with enzymes and yeasts; thermophilic anaerobes for the one-step hydrolysis and fermentation of cellulosics; packed columns containing live, immobilized yeast cells or both enzymes and yeast cells through which glucose solutions are passed; and recombinant-DNA techniques to develop new yeast strains for rapid conversion of starch to sugar. For example, packed columns of live saccharomyces yeast cells entrapped in carrageenan gel are reported to convert 20% aqueous glucose solutions containing nutrients to 12.8 vol % ethanol solutions in 2.5 hr. Biomass, such as pineapple, not normally used for alcohol production has also been evaluated for alcoholic fermentation. This plant species, which requires much less water to grow than sugarcane or cassava, was projected to yield higher ethanol quantities per acre than sugarcane or cassava.

Net energy production efficiency of ethanol fermentation processes is a controversial subject on which even commercial producers and their customers do not agree. Energy consumption in the fermentation ethanol plant is therefore a very important research area. Since distillation to separate ethanol consumes relatively large amounts of energy compared to consumption in other parts of the plant, several methods are being studied to try to improve postfermentation processing. Drying of the partially concentrated alcohol solution with dehydrating agents including corn and corn derivatives is reported to be effective for producing nearly anhydrous alcohol; the energy content of the ethanol is ten times that needed for dehydration. Other techniques for reducing energy consumption use azeotropic agents, low-energy distillation, and membrane filters. Another possible route to anhydrous ethanol is to use unleaded gasoline or other solvents for direct ethanol extraction. This technique is at the research stage.

Anhydrous alcohol is necessary because alcohol and gasoline separate when gasohol absorbs only a small amount of moisture. Several research groups are working on additive-treatment methods that would eliminate or reduce phase separation; it is possible that some of this work might lead to methods that permit the use of wet alcohol in gasohol blends. This would reduce energy consumption in the plant because the final drying steps in the production of anhydrous alcohol are the most energy-intensive. It is noteworthy that the use of neat alcohols as motor fuels would not require anhydrous liquids; some reports indicate that ethanol containing substantial amounts of water—up to 15–20 vol %—are effective motor fuels.

Little work is apparently being done on the thermochemical production of synthetic ethanol via hydration of ethylene derived from biomass.

But an interesting nonbiological method has been reported for the conversion to ethanol of furfural from rice hulls, corn cobs and material from the southern pine forests. Furfural undergoes ring cleavage and reduction in the presence of lithium metal and alkyl amine solvent, or lithium salts, amine solvent and gamma rays to form ethanol. These reactions are under laboratory study and could conceivably lead to new nonbiological routes to ethanol.

Methanol has not been produced in any appreciable yield by biological methods, and no practical route to fermentation methanol has been discovered. Methanol trapping agents have been evaluated for some biological processes in which methanol is believed to be a transitory intermediate, but to date, the derivatives have been isolated in only very small quantities. Direct oxidation of methane to methanol is known under very high pressures, but no practical application of this technique exists.

Today, almost all methanol is manufactured by conversion of synthesis gas, usually by the so-called low-pressure process originally developed by Imperial Chemical Industries (ICI) in the 1960s. Subsequent research to develop new methanol processes has usually been patterned after the ICI method which uses heterogeneous copper-based catalysts to reduce carbon monoxide. Pressures are of the order of 75–10 MPa and temperatures 220–270°C.

A recent discovery is the homogeneously catalyzed reduction of carbon monoxide to methanol and methyl formate at 132 MPa and 225–275°C in the presence of solutions of ruthenium complexes. This observation could be the forerunner of new catalytic systems for methanol manufacture.

Two factors could have a substantial negative impact on the future of fermentation ethanol. One is the use of neat alcohols as motor fuels. Currently, methanol is not marketed in gasoline blends even though methanol prices are significantly less than those of ethanol. Some feel that in blends of the gasohol type, ethanol is more competitive because of fewer phase-separation, performance and emissions problems. These factors plus the political pressures to market gasohol have established ethanol's position. But if neat alcohols were used, many of the perceived advantages for ethanol would disappear and cost per mile could determine the outcome. In that case, methanol would have the advantage. Fermentation ethanol from low-cost cellulosics would then be mandatory to make both alcohols competitive. Such competition might be healthy because it is conceivable that both methanol and ethanol could be marketed as motor fuels, and together, they would obviously have a better chance of eliminating our dependence on petroleum.

The other factor that could make fermentation ethanol obsolete is the development of processing technology for the manufacture of ethanol from synthesis gas. Coal and all forms of low-cost biomass could then be used as feedstocks. This possibility is more than a concept because it has been reported that the Institut Francais du Pétrole has developed such a process with a new type of catalyst. Typical product distributions are reported to be ethanol, 40%; methanol, 23%; n-propanol, 20%; n-butanol, 14%; and isopropanol, 3% [82]. This type of chemistry is reminiscent of mixed alcohol by-product production as reported in some of the older literature on Fischer-Tropsch conversion of synthesis gas to hydrocarbons. With some improvement in selectivity, inexpensive ethanol could result from low-cost feedstocks and displace fermentation ethanol in fuel applications.

The near-term large-scale use of ethanol and possibly methanol fuels seems asured at this time. Technological advances will determine their marketability over the long term.

SUMMARY AND CONCLUSIONS

The status of the technology on energy from biomass and wastes in the United States has been reviewed. Research and development on biomass production, gasification and liquefaction have continued to cover all facets of the technology. Commercial technology currently includes fermentation ethanol plants; wood and municipal solid waste combustion plants for heat, steam and electric power production; small-scale anaerobic digesters for farm wastes; air-blown wood gasification units for low-Btu gas manufacture; moderate-scale methane recovery systems in wastewater treatment plants and solid waste landfills; and one municipal solid waste gasification plant. Near-term commercial projects include a small-scale thermochemical waste gasification plant and a moderate-scale anaerobic digestion plant for manure. Several fermentation alcohol and waste combustion plants and landfill methane systems are either in the construction or advanced design stages. Overall, the number of small- and moderate-scale commercial energy recovery systems fueled with biomass and wastes is increasing, and the use of nonfossil renewable energy in the form of biomass and wastes is expected to grow as the costs of fossil fuels increase.

REFERENCES

1. Office of Technology Assessment, "Energy From Biological Processes," (1980).
2. "Energy Potential From Biomass High, OTA," *Chem. Eng. News* 58:11 (1980).

3. Klass, D. L. "Energy From Biomass and Wastes; 1979 Update," in Symposium Papers—Energy From Biomass and Wastes IV Sponsored by IGT, Lake Buena Vista, FL, January 21-25, 1980, pp. 1-41.

4. "Energy in the Developing Countries," World Bank, Washington, DC (1980), p. 38.

5. "French Renewables Plan Approved," *Financial Times European Energy Report,* No. 58:19 (1980).

6. Ranney, J. W., and J. H. Cushman. "Silvicultural Options and Constraints in the Production of Wood Energy Feedstocks," Proceedings, Bio-Energy 80 April 21-24, 1980, pp. 101-103.

7. Steinbeck, K., and C. L. Brown. "Short-Rotation Coppice Forestry Research in Georgia: A Status Report," Proceedings, Bio-Energy 80, April 21-24, 1980, pp. 106-108.

8. Lipinsky, E.S., and S. Knesovich. "Sorghums as Energy Crops," Proceedings, Bio-energy 80, April 21-24, 1980, pp. 91-93.

9. James N. I. "Potential of Sugarcane as an Energy Crop," Proceedings, Bio-Energy 80, April 21-24, 1980, pp. 94-95.

10. Doney, D. L. "Sugarbeet-Fodder as a Source of Alcohol Fuel," Proceedings, Bio-Energy 80, April 21-24, 1980, pp. 95-97.

11. Huang, H. T. "Useful Chemicals and Materials from Arid Land Plants," Proceedings, Bio-Energy 80, April 21-24, 1980, pp. 98-100.

12. Stauffer, M. D., B. B. Chubey and D. G. Dorrell. "Growth, Yield and Compositional Characteristics of Jerusalem Artichoke as It Relates to Biomass Production," Symposium on Fuels from Biomass, Division of Fuel Chemistry, American Chemical Society, Washington, DC, preprints of papers presented at Las Vegas, NV 25(4):193-203 (1980).

13. Nemethy, E. K., J. W. Otvos and M. Calvin. "Plants as a Source of High Energy Liquid Fuels," Symposium on Fuels from Biomass, Division of Fuel Chemistry, American Chemical Society, Washington, DC, preprints of papers presented at Las Vegas, NV 25(4):216-218 (1980).

14. Alexander, A. G. "Herbaceous Land Plants as a Renewable Energy Source for Puerto Rico," paper presented at the Symposium on Fuels and Feedstocks from Tropical Biomass, Sponsored by the Center for Energy and Environment Research, University of Puerto Rico, San Juan, PR, November 24-25, 1980.

15. Alexander, A. G. "Biomass Production in Puerto Rico," paper presented at the Symposium on Fuels and Feedstocks from Tropical Biomass, Sponsored by the Center for Energy and Environment Research, University of Puerto Rico, San Juan, PR, November 24-25, 1980.

16. Klass, D. L. "Fuels From Biomass," in *Encyclopedia of Chemical Technology, Vol. III,* 3rd ed. (New York: John Wiley & Sons, Inc., 1980), pp. 334-392.

17. "Nitrogen Fixation Research Advances," *Chem. Eng. News* 58:29-30 (1980).

18. Phillips, R. E. "No-Tillage Agriculture," *Science* 208:1108-1113 (1980).

19. U.S. Dept. of Agriculture, "United States 1979 Crop Production Was Record High," Dept. of Agricultural Statistics, Purdue University, West Lafayette, IN, (1980).

20. Klass, D. L. "The Potential of Biomass Energy for Indiana," paper presented to Indiana Energy Resource Development Board, Indianapolis, IN, August 12, 1980.

21. Benemann, J. R. "Problems and Potential of Land-Based Aquatic Biomass Energy Systems," Proceedings, Bio-Energy 80, April 21-24, 1980, pp. 109-114.
22. Oswald, W. J., and D. M. Eisenberg. "Biomass Generation Systems as an Energy Source," Proceedings, Bio-Energy 80, April 21-24, 1980, pp. 123-126.
23. Tornabene, T.G. "Microbial Production of Hydrocarbons," Proceedings, Bio-Energy 80, April 21-24, 1980, pp. 491-492.
24. Wayman, M. Paper presented at the American Association for the Advancement of Science Annual Meeting, Toronto, Canada, January 3-8, 1981.
25. *Chicago Tribune* (January 5, 1981), p. 1.
26. Pirt, S. J., Y. K. Lee, A. Richmond and M. W. Pirt. "The Photosynthetic Efficiency of *Chlorella* Biomass Growth with Reference to Solar Energy Utilization," *J. Chem. Technol. Biotechnol.* 30:25-34 (1980).
27. Wolverton, B.C., and R. C. McDonald. "Vascular Plants for Water Pollution Control and Renewable Sources of Energy," Proceedings, Bio-Energy 80, April 21-24, 1980, pp. 109-114.
28. "Hyacinths Are Being Used to Treat Sewage Water," *Solar Energy Digest* 15:(1):3 (1980).
29. Pratt, D. C., and N. J. Andrews. "Wetland Energy Crops," Proceedings, Bio-Energy 80, April 21-24, 1980, pp. 115-119.
30. Ryther, J. H. et al. "Studies on Biomass and Biogas Production by Aquatic Macrophytes," Proceedings, Bio-Energy 80, April 21-24, 1980, pp. 130-133.
31. Flowers, A. et al. "Ocean Farms," Proceedings, Bio-Energy 80, April 21-24, 1980, pp. 464-486.
32. Flowers, A., and J. R. Frank. "Marine Biomass Energy Systems for the Production of Methane," Conference Proceedings, National Conference on Renewable Energy Technologies December 7-11, 1980, Honolulu, pp. 3-36, 3-37.
33. Berry R. I. "Ancient Fuel Provides Energy for Modern Times," *Chem. Eng.* 86:73-76 (1980).
34. "A Harvest of Trees That Will Fuel a Factory," *Business Week* (2624):110L (1980).
35. "PG&E Plugs into World's Largest Biomass Plant," *Energy Daily* (July 29, 1980), p. 4.
36. Hendon, J. "Biomass and Co-generation Are More Than Talk in California," *Oil Daily* (September 9, 1980).
37. Budiansky, S. "Bioenergy: The Lesson of Wood Burning," *Environ. Sci. Technol.* 14(7):769-771 (1980).
38. "Garbage Gets Hot as a Fuel Once Again," *Business Week* (May 19, 1980), pp. 44N,P, U.
39. Deutsch, D. J. "Energy from Garbage Tempts CPI Firms," *Chem. Eng.* 86:79,81 (1980).
40. Ganotis, C. G., Wheelabrator Fry Inc. Personal communication (January 5, 1981).
41. Klass, D. L. "Anaerobic Digestion for Methane Production—A Status Report," Proceedings, Bio-Energy 80, April 21-24, 1980, pp. 143-149.
42. Ghosh, S., M. P. Henry and D. L. Klass. "Bioconversion of Water Hyacinth-Coastal Bermuda Grass-MSW-Sludge Blends to Methane," Proceed-

ings of the 2nd Symposium on Biotechnology in Energy Production and Conservation, Gatlinburg, TN, October 3-5, 1979, Biotech. and Bioengr. Symposium No. 10 (New York; John Wiley and Sons, Inc., 1980), pp. 163-187.

43. Klass, D. L., and S. Ghosh. "Methane Prduction by Anaerobic Digestion of Water Hyacinth (*Eichhornia crassipes*)," Preprints, ACS Div. Fuel Chem., 25(4):221-232 (1980).

44. van den Berg, L., and C. P. Lentz. "Effects of Film Area-to-Volume Ratio, Film Support Height and Direction of Flow on Performance of Methanogenic Fixed Film Reactors," paper presented at the Seminar on Anaerobic Filters, U.S. Department of Environment, Harvey-in-the-Hills, FL, January 9-10, 1980.

45. van den Berg, L., and C. P. Lentz. "Anaerobic Waste Treatment Efficiency Comparisons between Fixed Film Reactors, Contact Digestors and Fully Mixed Continuously Fed Digestors," paper presented at the 35th Purdue Industrial Waste Conference, May 13-15, 1980.

46. van den Berg, L., and C. P. Lentz. "Performance and Stability of the Anaerobic Contact Process as Affected by Waste Composition, Inoculation and Solids Retention Time."

47. Jewell, W. J. "Development of the Attached Microbial Film Expanded Bed Process for Aerobic and Anaerobic Waste Treatment," University of Manchester Institute of Science and Technology, Hertfordshire, England, April 14-17, 1980.

48. "Fuel From Biomass and Wastes," *IGT* Gas Scope (51) (1980).

49. Wright, J. "Energy From Biomass," *Brown Caldwell Newsl.* 9(1):9 (1980).

50. Bode, D. "Cutting Costs With Digester Gas," *Diesel Gas Turbine Prog.*" (February 1980), pp. 42-43.

51. "Landfill Gas Development Offers Significant Supplemental Supply," *Gas Ind.* (October 1980), pp. 31-32,34.

52. Wander, T. J., and R. V. Griffin. "The Gas Industry Perspective on Methane From Renewable Sources," Proceedings, Bio-Energy 80, April 21-24, 1980, pp. 238-241.

53. Martinell, R. "VRYMETHANE, Bacteriological In Situ Gasification of Peat to Methane," Vyrmetoder AB, Sweden (1980).

54. "Power Generation Progress," *Diesel Gas Turbine Prog.*, XLVI(5):15 (1980).

55. Varani, F. T. "Materials Handling in Anaerobic Digestion Systems," Proceedings, Bio-Energy 80, April 21-24, 1980, pp. 140-142.

56. "Fuel-From-Manure Firm Shifts to Animal Feed," *Chem. Eng. News* 58:6 (1980).

57. Hashimoto, A. G., Y. R. Chen and R. P. Prior. "Thermophilic Anaerobic Fermentation of Beef Cattle Residue," in Symposium Papers, Energy from Biomass and Wastes sponsored by IGT, August 14-18, 1978, Washington, DC, pp. 379-402.

58. van der Meer, R. R. et al. "The Upflow Reactor for Anaerobic Treatment of Wastewater Containing Fatty Acids," 34rd International Congress on Industrial Wastewater and Wastes, Stockholm, Sweden, February 1980.

59. Sax, R. I., M. Holtz and R. C. Pette. "Production of Biogas From Wastewaters of Food Processing Industries," 2nd Annual Conference on Industrial Energy Conservation Technology, Houston, Texas, April 13-15, 1980.

60. "Biogas From Your Wastewater," Joseph Oat Corp., Camden, NJ (1980).

61. Jewell, W. J. et al. "Anaerobic Fermentation of Agricultural Residue Potential for Improvement and Implementation," Final Report for DOE Under Grant EY-76-S-02-2981 (1978).
62. Pitt, W. W., and R. K. Genung. "Energy Conservation and Production in a Packed-Bed Reactor," in Symposium Papers, Energy From Biomass and Wastes Sponsored by IGT, January 21–25, 1980, Lake Buena Vista, FL, pp. 451–472.
63. Mitchell, D. H. et al. "Methane/Methanol by Catalytic Gasification of Biomass," *CEP* (September 1980), pp. 53–57.
64. Beck, S. R., and M. J. Wang. "Wood Gasification in a Fluidized Bed," *Ind. Eng. Chem. Process Des. Devel.* 19(2):312–317 (1980).
65. Raman, K. P., W. P. Walawender and L. T. Fan. "Gasification of Feedlot Manure in a Fluidized Bed," Preprints, ACS, Div. Fuel Chem., 25(4); 233–244 (1980).
66. "Coal/Biomass Gasifier Lab Tests Are A Success," *Chem. Eng. News* 58:28 (1980).
67. Mark, Jr., S. D. "Development and Commercialization of Slagging Pyrolysis Process for Conversion of Solid Waste to Energy," Proceedings, Bio-Energy, April 21–24, 1980, pp. 177–180.
68. "Wood Gasification System, Northwest Georgia Regional Hospital," Press Kit, Applied Engineering Co., Orangeburg, SC, October 23, 1980.
69. Bozdech, S. L. "Use of Corn Cobs for Seed Drying Through Gasification," Preprints, ACS Div. Fuel Chem. 25(4):251–256 (1980).
70. Bailie, R. C. "Economic Justification for Small-Scale Biomass Fueled Electrical Generating Systems," Proceedings, Bio-Energy 80, April 21–24, 1980, pp. 186–190.
71. *Biomass Refining Newsletter* (Summer 1980) pp. 19–24, 32, 43–44.
72. Hansen, S. M., and G. C. April. "Chemical Feedstocks From Wood: Aqueous Organic Alcohol Treatment," Preprints ACS Div. Fuel Chem. 25(4):319–326 (1980) August 24–29.
73. "Lignin Conversion Process Shows Promise," *Chem. Eng. News* 58:35–36 (1980).
74. Soltes, E. J. "Pyrolysis of Wood Residues," *Tappi* 63(7):75–77 (1980).
75. "Processes Promising for Cellulose Pyrolysis," *Chem. Eng. News,* 58:26–28 (1980).
76. Atterbury, A., County of San Diego, Dept. of Sanitation and Flood Control, Personal communication (January 5, 1981).
77. Elliott, D. C. "Process Development for Biomass Liquefaction," Preprints, ACS Div. Fuel Chem. 25(4):257–263 (1980).
78. "New Zealand's Plan to Turn Gas to Gasoline," *Business Week* (July 14, 1980) p. 50H.
79. Kam, A. Y. "Hydrocarbon Liquids and Heavy Oil From Biomass: Technology and Economics," in Symposium Papers, Energy From Biomass and Wastes IV Sponsored by IGT, Lake Buena Vista, FL, January 21–25, 1980, pp. 589–615.
80. "Plant Materials Tested as Hydrocarbon Source," *Chem. Eng. News* 58:43 (1980) September 15.
81. Klass, D. L. "Alcohol Fuels For Motor Vehicles: An Overview," *Energy Topics* (April 14, 1980).
82. Kampen, W. H. "Engines Run Well on Alcohol," *Hydrocarbon Processing* (February 1980), pp. 72–75.

SECTION 1

BIOMASS PROCUREMENT AND PRODUCTION

CHAPTER 2

PROCUREMENT PROBLEMS AND
THEIR SOLUTIONS FOR LARGE-SCALE
WOOD-FUELED FACILITIES

J. Phillip Rich and Sandy Thomas
  J. P. R. Associates, Inc.
  Stowe, Vermont

When a facility that has not previously used harvested wood begins to use wood on a large scale, it has very different procurement needs and requirements from an established wood user. As industry looks to wood for fuel, procurement of the required amount is a major consideration. Five areas stand out as crucial in planning for harvested wood fuel use as well as in the actual procurement of the fuel.

BIOMASS INVENTORY

As with any new industry, a survey of the availability of raw material (in this case, the fuel) is necessary. This presents problems for any prospective user of forest biomass.* Forest inventories have historically been measured in merchantable bole material, which is to say the trunks or stems of certain species of trees used in extant manufacturing processes. The U.S. Forest Service continues to measure forests in this way. With the advent of the fuel market, trees that were previously overlooked took

---

*Forest biomass, when used in relation to wood fuel use, ordinarily refers to the trees or parts of trees that are unmerchantable, but that can be removed for use as fuel. Included are thinnings, culls, rough and rotten trees, and unmerchantable species, as well as tops and limbs from conventional harvest operations.

on a value. Data on the amount of material in nonmerchantable forest biomass and its growth rates are sadly lacking in all parts of the world.

HARVEST LANDS

Established industrial plants using large quantities of wood have ordinarily acquired sufficient landhold from which to draw the majority of their fiber, or at least enough to even out periods when other harvest lands may not be available or suitable for one reason or another. A controlled landhold also allows for continuing forest management and maximum utilization. A new user is not likely to possess large areas of forest lands.

FORESTRY SUPERVISION

Pulp manufacturers and other large wood users have a full-time forestry staff that not only manages their own lands, but makes sure that there are other lands available for harvest and that there are loggers to work on them. Most manufacturing plants or power plants using coal, oil or gas have only to lift the phone when supplies of fuel get low, and the fuel arrives in a short time. This is not the case with wood. Supply has to be assured by careful planning, management, and storage.

HARVEST CONTRACTORS

A new user will not have an established network of loggers that could be encouraged to expand production or go into whole-tree chip harvesting. The new user will have to plan carefully to assure an orderly startup and a continuing and sufficient supply of fuel.

PUBLIC OPINION

In addition to the more or less mechanical problems assuring supply, public resistance to harvesting forest biomass, particularly in areas where logging has not been a visibly established practice, is sure to appear. For the new or prospective wood fuel user, this is a serious problem, but it can be controlled with careful and early attention. Public opinion must be taken into consideration even before the project is on the drawing boards and must be of paramount concern in planning and procurement.

METHODS

During the past six years, we have worked with the Vermont state government, the U.S. Department of Energy (DOE), Rust Engineering, the Burlington Electric Department (Vermont) and several other private industrial concerns on the study of and planning for large-scale use of wood for fuel, primarily in New England. After the first flush of excitement passed, the problems in planning for and using this "renewable local resource" had to be faced.

A thorough and continuing review of relevant literature has been the main source of information in planning large-scale wood-fuel use. In addition, visits to facilities already using large volumes of wood (primarily pulp and paper companies), onsite inspections of harvest operations, attendance at conferences and demonstrations, and close contact with people working in the field of forest biomass for energy have provided the most up-to-date information and thinking on the use of wood for energy and related subjects, as well as the cross- pollination of ideas necessary to deal with the problems of developing a new use for an unused resource.

RESULTS AND DISCUSSION

Biomass Inventory

The first question that invariably arises when the use of wood for fuel on a large scale is proposed is whether there is enough wood. This has turned out to be one of the most difficult questions to answer with any degree of accuracy.

At the present cost of competitive fuels, a harvest radius of about 50 miles would appear to be maximum. This will increase or decrease, depending on competing prices as well as the cost of the transportation itself. When the viable harvest radius is estimated, a second estimate of forest land within the radius is required.

The U.S. Forest Service surveys have acreage figures for forest lands on a county or regional basis for all states. However, not all of the commercial land is physically accessible for harvest, particularly if mechanized removal of whole-tree chips is the primary harvest method planned. Road access, slope, wetness, soil types and rockiness can be limiting factors to an economically feasible harvest of wood for fuel. If the total acreage is marginal within the harvest area, or if in further development of biomass estimates, it appears there might not be enough fuelstock available, a careful study of the profile of the potentially available forest acreage will have to be undertaken. (To date, in Maine, New Hampshire and Vermont, where facilities have been proposed, this has not been a serious problem because of the vast acreage and biomass potential.)

Having ascertained the amount of forest land available, the next step is to estimate the standing forest biomass. This is a highly conjectural process. As has previously been stated, the U.S. Forest Service surveys are primarily concerned with merchantable bole material, although they do include inventories of rough, rotten and cull trees. Unfortunately, the only comprehensive inventories of commercial forest lands are those done by the U.S. Forest Service. To determine biomass from conventional forest surveys, it is necessary to resort to whole-tree weight tables such as those developed for Maine [1] and New York State [2]. In addition to the amount of forest biomass that is not included in the inventories, the U.S. Forest Service surveys are scheduled only once every ten years, and it is ordinarily two or three years after the inventories that the results are published. With widespread, large-scale use of wood for fuel looming on the horizon, these surveys are becoming less and less useful.

From these barely adequate (for the purposes of wood fuel estimation) surveys, it is possible to extrapolate gross biomass estimates. It has been our fortunate experience to find these estimates useful, but only because of the plethora of material they turned up. For instance, a study of biomass availability for the Burlington Electric Department showed enough standing rough and rotten within 50 miles of Burlington to provide fuel to the plant for 56 years, and enough annual growth to fuel 10 such plants using 0.5 million tons/yr [3]. If there is any question of the amount of biomass available from gross estimates, the feasibility of using wood for fuel must seriously be reconsidered. Achieving accurate biomass assessments with the present forest inventory and biomass estimating procedures is simply impossible. The cost of achieving reasonably accurate state or regional biomass inventories is probably prohibitive without substantial government funding.

In addition to standing biomass, it is necessary to estimate growth. Modern forestry management practices (and public demands) require a "sustained yield" approach to wood harvesting, which is to say that no more is removed from the forest than can be expected to grow before the next harvest is undertaken. Therefore, estimates of biomass growth are necessary to determine the continuing availability of fuel. Again, there are no really good estimates of biomass growth, although the U.S. Forest Service supplies growth figures for certain portions of U.S. forests. Unless a study has been conducted on biomass growth rates for a particular area, such as the studies for New Hampshire [4], Maine [5], and the oak-pine forest in Brookhaven, Long Island [6], it will be necessary to "massage" the U.S. Forest Service inventories to achieve gross estimates, which may be used to satisfy questions raised by both the public and the government offices involved in granting permits. If there is any question of the amount available, money is going to have to be spent and on-the-ground research conducted. Also, other current and projected uses for the biomass have to be taken into consideration in planning, both in terms of price and availability.

Harvest Lands

Ordinarily, a new wood user does not possess forest lands in sufficient quantity to supply even a relevant portion of fuel needs. There are two ways of approaching this problem: purchase suitable lands, or solicit harvest sites from private or public lands. It is possible to combine the two approaches, both of which have advantages and disadvantages.

The obvious disadvantage of land purchase is the capital investment involved. When this is added to the investment required for the facility, total investment can become prohibitive. Additionally, there are the ongoing costs of management and taxes. On the other hand, when there is no or little landhold, harvest site procurement is entirely dependent on public willingness to allow harvesting, and this requires careful supervision of harvesting operations. Traditionally, the large landholdings of wood users have allowed them to manipulate prices paid for wood products removed.

If a company is able to purchase sufficient lands, and lands are available, they have solved their stumpage procurement problem. If this is not the case (a far more likely possibility), the prospective user is going to have to launch an extensive campaign to assure landowners of their intent to improve the forests, rather than decimate them, and to encourage forest management. The use of a tree farm system can be an effective approach, as can a landowner assistance program, which differs from the tree farm system in that the landowner makes a commitment to sell to the sponsoring wood user. It would appear that the commitment is unnecessary in securing the product [7,8] and might even prove a deterrent to landowner cooperation.

Once the harvest is underway, the using industry will have to supervise all harvest operations and assure that they are within acceptable parameters for conservation, esthetics, wildlife habitat, recreational use or whatever other values are held by the landowners and users of the lands. The Burlington Electric Department, for instance, has stated unequivocally to its harvested wood suppliers that their jobs will be inspected at least weekly and that if they do not meet the foresters' approval, the chips will not be purchased by Burlington Electric. This is a strong revolutionary position that is only possible because Burlington Electric is the only user of whole-tree chips in the immediate area. However, other users will have to make similar commitments and find ways of enforcing them. This will become more and more vital as pressure on the resource increases and signs of harvesting become more visible.

Forestry Supervision

To have sufficient harvest lands and to supervise harvest operations, a staff of trained forestry personnel is required. If a facility is going to use wood that is harvested from the forest, the foresters are going to have to assume responsibility for bringing it in. Foresters will have to negotiate with landowners, support and consult with loggers, mark timber, prepare road designs, plan harvest sites months in advance, and take climatic conditions and fluctuating supplies into account. They will be responsible for the flow of wood into the facility and for the maintenance of good relations with landowners, the public and the wood harvesters. The size of the forestry staff is dependent on the amount of fuel needed, the type of wood that is harvested and the amount of work required to get sufficient harvest acreage and fuel into the plant. For whole-tree chip harvesting, a rule of thumb for a new operation is one forester per whole-tree chipper.

Harvest Contractors

Sufficient harvest contractors must be in operation to supply plant needs on a year-round steady basis. Since New England, for example, has areas where harvesting is not possible from four to eight or even more weeks out of the year, it is necessary for the plant to have sufficient storage space to sustain the plant during these periods.

The wood fuel that is most economical to harvest, transport and handle is whole-tree chips. If this is the fuel that is selected, it is necessary to have harvest contractors who can produce it. This requires considerable capital investment ($200,000–500,000) for a mechanized system, and from six months to a year to reach full production [9,10]. This means the wood fuel user has to be prepared to aid in securing financing and in providing a market for some production of chips before the wood-fueled plant begins operation. Another possibility is to begin operating at less than full capacity until chip production becomes sufficient to operate at full capacity. Unlike many conventional loggers who operate with a chainsaw, venerable skidder and aged log truck, the whole-tree chip operator has a considerable investment in equipment, so it is imperative that the whole-tree chip operation run full tilt on a daily basis, and that there be a market for the production at a fair price. There is little latitude to manipulate the market for whole-tree chips and keep a reliable stable of harvest operators.

This is an extremely sensitive area where the forestry staff must function cooperatively and fairly, carefully balancing the requirements of the wood-fueled facility for the most inexpensive fuel supply with the requirements of the wood harvester for a reasonably profitable operation. The question of the viability of whole-tree chip harvesting has at least been settled to some extent by the fact that there are now several successful independent operators in the Northeast who have been working for several years.

Public Opinion

There is probably no new industry or building contemplated that does not encounter "public opinion," both favorable and unfavorable. This is particularly true of a new wood-fueled facility, and unfortunately the first response tends to be negative. Joyce Kilmer has planted his seedlings in the hearts of the American public, and the tree has become nearly as sentimental a symbol as motherhood, hot dogs and apple pie. It appears that a massive education campaign is required to demonstrate that cutting down some trees allows others to grow better and that trees do grow. As simple and obvious as these two statements may seem, the task of convincing an environmentally aroused public is neither simple nor obvious.

There is the ever-present and not entirely unfounded fear that the ground and a barren landscape will appear overnight. It is the responsibility of the prospective wood users to ensure that this will not happen, and to convince the public that they will live up to the assurances. It is important that this education process begin before rumors start. Once the word is out that wood is going to be cut for fuel, the fear has already set in. It is far easier to begin with a positive and strong statement of intent than to have to respond to public criticism that has already begun to snowball. A forester with a strong conservation ethic and a good public image should be hired at the onset of the project to work with the public and with the prospective user in planning wood procurement. This is considerably less expensive than having to fight a rearguard action against an organized resistance.

CONCLUSIONS

Any facility planning to use wood for fuel must meet certain criteria in the planning process. The owner-operator must first ascertain whether the necessary wood is available, and then determine whether it can be obtained. In addition, a prospective wood user must take into consideration public opinions and attitudes and must begin as early as possible to influence the public in a positive manner toward the proposed facility.

Unlike distant fuel supplies, the "mining" of wood occurs in the backyard of the user, and it is vital that the removal be undertaken responsibly and carefully. Supervision by trained forestry personnel is essential.

REFERENCES

1.   Young, H. E., L. Strand and R. Altenberger. "Preliminary Fresh and Dry Weight Tables for Seven Tree Species in Maine," Maine Agricultural Experiment Station Technical Bulletin No. 12, University of Maine (1964).
2.   Monteith, D. B. "Whole Tree Weight Tables for New York," AFRI Research Report No. 40, Applied Forestry Research Institute, State University of New York, Syracuse, NY (1979).
3.   Hughes, G. A., "The Potential Average Yield of Wood Fiber per Harvested Acre within 75 Miles of Burlington, Vermont," report prepared for J.P.R. Associates Inc., Stowe, VT (1979).
4.   Yaeger, H., Westvaco, Covington, VA. Personal communication (July 12, 1978).
5.   Young, H. E., J. H. Ribe and D. C. Hoppe. "A Biomass Study of the Thinning Potential and Productivity of Immature Forest Stands in Maine," Bulletin No. 758, University of Maine, Orono, ME (1979).
6.   Whittaker, R. H., and G. M. Woodwell. "Dimensions and Production Relations of Trees and Shrubs in the Brookhaven Forestry, N.Y.," *J. Ecol.* 56:1–25 (1968).
7.   Clement, D. B. and S. D. Forester, S. D. Warren Company, Westbrook, ME. Personal communication (August 2, 1979).
8.   Lewis, R., National Tree Farm System Program, American Forest Institute, Washington, DC. Personal communication (January 9, 1980).
9.   Brown, S. A., and N. Liggett. Mead Paper Company, Chillicothe, OH. Personal communication (July 10, 1979).
10. Whittaker, R. H., F. H. Bormann, G. E. Likens and T. G. Siccama. "The Hubbard Brook Ecosystem Study: Forest Biomass and Production," *Ecol. Monog.* 44:223–254 (1974).

CHAPTER 3

A COMPARISON OF THE ENERGY EFFICIENCY OF INTENSIVE AND EXTENSIVE HYBRID POPLAR PRODUCTION SYSTEMS

Dietmar W. Rose, Barbara A. Walker, Karen Ferguson and David Lothner
College of Forestry
University of Minnesota and
North Central Forest Experiment Station
U.S. Department of Agriculture Forest Service
Duluth, Minnesota

Will the amount of energy invested in hybrid poplar plantations exceed the amount of energy they produce? How do energy output/input ratios for extensive management systems compare to the ratios for intensive management systems? Previously, we examined the economic attractiveness of hybrid poplar cultures for industrial users and found irrigated, intensive cultures economically unattractive and nonirrigated crops only marginally attractive [1]. Due to constantly changing economic conditions, i.e., cost for production inputs and prices for energy and for wood, the economic attractiveness of such systems is subject to change in the future. Another factor is the energy balance between production inputs and outputs. This study examines the energy efficiency of, and draws comparisons between, intensive and extensive fiber production systems.

STUDY DESCRIPTION

Methods

Specific production systems were identified through consultation with U.S. Forest Service researchers at the North Central Forest Science Laboratory, Rhinelander, WI, agriculture extension specialists and M. Morin, the forester in charge of what is currently the only industrial large-scale hybrid poplar intensive culture. (Packaging Corporation of America, Filer City, MI). Energy inputs and outputs of these production systems were estimated and evaluated using cashflow techniques and two performance measures. The first was the net present energy value (NPEV) produced by a production system, i.e., the difference between discounted energy outputs and inputs; four discount rates from 0 to 15% were used for this measure. The rationale for discounting will be explained below. The second measure was the ratio of (discounted) energy yields and energy inputs, or the benefit/cost ratio (B/C). The sensitivity of these two measures to changes in energy inputs and/or outputs was evaluated.

Description of Alternative Production Systems

The study compared four intensive production systems and one extensive system. The intensive alternatives were chosen to represent a likely range of spacings, rotations and cultural practices. Two spacings [1.2 × 1.2 m (4 × 4 ft) and 2.4 × 2.4 m (8 × 8 ft)] and three rotations (5, 10 and 15 years) were chosen; irrigation, fertilization and drying were treated as options (Table I).

The size of operation for each intensive alternative was 405 ha (1000 ac) of cleared, marginal agricultural land arranged in 10 tracts of 32–48 ha (80–120 ac) each. Site preparation for these hypothetical plantations included plowing, disking and preplanting herbicide treatment of Round-up and Simazine. All these activities were assumed to take place the late summer or fall prior to spring planting. Postplanting establishment activities were cultivation (three times) and two more applications of Simazine. The method of irrigation chosen was a traveling gun system (one system per tract) applying 10 effective inches per acre annually. Fertilization included only nitrogen additions and was applied at an annual rate of 247 lb/ha (100 lb/ac). Harvesting methods were whole-tree chipping for all rotations greater than five years, and forage harvesting for the five-year coppice rotations.

Table I. Description of Wood Production Alternatives

Alternative	Rotation	Metric ton/ ha-yr	Dry ton/ ac-yr
1.2 × 1.2 m spacing (4 × 4 ft spacing) irrigated, fertilized	10 yr for orig. stock plus four 5-yr coppicings	6.30 7.20 each	14.12 16.14
2.4 × 2.4 m spacing (8 × 8 ft spacing) irrigated, fertilized	15 yr for orig. stock plus one 15-yr coppice	3.20 3.60	7.17 8.07
1.2 × 1.2 m spacing (4 × 4 ft spacing) nonirrigated, fertilized	10 yr for orig. stock plus four 5-yr coppicings	3.15 3.60 each	7.06 8.07
2.4 × 2.4 m spacing (8 × 8 ft spacing) nonirrigated, fertilized	15 yr for orig. stock plus one 15-yr coppice	1.60 1.80	3.59 4.03
1.2 × 1.2 m spacing (4 × 4 ft spacing) nonirrigated, nonfertilized	10 yr for orig. stock plus four 5-yr coppicings	2.52 2.88 each	5.65 6.46
2.4 × 2.4 m spacing (8 × 8 ft spacing) nonirrigated, nonfertilized	15 yr for orig. stock plus one 15-yr coppice	2.02 2.30	4.53 5.15
Extensive aspen management (Site Index 60)	45 yr	1.41	3.16

Yields of irrigated and fertilized systems are based on Ek and Dawson [2]. Yields in the no-irrigation, fertilization alternative are assumed to be 50% of the irrigation-fertilization alternatives. For the no-irrigation, no-fertilization alternatives, yields were reduced by 20% over the nonirrigation, fertilization alternatives.

The extensive management alternative was based on management of natural aspen stands under lake states conditions. Energy inputs are incurred only in site preparation (heavy disking), harvesting, hauling and drying of the wood. Site index was assumed to be 60, and one harvest occurs at age 45 with a yield of 235 cord/ha (95 cord/ac) using 50% higher values than reported [3] to account for full-tree harvesting.

Assumptions used in estimating the energy inputs and outputs are listed in Table II. The gross heat value of hybrid poplar is reported [4] to be 19,525 MJ/metric ton (16.8 MBtu/dry ton), a weighted average of the heat content for stem and branch wood. The gross heat value of wood chips assumed at 100% moisture content (dry basis) was reduced by 65%

Table II(A).   Derivation of Energy Inputs: Site Preparation

Alternatives	Year	Operation	Production Rate[a] ha/hr	ac/hr	Total Hours[b]	
Intensive	0	Plow	2.65	6.55	137.4	
	0	Disk	4.06	10.04	89.6	
	0	Round-up & Simazine	5.74	14.18	63.5	
	1	Cultivation (3×)	3.68	9.09	297.0	(3×)
	1	Simazine	5.74	14.18	63.5	
	2	Simazine	5.74	14.18	63.5	
Extensive	0	Heavy disk	0.81	2.00	450[c]	

Table II(B).   Derivation of Energy Inputs: Planting

Spacing	Equipment	Production Rate[a] ha/hr	ac/hr	Total Hours[d]
1.2 × 1.2 m (4 × 4 ft)	Three 225-hp tractors, six planters	0.30	0.75	1200
2.4 × 2.4 m (4 × 4 ft)	Two tractors, four planters	0.81	2.00	450

Table II(C).   Derivation of Energy Inputs: Harvesting[e]

Machine	Production Rate (green metric ton/hr) 1.2 × 1.2 m	2.4 × 2.4 m	Fuel Consumption (liter/hr)	liter/green metric ton 1.2 × 1.2 m	2.4 × 2.4 m
Med. Skidder		13.3	12.5		0.8
Two Small Skidders	11.2		15.9	1.3	
Feller Buncher	19.2	63.8	17.8	0.8	0.2
Chipper Baler	11.2	13.3	29.0	2.3	2.0
Total				4.4	3.1
(liter/dry metric ton)				8.9	6.1
(gal/dry ton)				2.4	1.6

Table II(D).   Derivation of Energy Inputs: Hauling[f]

40-km haul, loaded 16 km of class II county roads × 0.647 liter/km =	10.4 liters
One way, unloaded 16 km of class II county roads × 0.257 liter/km =	4.2 liters
Loaded, 24 km of class I paved road × 0.526 liter/km =	12.7 liters
Unloaded, 24 km of class I paved road × 0.284 liter/km =	6.8 liters
Total	34.1 liters

34.1 liter/10.9 metric tons per van = 3.1 liter/dry metric ton (0.75 gal/dry ton)

Table II(E).   Derivation of Energy Inputs: Irrigation

Equipment Used: one traveling gun system per 32- to 48-ha tract and 30.5-m well; 2127 liter/min sprinkler system; turbine pump; diesel power unit

Energy Consumption: 636.1 liter/ha-yr (68 gal/ac-yr) of diesel fuel; lubrication and repairs included as percentages of fuel costs and initial investment costs, respectively[g]

[a]From Benson [6].
[b]364 workable ha/production rate (900 workable ac/production rate); fuel consumption per hour = 0.16 liter/hp-hr × 225 hp = 37 liter/hr (9.9 gal/hr).
[c]0.16 liter/hp-hr × 100 hp = 16.6 liter/hr (4.4 gal/hr).
[d]Total hours for each equipment group. Each equipment group (tractor and two planters) works simultaneously; total hours is the additive number of hours for each group.
[e]From Mattson [7].
[f]From Aube [8].
[g]From Reference 9.

to account for the boiler efficiency in burning green wood [5]. For dried wood (20% moisture content), an 80% boiler efficiency was assumed.

Only direct energy expenditures (fuel and chemicals) were considered in the energy analysis. Energy expended in manufacturing equipment and in labor was not considered as it is an insignificant part of the total energy picture.

Energy inputs for site preparation and establishment include fuel and herbicides. Fuel energy was calculated from the total hours spent and the fuel consumption rates (see Table II), using a conversion factor of 38.44 MJ/l (138,000 Btu/gal) of diesel. Herbicides were estimated to contain 102 MJ/kg (11,000 kcal/lb) [10].

Fuel was the only energy input accounted for in planting and irrigation, calculated in the same way as above, using information from Table II.

Energy content of nitrogen fertilizer is estimated to be 77 MJ/kg (8400 kcal/lb), or for 112 kg/ha (100 lb/ac), 8676 MJ/ha (3.33 MBtu/ac).

Harvest fuel consumption/dry metric ton was estimated using information from the simulation of a whole tree harvesting system for intensively grown poplar [7]. For lack of a better estimate, it was assumed that forage harvesting would take the same amount of energy (see Table II).

An 80-km (50-mile) round trip over 16 km (10 miles) of good gravel road and 24 km (15 miles) of average paved road using a 40-ft van holding 10.9 metric tons (12 dry tons) of chips was used to estimate fuel consumption/dry ton of hauling (see Table II).

An estimated 3700 MJ/metric ton (3.184 MBtu/dry ton) is used to dry wood chips [11].

RESULTS

The discussion will focus first on the performance of the alternatives in terms of NPEV and B/C under the stated energy input and output assumptions. Second, changes in these two measures due to a change in energy inputs and/or outputs will be discussed utilizing results of sensitivity analyses. Results of both these analyses are shown together in one table each for the four discount rates utilized (Tables III to VI).

It is apparent that all alternatives produce positive net energy returns (NPEV). Irrigated systems produce higher net energy returns than nonirrigation alternatives. For the nonirrigated alternatives, the ones with fertilizer application resulted in higher NPEV than the ones without fertilizer application. The smaller spacing [1.2 × 1.2 m (4 × 4 ft)] in all cases is superior to the wider spacing [2.4 × 2.4 m (8 × 8 ft)] in terms of NPEV. The extensive alternative has the lowest NPEV.

These results are almost exactly reversed for the B/C ratios. The least intensive alternatives have the highest B/C ratios. All production systems have positive B/C ratios, indicating that they produce more energy than they consume. Extensive, nonirrigated and nonfertilized alternatives are preferable on the basis of the B/C ratio.

Results change when discounting is introduced. Our rationale for discounting energy flows is that it provides a means for comparing the timing and risk involved in using energy inputs of known practical value (petroleum, electricity) for different energy production schemes. For example, a barrel of oil can be invested today into producing more energy in the form of equipment for mining or for growing trees. Certainly the timing and risks of energy outputs for the same energy inputs are different and need a common basis for comparison. For those who do not agree with this rationale, energy inputs and outputs without discounting should be compared.

At relatively low discount rates, the results remain about the same as before, i.e., the more intensive alternatives have the higher NPEV and

Table III.   Energy Performance (MJ/ha) of Wood Production Alternatives, and Sensitivity of Discounted Energy Inputs and Yields to Changes in Inputs and Yields[a]

	Alternative						
	1.2 × 1.2 m (4 × 4 ft)[b]	2.4 × 2.4 m (8 × 8 ft)[b]	1.2 × 1.2 m (4 × 4 ft)[c]	2.4 × 2.4 m (8 × 8 ft)[c]	1.2 × 1.2 m (4 × 4 ft)[d]	2.4 × 2.4 m (8 × 8 ft)[d]	Extensive
NPEV[e]	4,660,268	4,221,344	2,561,163	1,343,356	2,255,660	2,082,013	1,851,515
B/C[f]	4.79	4.65	7.68	7.79	23.53	31.25	47.81
NPGEV[g]	5,889,237	5,377,130	2,944,619	2,688,565	2,355,757	2,150,847	1,890,963
NPEI[h]	1,228,970	1,155,785	383,455	345,209	100,098	58,833	39,549
Change in Discounted NPGEV and/or NPEI due to a 10% Change in Input Factors and Yield							
Site Preparation	339	287	339	287	339	287	78
Planting	417	156	417	156	417	156	
Irrigation	73,002	73,002					
Fertilization	26,027	26,027	26,027	26,027			
Whole-Tree Harvesting	5,341	11,021	2,657	5,497	2,136	4,403	1,928
Forage Harvesting	12,193		6,097		4,872		
Hauling	5,575	5,106	2,788	2,553	2,241	2,032	1,954
All Inputs	122,894	115,599	38,325	34,520	10,005	6,878	3,960
Yield	588,913	537,718	294,457	268,846	235,576	215,098	189,096

[a] Discount rate, 0%; boiler efficiency, 65%.
[b] Irrigated and fertilized.
[c] Nonirrigated and fertilized.
[d] Nonirrigated and nonfertilized.
[e] NPEV = discounted net energy (NPGEV − NPEI).
[f] B/C = benefit/cost ratio (NPGEV/NPEI).
[g] NPGEV = discounted gross energy yield adjusted for boiler efficiency.
[h] NPEI = discounted energy inputs.

Table IV.   Energy Performance (MJ/ha) of Wood Production Alternatives, and Sensitivity of Discounted Energy Inputs and Yields to Changes in Inputs and Yields[a]

	Alternative						
	1.2 × 1.2 m (4 × 4 ft)[b]	2.4 × 2.4 m (8 × 8 ft)[b]	1.2 × 1.2 m (4 × 4 ft)[c]	2.4 × 2.4 m (8 × 8 ft)[c]	1.2 × 1.2 m (4 × 4 ft)[d]	2.4 × 2.4 m (8 × 8 ft)[d]	Extensive
NPEV	1,904,902	1,346,289	1,069,131	791,401	960,540	738,955	205,302
B/C	4.10	3.37	6.62	5.76	21.49	28.19	40.80
NPGEV	2,518,462	1,915,323	1,259,244	957,675	1,007,411	766,129	210,460
NPEI	613,560	569,034	190,112	166,221	46,870	27,174	5,159
Change in Discounted NPGEV and/or NPEI due to a 10% Change in Input Factors and Yield							
Site Preparation	339	287	339	287	339	287	78
Planting	417	156	417	156	417	156	
Irrigation	37,413	37,413					
Fertilization	13,340	13,340	13,340	13,340			
Whole-Tree Harvesting	3,283	3,934	1,641	1,954	1,303	1,563	208
Forage Harvesting	4,221		2,110		1,693		
Hauling	2,397	1,824	1,198	912	964	729	208
All Inputs	61,410	56,954	19,045	16,649	4,716	2,735	498
Yield	251,859	191,545	125,917	95,773	100,749	76,623	21,051

[a]Discount rate, 5%; boiler efficiency, 65%.
[b]Irrigated and fertilized.
[c]Nonirrigated and fertilized.
[d]Nonirrigated and nonfertilized.

Table V. Energy Performance (MJ/ha) of Wood Production Alternatives, and Sensitivity of Discounted Energy Inputs and Yields to Changes in Inputs and Yields[a]

			Alternative				
	1.2 × 1.2 m (4 × 4 ft)[b]	2.4 × 2.4 m (8 × 8 ft)[b]	1.2 × 1.2 m (4 × 4 ft)[c]	2.4 × 2.4 m (8 × 8 ft)[c]	1.2 × 1.2 m (4 × 4 ft)[d]	2.4 × 2.4 m (8 × 8 ft)[d]	Extensive
NPEV	874,746	458,568	507,652	301,048	470,135	305,451	24,568
B/C	3.38	2.35	5.48	4.08	18.71	23.42	18.79
NPGEV	1,241,684	797,706	602,855	398,853	496,684	319,077	25,949
NPEI	366,937	399,138	113,202	97,805	51,169	13,636	1,381
Change in Discounted NPGEV and/or NPEI due to a 10% Change in Input Factors and Yield							
Site Preparation	399	287	339	287	339	287	78
Planting	417	156	417	156	417	156	
Irrigation	22,927	22,927					
Fertilization	8,181	8,181	8,181	8,181			
Whole-Tree Harvesting	2,058	1,641	1,016	808	834	651	26
Forage Harvesting	1,641		808		651		
Hauling	1,172	756	599	391	469	313	26
All Inputs	36,735	33,948	11,360	9,823	2,710	1,407	130
Yield	124,171	79,776	62,085	39,888	49,658	31,916	2,579

[a] Discount rate, 10%; boiler efficiency, 65%.
[b] Irrigated and fertilized.
[c] Nonirrigated and fertilized.
[d] Nonirrigated and nonfertilized.

Table VI. Energy Performance (MJ/ha) of Wood Production Alternatives, and Sensitivity of Discounted Energy Inputs and Yields to Changes in Inputs and Yields[a]

	Alternative						
	1.2 × 1.2 m (4 × 4 ft)[b]	2.4 × 2.4 m (8 × 8 ft)[b]	1.2 × 1.2 m (4 × 4 ft)[c]	2.4 × 2.4 m (8 × 8 ft)[c]	1.2 × 1.2 m (4 × 4 ft)[d]	2.4 × 2.4 m (8 × 8 ft)[d]	Extensive
NPEV	427,981	139,230	261,994	119,064	253,813	140,038	2,579
B/C	2.71	1.60	4.40	2.79	15.56	17.74	3.74
NPGEV	678,094	371,002	339,034	185,501	271,243	148,401	3,517
NPEI	250,114	231,772	77,040	66,436	17,430	8,363	938
Change in Discounted NPGEV and/or NPEI due to a 10% Change in Input Factors and Yield							
Site Preparation	339	287	339	287	339	287	78
Planting	417	156	417	156	417	156	
Irrigation	15,971	15,971					
Fertilization	5,706	5,706	5,706	5,706			
Whole-Tree Harvesting	1,329	756	651	391	521	313	0
Forage Harvesting	703		339		287		
Hauling	651	339	313	182	261	130	0
All Inputs	25,116	23,215	7,765	6,722	1,825	886	78
Yield	67,817	37,100	33,896	18,550	27,122	14,824	365

[a] Discount rate, 15%; boiler efficiency, 65%.
[b] Irrigated and fertilized.
[c] Nonirrigated and fertilized.
[d] Nonirrigated and nonfertilized.

the lower B/C ratios and vice versa for the more extensive alternatives. The major change in the rankings with higher interest rates is an improved performance in terms of NPEV of the alternatives with early and frequent energy harvests, i.e., the small spacing alternative.

As will be discussed later, results are most sensitive to energy yields. Therefore, the timing of the yield(s) becomes more and more important with higher discount rates and differences among alternatives become greater. The most dramatic changes occur in the extensive alternative.

In the previous discussion, wood was assumed to be fired green with a resulting boiler efficiency of 65%. Burning of predried wood has been recommended by some. However, drying can be a significant energy expenditure in both intensive and extensive systems. The energy flow analyses were performed with and without drying to determine the significance of drying. The chips that were not dried were assumed to have a moisture content of 100% on a dry basis while those that were dried were assumed to contain only a moisture content of 10% on a dry basis.

Considering that dried and undried chips burn at different temperatures, their gross heat energy yields cannot be directly compared. Instead, the different boiler efficiencies can be applied to achieve values suitable for comparison. In our example, a boiler efficiency of 80% and a boiler efficiency of 65% were used for dried and green wood respectively [5].

Table VII illustrates the results when drying is included in the analysis. It reflects energy inputs for drying, and also the increased energy efficiency of burning the wood chips in a boiler.

The results indicate that drying inputs are not offset by increased usable energy. For this reason, in the further discussion, alternatives without drying only will be considered. It should be noted that the rankings in terms of NPEV and B/C ratios remained the same as for the alternatives without drying. Results for other discount rates are not shown because the comparisons of systems with and without drying produced parallel results.

SENSITIVITY ANALYSIS

Sensitivity analysis is a valuable tool to predict the potential effect of changes in estimates of inputs and outputs. Estimates to which results are sensitive and which also are uncertain must receive the greatest attention. The three most uncertain estimates in the analyses are the yields that can be obtained in various production systems, irrigation energy inputs and energy use of various harvesting systems. Sensitivity analysis of two mea-

Table VII.   Energy Performance (MJ/ha) or Wood Production Alternatives, and Sensitivity of Discounted Energy Inputs and Yields to Changes in Inputs and Yields[a]

	Alternative						
	1.2 × 1.2 m (4 × 4 ft)[b]	2.4 × 2.4 m (8 × 8 ft)[b]	1.2 × 1.2 m (4 × 4 ft)[c]	2.4 × 2.4 m (8 × 8 ft)[c]	1.2 × 1.2 m (4 × 4 ft)[d]	2.4 × 2.4 m (8 × 8 ft)[d]	Extensive
NPEV	4,302,214	3,894,373	2,382,124	2,179,844	2,112,418	1,951,199	1,736,440
B/C	2.46	2.43	2.92	2.93	3.68	3.80	3.94
NPGEV	7,248,292	6,618,006	3,624,146	3,309,003	2,899,390	2,647,192	2,327,333
NPEI	2,946,078	2,723,633	1,242,022	1,129,159	786,972	695,993	590,893
Change in Discounted NPGEV and/or NPEI due to a 10% Change in Input Factors and Yield							
Site Preparation	339	286	339	286	339	286	78
Planting	417	156	417	156	417	156	
Irrigation	73,002	73,002					
Fertilization	26,027	26,027	26,207	26,207			
Whole-Tree Harvesting	5,341	11,021	2,657	5,497	2,136	4,403	1,928
Forage Harvesting	12,193		6,096		4,872		
Hauling	5,575	5,106	2,788	2,553	2,241	2,032	1,954
Drying	171,719	156,790	85,846	78,395	68,677	62,711	55,129
All Inputs	294,613	272,389	124,171	112,916	78,681	69,589	59,089
Yield	724,834	661,811	362,404	330,905	289,949	264,730	232,606

[a] Discount rate, 0%; boiler efficiency, 80%.
[b] Irrigated and fertilized.
[c] Nonirrigated and fertilized.
[d] Nonirrigated and nonfertilized.

sures of energy efficiency (NPEV and B/C) to changes in the energy flows was used to identify critical factors.

The sensitivity tables (Tables III to VI) indicate how much the discounted energy yields or inputs would change if an input activity or yield changed by 10%. An increase of an input would increase the total discounted energy inputs by the amount specified in the table; a decrease vice versa would decrease the total input by that amount. Changes in energy yields would change total discounted energy yield by the amount specified in the table. A 20% change would result in changes two times the values in the sensitivity table and, in general, an X% change would result in a change equal to the table value multiplied by the ratio of X/10.

Any change in discounted energy yields or energy inputs due to changed input or output assumptions results in a new NPEV and B/C which can easily be calculated from the values provided in the summary tables. Effects of changes in several activities are additive, i.e., a 10% reduction in all energy inputs would reduce total discounted energy inputs by the sum of the table values for these activities.

Both measures of energy efficiency are most sensitive to changes in yield. For both intensive and extensive fiber production systems, yields can be highly variable. For hybrid poplar cultures in the lake states, yield data have been recorded only for carefully tended research plots grown for short periods. In long-term operations, yields could therefore be lower than predicted. Yields for nonirrigated and nonfertilized cultures are also very speculative. Since there are no data on this type of culture, we assumed the yield was 40% of irrigated and fertilized cultures. On the other hand, the lack of irrigation and fertilization could make the difference between the success or failure of a crop and not just reduce growth, as we assumed. Another possible cause for reduced yields is that both intensive and extensive cultures are susceptible to risks from insects and diseases which may destroy an entire crop. Thus, uncertainty about yields can have a substantial impact on wood energy and should receive attention before wood energy production is contemplated. Generally, yields would have to be reduced more than 78-98% before NPEV would become zero or the B/C ratio would become one when a zero discount rate is used (Table VIII). Even with a 15% discount rate, yields would have to decline by 37-94%.

Irrigation is not only one of the largest energy inputs into an intensive wood fiber production system, but it is also the most unpredictable. The amount of energy consumed during irrigation is dependent on:

- soil type,
- climate,
- crop requirements,

Table VIII. Necessary Percent Reduction in Energy Yield to Make NPEV = 0 and B/C = 1

Discount Rate (%)	1.2 × 1.2 m[a]	2.4 × 2.4 m[a]	1.2 × 1.2 m[b]	2.4 × 2.4 m[b]	1.2 × 1.2 m[c]	2.4 × 2.4 m[c]	Extensive
0	79	78	87	87	96	97	98
5	76	70	85	83	95	96	97
10	70	57	82	75	95	96	95
15	63	37	77	64	94	94	71

[a] Irrigated and fertilized.
[b] Nonirrigated and fertilized.
[c] Nonirrigated and nonfertilized.

- type of fuel used,
- type of equipment used,
- well depth and capacity, and
- unforeseen drought or flooding.

A large reduction in irrigation energy expenditures can be accomplished by switching energy sources. A diesel fuel irrigation system consumes almost four times the energy that an electric system does. One drawback to the use of electricity as a power source is the lack of power lines available near remote sites.

Another determinate of energy consumption is the amount of water applied. The better the site quality and the moister the environment, the less water would be required to irrigate the site. Also, if less water is used in the site, more acreage can be covered by one system. The depth and capacity of a well also influence the amount of energy consumed. For instance, the deeper the well, the more energy that must be expended to get the water to the surface. The rate at which water is pumped can be adjusted to fit a well's capacity and the crop needs. Other possible reductions in energy inputs to irrigation systems would be to decrease the amount of water supplied to each parcel, e.g., by irrigating during the critical establishment period or by introducing new technology such as trickle irrigation. Irrigation is a very energy-intensive process with energy inputs often exceeding the benefits. Because the range in energy inputs in irrigation systems varies greatly with the type of irrigation system, type of fuel used, amount of water required and site quality, the decision to irrigate requires careful planning. Under some circumstances, irrigation can be a feasible alternative to increase yields. Under other circumstances, the use of irrigation in plantations could require large enough amounts of energy to lead to a low energy output/input ratio. Let us examine two extreme cases to illustrate this point.

For the first case, the plantation is located in a dry climate with a low water table and no convenient source of fuel, like electricity, close to the site. This plantation would require large energy inputs (67,739 MJ/ha-yr or 26 MBtu/ac-yr) to effectively irrigate the site (Table IX). This input is 278% higher than assumed in our analysis and would cause a reduction in the B/C ratios to 1.89 and 1.50 for alternatives 1 and 2, respectively (Table III).

The second case takes the opposite extreme. This system is located in a moister climate with a high water table and a readily accessible source of inexpensive fuel. In this case, the energy inputs would be quite low (2345 MJ/ha-yr or 0.9 MBtu/ac-yr) due to such factors as the short distance from the water table to the surface, the larger area covered by one system, and the lower amount of water required (Table IX). Irrigation

Table IX.   Energy Inputs for Two Extreme
Irrigation Alternatives [9]

	Lowest Energy Expenditure	Highest Energy Expenditure
Type	Electric	Diesel
Rate (liter/min)	2270	3407
Available H$_2$O (m)	20.3 total	61.0 total
Energy Consumed (MJ)	173,137	3,412,833
Hours of Operation	600	1200
Lift (m)	6.1	76.2
Acreage Irrigated (ha)	72.8	48.6
Water/acre (ha-cm/hr)	1.37	2.06
Effective Water (cm)	12.7	38.1
Energy Units	49,099.2 kWh	81,686 liters

energy inputs would be 90% below the level assumed in our study and cause an increase in the B/C ratios to 10.3 and 10.8 for alternatives 1 and 2, respectively (Table III).

Other energy inputs in site preparation, planting, harvesting and hauling do not significantly affect energy efficiencies (Tables III to VI). Increases of all energy input requirements of 365–4675% would be necessary before NPEV would become zero and B/C would become equal to one (Table X). It is apparent that the more extensive alternatives would require substantially larger changes in energy inputs for this to happen. Such changes are extremely unlikely, especially since these alternatives do not involve highly uncertain energy input activities such as irrigation. The sensitivity of NPEV and B/C to total cost increases is much more pronounced at the higher discount rates. With a 15% discount rate, the irrigation alternatives would require relatively smaller energy input increases to make these alternatives unattractive for energy production, i.e., NPEV equal to zero and B/C equal to one. Nonetheless, even changes of the magnitude indicated are highly unlikely.

CONCLUSIONS

Production of energy in intensive or extensive cultures of poplars is feasible from an energy budget view, i.e., all systems examined produced more energy than they used as production inputs. Intensive production systems produce higher net energy yields than more extensive systems, but produce considerably fewer energy units per unit of energy input. The more intensive systems also are relatively more sensitive to changes in energy input requirements. Irrigation input is the most uncertain and

Table X. Necessary Percent Reduction in Total Energy Yield to Make NPEV = 0 and B/C = 1

Discount Rate (%)	1.2 × 1.2 m[a]	2.4 × 2.4 m[a]	1.2 × 1.2 m[b]	2.4 × 2.4 m[b]	1.2 × 1.2 m[c]	2.4 × 2.4 m[c]	Extensive
0	380	365	668	679	2255	3027	4675
5	310	236	561	475	2037	2701	4147
10	238	135	447	306	1644	2171	1886
15	170	60	337	177	1390	1581	330

[a] Irrigated and fertilized.
[b] Nonirrigated and fertilized.
[c] Nonirrigated and nonfertilized.

is also the input to which energy performance measures are most sensitive. Yield is the single most important factor to which net energy yield and B/C ratios are most sensitive. On the basis of this study and a previous economic analysis [1], nonirrigated, nonfertilized hybrid poplar plantations at close spacings appear to be most promising.

REFERENCES

1.   Rose, D. W., K. D. Ferguson, D. Lothner and J. Zavitkovski. "Hybrid Poplar Plantations in the Lake States—A Financial Analysis" *J. For.* (in press).
2.   Ek, A. R., and D. H. Dawson. "Yields of Intensively Grown *Populus*—Actual and Projected," U.S. Forest Service General Technical Report NC-23 (1976), pp. 5–9.
3.   Kittredge, J., and S. R. Gevorkiantz. "Forest Possibilities of Aspen Lands in the Lake States," University of Minnesota Agricultural Experiment Station Technical Bulletin 60 (1929).
4.   Zavitkovski, J. "Energy Production in Irrigated, Intensively Cultured Plantations of *Populus* 'Tristis #1' and Jack Pine," *For. Sci.* 25:383–392 (1979).
5.   Vanelli, L. S., and W. B. Archibald. "Economics of Hog Fuel Drying," FPRS Proceedings, Energy and the Wood Products Industry, p-76-14 (1976), pp. 55–62.
6.   Benson, F. J. "Machinery Cost Estimates," Agricultural Extension Service, University of Minnesota. Unpublished results (1979).
7.   Mattson, J. A. "Harvesting Research for Maximum Yield Systems," U.S. Forest Service, North Central Forestry Experiment Station, Forest Engineering Laboratory. Unpublished results (1976).
8.   Aube, P. J. "A Cost Analysis of a 25 Megawatt Wood-Fueled Power Plant in Minnesota's Inventory Region 2," MS Thesis, University of Minnesota, College of Forestry (1980).
9.   "Water Sources and Irrigation Economics," Misc. Report 150-1978, Agricultural Experiment Station, University of Minnesota (1978), p. 76.
10. Pimentel, D., L. E. Hurd, A. C. Belotti, M. J. Forster, I. N. Oka, O. D. Sholes and R. J. Whitman. "Food Production and the Energy Crisis," *Science* 182:443–449 (1973).
11. Blankenhorn, P. R., T. W. Bowersox and W. K. Murphey. "Recoverable Energy from the Forests, an Energy Balance Sheet," *Tappi* 61(4):57–60 (1978)

CHAPTER 4

ENERGY OUTPUT/INPUT RATIOS FOR SHORT-ROTATION GROWTH OF AMERICAN SYCAMORE

Klaus Steinbeck
School of Forest Resources
University of Georgia
Athens, Georgia

WOOD AS AN ENERGY SOURCE

Wood is regaining importance as an energy source in the United States. In its 1979 report, the Council on Environmental Quality estimated that in 1977 about 2.3% of the country's total energy consumption was derived from wood [1]. Most of this was generated and used in-house by forest-based industries. The pulp and paper industry, for example, is among the top ten industrial energy users in the nation, and estimates that it is meeting about half of its energy requirements with woody materials. Logging and mill residues, thinnings, and entire stands which lack traditional markets are among the various types of "hogfuel" which fire boilers. A task force of the Society of American Foresters reported that the forest biomass potentially available from the commercial forest acreage in 1979 was the equivalent of 9.5 quadrillion Btu, or about five times more than was actually generated from wood [2]. They estimated that the net annual growth on these lands was about 530 million dry tons of wood, and that only 200 million tons were being used for traditional products like lumber and paper by the forest industry.

All kinds of biomass share certain advantages and disadvantages as energy sources. Among the most important positive, long-term aspects are that they are renewable indefinitely, cause no net change in atmospheric carbon dioxide levels and contain relatively few pollutants. Biomass is a versatile source of energy in that it can be converted into a variety of liquid or gaseous fuels or directly into heat. Disadvantages of biomass include the removal of nutrients and organic matter from soils as well as high harvest and transport costs because the resource generally is scattered, bulky and high in moisture.

Forest trees offer several advantages as energy crops in comparison with agricultural crops or other herbaceous plants. The perennial habit of trees offers flexibility in harvest timing and the option to accumulate biomass in the field. Trees make low nutritional demands on soils and protect them from erosion. Broad-leaved species sprout from the stump or rootsystem, obviating the need for additional site preparation or planting and thus reducing energy inputs after the initial establishment. Also, woody materials are less subject to deterioration in post-harvest storage than the more succulent, herbaceous plant tissues.

The increasing use of trees for energy is causing some concern among forest-based industries about the effects of a potentially major competitor for its raw materials base. At the same time some of the same firms see opportunities for diversification into the energy business. Modern tree harvesting and wood processing methods leave less and less waste. In some parts of the country, traditional users of trees already compete rather strenuously for wood, and major efforts have been expended to increase the raw material production per unit of land area and time. The short-rotation forestry system is one such effort.

SHORT-ROTATION FORESTS

Short-rotation forests are plantations of closely spaced, broad-leaved trees which are harvested repeatedly on cycles of less than ten years. The key to higher production than that achieved under more conventional management systems lies in the rootstocks. Once established, they remain in the ground after each harvest of the aboveground tree portions, and resprout. The emerging shoots grow rapidly because the roots already have established access to soil water and nutrients and also contain stored carbohydrates which sustain rapid regrowth.

The Southeastern United States, with its long growing season and moist climate, is well suited for high biomass production rates, and several tree species are being screened for their suitability for short-rotation forestry [3]. American sycamore (*Platanus occidentalis*), sweetgum (*Liquidambar styraciflua*), red maple (*Acer rubrum*), black locust (*Robinia pseudoacacia*) and European black alder (*Alnus glutinosa*) are included in the screening program of the University of Georgia. The latter two species fix atmospheric nitrogen and have been planted in both pure plots and mixed with sycamore and sweetgum. Their ability to symbiotically incorporate fixed nitrogen could reduce the fertilizer (and energy) inputs needed to sustain high productivity.

Biomass yields of short-rotation forests depend on many factors, among them choice of tree species, site quality, length of harvesting cycle (rotation) and cultural practices. Yields should be evaluated in terms of energy inputs needed to achieve various levels of biomass or energy outputs. Site preparation, seed or seedling production, planting, cultivation, fertilization, harvest, transport and conversion of the crop require energy expenditures. These inputs can include energy needed to manufacture machinery and that needed to run it in the various production and utilization processes, to produce fertilizers, and to feed the labor force. Energy inputs may also include environmental sources like solar radiation, and rain and wind energies. Outputs also can be measured at several stages, such as before or after harvest, transportation, pretreatment (chipping, pelleting, drying) or final conversion.

ENERGY INPUTS

This study assesses the direct energy inputs after site preparation and outputs "on the stump" (before actual harvest) for short-rotation forests of American sycamore. Even within this limited scope, it will be necessary to combine the energy inputs of a planting established in 1978 with the outputs of another one planted in 1967. Both are very similar and are located in the rolling terrain of the Piedmont province of Georgia and, for the most part, on the same soil series. The reader interested in other energy inputs and outputs necessary for equipment manufacture, seedling production, labor sustenance, harvesting, drying and transport can find such information in other articles [4–6].

One-year-old, nursery-grown sycamore seedlings were machine planted at a 1.2 × 2.4 m spacing (3470 seedlings/ha) on a clean-cultivated site. Details of soils, site and equipment used have been reported previously [7]. The site had supported a mixed pine-hardwood stand which had been clearcut commercially. Site preparation consisted of piling the unmerchantable trees and brush, pushing up of stumps and turning the soil twice with a heavy wildland harrow. This land clearing and site preparation required 13 times more energy than all subsequent operations (Table I), and probably represents the high end of the range of what would normally be expected.

Table I.   Diesel Fuel Inputs Needed for the Establishment
and Cultivation of a Short-Rotation Plantation

Operation	Frequency	Liters Fuel/ha	10^6 g-cal/ha[a]
Heavy land clearing, root raking and			
harrowing	1	813.8	7328.6
First Year			
Planting	1	37.5	337.7
Herbicide spray	1	2.6	23.4
Fertilizer spreading	1	3.6	32.4
Disking	2	6.3	56.7
Second Year			
Disking	1	3.3	29.7
Fertilizer spreading	1	3.6	32.4
Fifth Year			
Fertilizer spreading	1	3.6	32.4
Total		874.3	7873.3

[a]Diesel fuel (kerosine) heat of combustion at 11,006 g-cal/g [8] or 9,005,438 g-cal/l or
37.65×10^6 J/l.

Once the site had been prepared, machine planting accounted for about two-thirds of the remaining energy inputs needed to establish and maintain this planting. Once planted, such plantations typically receive an herbicide spray, one fertilizer application, and two diskings during the first year.

All values in Table I pertain to one site and to equipment which was not necessarily of optimum size for the operations. Once the field crews had measured the fuel being poured into their equipment for a while, they became conscious of fuel consumption and began to experiment with different tractor rpm and gear combinations. Later, this often resulted in significant fuel savings (not reported here). Even then, the 545 $\times$ 10^6 g-cal/ha used to prepare the site and plant and maintain this short-rotation plantation was 30% less than the 770 $\times$ 10^6 g-cal/ha estimated by Blankenhorn et al. [4] for roughly the same operations.

The energy needed to manufacture the fertilizers and herbicides applied to the plantation during the first five years of its existence totaled 5723 $\times$ 10^6 g-cal/ha (Table II). About 95% of this was accounted for by nitrogen alone. It is hoped the plantings containing nitrogen-fixing tree species will reduce the energy inputs for this vital element. Estimates of 25,084 $\times$ 10^6 g-cal/ha fertilizer input for an intensive biomass production system managed on a 10-yr growth and harvest cycle have been

Table II. Fertilizer and Herbicide Energy Inputs Used
in the Maintenance of a Short-Rotation Plantation

Timing	Elemental or Active Ingredient (kg/ha)	10^6 g-cal/ha[a]
First Year		
N	89.7	1661.1
P	17.4	58.3
K	16.6	38.4
Herbicide	4.0	0.1
Second Year		
N	147.9	2738.9
Fifth Year		
N	56.0	1037.0
P	24.4	81.8
K	46.5	107.6
Total		5723.2

[a]Conversion factors from Pimentel et al. [5].

reported [4]. An estimate of 29,331 $\times$ 10^6 g-cal/ha for the manufacture, transport and application of fertilizer to *Populus* "Tristis #1" plantations grown on 10-yr rotations has also been reported [6]. These two estimates are more than double the fertilizer inputs used in this study. Several reasons for this difference can be advanced, chiefly, differences in soils and nutrient requirements of various tree species. Lack of knowledge of the actual nutrient needs of the American sycamore certainly decreased the efficiency with which the dosage and timing of fertilization were determined in this study. Direct energy input in fuel and fertilizer totaled about 13,600 $\times$ 10^6 g-cal/ha for this sycamore plantation.

ENERGY OUTPUTS

The biomass yields and their equivalent energy outputs were determined for three different harvesting cycles of coppiced sycamore. In each harvesting option, the trees were initially cut two years after field planting and in the ensuing four years, some plots were cut annually, others every other year, and still others only once at the end of the four-year period. Trees and sprouts were cut 15 cm above groundline during the winters, and the seedling [9] and coppice [10] yields have been reported.

The annually coppiced plot produced the lowest total output, at 167 × 10^9 g-cal/ha in 6 yr (Table III). Under the previously stipulated conditions, this represents an energy output/input ratio of about 12. The plots harvested only at age two and then after four years of sprout growth produced 17.6 g-cal for every g-cal of input. All harvesting options averaged 15.3 g-cal output for every g-cal input. This compares very well with the on-the-stump ratio of 16 reported for intensively cultivated, nonirrigated *Populus* "Tristis #1" harvested on 10-yr cycles [6]. Obviously, harvesting, transport, drying and conversion energy expenditures should be deducted from the energy outputs before an overall energy balance can be calculated.

Table III.   Preharvest Energy Outputs of American
Sycamore Coppiced on Different Rotations[a] (g-cal/ha)

| Rotation (yr) | Plantation Age (yr in field) | | | | | Total |
	2[b]	3	4	5	6	
1	31,888	20,470	39,987	39,511	35,207	167,063
2	31,888		72,351		112,391	216,630
4	31,888				207,492	239,380

[a] Derived from yields presented by Steinbeck and Brown [10], bark and wood content of various ages of coppice material by Sockwell and Presnell [11] and caloric content of sycamore bark at 4779 g-cal/g and wood at 4756 g-cal/g [12]. All trees were cut during the winter; therefore, no foliage is included in these outputs.
[b] This initial harvest [9] is of seedling origin, and all subsequent ones are sprout material.

CONCLUSIONS

This study represents at best a first approximation of the energy balance of short-rotation forestry in Georgia. Opportunities to improve that balance exist. Genetic selection may improve biomass yields by 20% or more after one selection cycle. Nitrogen-fixing species may reduce the need for fertilizer inputs significantly, and equipment specifically designed to operate in short-rotation forests should require less fuel.

REFERENCES

1.   "The Good News about Energy," Council on Environmental Quality, Washington, DC (1979).

2. "Forest Biomass as an Energy Source," Society of American Foresters, Washington, DC (1979).
3. Evans, R. S., "Energy Plantations: Should We Grow Trees for Power Plant Fuel?" Canadian Forest Service Report No. VP-X-129, Western Forest Products Laboratory, Vancouver, BC (1974).
4. Blankenhorn, P. R., T. W. Bowersox and W. K. Murphey. "Recoverable Energy from the Forests," *Tappi* 61(4):57–60 (1978).
5. Pimentel, D., L. E. Hurd, A. C. Belotti, M. J. Forster, I. N. Oka, O. D. Sholes and R. J. Whitman. "Food Production and the Energy Crisis," *Science* 182:443–449 (1973).
6. Zavitkovski, J. "Energy Production in Irrigated, Intensively Cultured Plantations of *Populus* 'Tristis #1' and Jack Pine," *For. Sci.* 25:383–392 (1979).
7. Steinbeck, K. "Increasing the Biomass Production of Short-Rotation Coppice Forests," in *Proceedings of the Third Annual Biomass Energy Systems Conference* (Golden, CO: Solar Energy Research Institute, 1979).
8. *Handbook of Chemistry and Physics,* 44th ed. (Cleveland, OH: CRC Press, 1961), p. 1936.
9. Steinbeck, K., and J. T. May. "Productivity of Very Young *Plantanus occidentalis* L. Plantings Grown at Various Spacings," in *Forest Biomass Studies,* Misc. Pub. 132, Life Science and Agriculture Experiment Station, University of Maine, Orono, ME (1971).
10. Steinbeck, K., and C. L. Brown. "Yield and Utilization of Hardwood Fiber Grown on Short Rotations," *Appl. Polymer Symp.* 28:393–401 (1976).
11. Sockwell, T. R., and R. F. Presnell. "Age and Spacing Effects on the Yields of American Sycamore Coppice," Unpublished, University of Georgia, Athens, GA (1973).
12. Neenan, M., and K. Steinbeck. "Caloric Values for Young Sprouts of Nine Hardwood Species," *For. Sci.* 25:455–461 (1979).

CHAPTER 5

GROWTH, YIELD AND COMPOSITIONAL CHARACTERISTICS OF JERUSALEM ARTICHOKE AS THEY RELATE TO BIOMASS PRODUCTION

M. D. Stauffer

International Harvester Co.
Hillside, Illinois

B. B. Chubey

Agriculture Canada
Morden, Manitoba   Canada

D. G. Dorrell

Agriculture Canada
Winnipeg, Manitoba   Canada

The Jerusalem artichoke (*Helianthus tuberosus* L.) has shown excellent potential as a carbohydrate-rich crop. It has been grown and investigated in Europe [1–3], the United States [4–7] and Japan [8]. Interestingly, some of the early research on Jerusalem artichoke proposed production systems for processing either alcohol or crystalline sugar. However, yields, agronomic technology and economic structures at the times of investigation were generally unfavorable and therefore, research was abandoned. More recently, Jerusalem artichoke has received renewed interest as a vegetable (popular press), as a source for fructose sugar [3,9] and as a potential biomass crop for alcohol production [10].

In the past decade, fructose and high-fructose corn syrups have gained an increasing portion of the sweetener market. Fructose is the sweetest naturally occurring sugar, rated as 1.5–1.7 times sweeter than sucrose. The lower caloric-to-sweetness ratio of fructose is considered as a beneficial alternative in many foods. Fructose is produced commercially by two methods. One is by acidic or enzymatic hydrolysis of sucrose and isomerization of the glucose to fructose; the other method is enzymatic processing and isomerization of cornstarch.

Alternative sources of fructose are plants which produce fructofurano-sides as their storage carbohydrate. Among the inulin-storing plants [11] are Jerusalem artichoke, chicory, dahlia, dandelion, Canada thistle, goldenrod and wild onion. Jerusalem artichoke tubers, probably the highest yielding source, contains a wide range of fructosans. For conven-ience, they can be divided into two groups: oligosaccharides, which are soluble in 80% ethanol and contain up to eight fructose molecules, and inulin, where the number of fructose molecules varies from 9 to 35 [12]. Fructose content varies from 70 to 87% of the total reducing sugar content in Jerusalem artichoke, depending on genetic strain and harvest time [9]. During storage, a decrease in the high-molecular-weight frac-tion and a concurrent increase in the low-molecular-weight fraction occurs [13,14]. No change in free fructose level was observed during that time; however, the fructose-to-glucose ratio increased. This suggests that the fructose was released from the high-molecular-weight fraction. The free fructose may then be added onto oligofructosans, or isomerized to glucose and synthesized into sucrose. Several enzymes are likely in-volved [15]. Jerusalem artichoke extracts were found to contain three beta-D-fructo-furanosidases, two of which are hydrolases and probably responsible for depolymerizing the high-molecular-weight fraction. The third enzyme is an invertase, which acts on the sucrosyl linkage. The activities of these enzymes have been monitored throughout the growing season [16]. Invertase activity was high during tuber formation and again when tubers were sprouted. High activity of the two hydrolyases oc-curred during tuberization and for some weeks during dormancy. Pre-viously, it was reported that invertase activity could be developed in "aged" tuber disks [17]. This enzyme is apparently bound to the cell wall since attempts to solubilize it failed.

Studies in western Canada show that high yields of Jerusalem arti-choke are obtainable. The yield of sugar under prairie conditions would be 6.2–8.6 metric ton/ha compared to 4.9 metric ton/ha from sugarbeets [10]. In the Netherlands, tuber yields of 45.8 metric ton/ha (11.1 metric ton/ha DM) have been reported [3]. Sugar yield was 8 metric ton/ha, which compares favorably with yields from sugarbeets under those conditions. High yields are obtained on more fertile soils, although Jeru-salem artichoke will produce satisfactorily on less fertile lands [18].

Investigations showed that dwarf cultivars having less leaf area per plant could produce tuber yields equivalent to or greater than taller, more leafy types by increasing plant populations to more fully utilize the soil and environment [3]. However, their investigations showed duration of the tuberization period to be more important in developing tuber yield.

Jerusalem artichoke forage quality has not been well documented.

Generally, it is characterized as a roughage lacking palatability. When harvested after frost, it was reported to be comparable to corn stover [5]. At that stage, leaf loss was significant. Crude fiber content has been found to remain constant at approximately 24% up to the time where leaves were dropped [6]. More recently, however, it was reported that in the United Kingdom, foliage can be made into good hay or silage [19].

Our initial investigations determined the reducing sugar content as an estimate of inulin content and tuber yields, and the variability that exists among different lines. However, when additional studies showed that good-quality pulp remained after inulin extraction and high forage yields were obtainable, the scope of our investigations was broadened to assess utilization of the total plant. Plant growth, yield and compositional characteristics of Jerusalem artichoke as they relate to biomass production are described in this study.

PLANT DESCRIPTION AND CHARACTERISTICS

Jerusalem artichoke is native to temperate North America and adapted to the region circumscribed by the agricultural region contiguous with the northern shore of the Great Lakes, the Red and Mississippi Rivers on the west, as far south as Arkansas, eastward to the Piedmont coastal plain of the eastern seaboard, extending from Georgia north to Quebec (Figure 1). However, it is now being successfully grown in western regions of both the United States and Canada. European production of Jerusalem artichoke is also known. Although often referred to as wild sunflower, it differs from other native *Helianthus* species by producing perennial, fleshy tubers. Plants grow tall and upright, having either a branching or nonbranching form of growth (Figures 2,3). Small yellow ray and disk florets borne at the end of main stems and branches may produce small, hard seeds, although seed production is often poor. Tuber shape, size and display vary from round, knobby clusters to long, smooth, single tubers (Figure 4).

Growth begins from tubers early in the season. Maturity is reached within 100–130 days in southern Manitoba. The length of the growing season throughout the northern half of the United States has been reported to range between 125 and 213 days [6]. The moderate tolerance of Jerusalem artichoke to spring and fall frosts extends its growing season beyond that of conventional field crops in areas having short growing seasons. This characteristic aids some experimental lines in achieving high tuber yields. Late-flowering genotypes, however, do not mature.

Pest and disease problems are few. Of the diseases, Sclerotinia wilt (*Sclerotinia sclerotiorum*) is the most efficacious; however, rust, Septoria

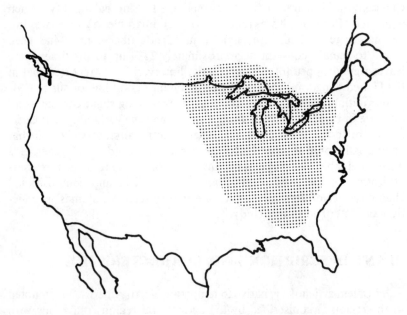

Figure 1.   Native region of Jerusalem artichoke adaptation in North America.

leaf spot and downy mildew are potential problems [20]. White mold and soft rot tuber diseases may occur in storage. Unharvested tubers winter well in the soil and remain healthy until their removal in the spring. In all respects, Jerusalem artichoke is adaptable and remarkably resilient to damage, which explains its high yield potential.

MATERIALS AND METHODS

Experiment 1: Growth Functions

Ontogenic studies of Jerusalem artichoke (experimental line NC10-69) tuber growth and the relationship between tuber growth and top growth were conducted in field plots in 1977 and 1978 at Morden, Manitoba, Canada. Plots are located in an Eigenhof orthic black soil. Seed tubers, weighing 60 ± 5 g, were planted in 1-m-wide rows at 0.3-m spacings within the row. Plots consited of four 7.5-m rows, and were arranged in a random complete block design. Three replications were used. Planting dates were April 22, 1977, and May 4, 1978. Nitrogen, phosphorus and

CHARACTERISTICS OF JERUSALEM ARTICHOKE 83

Figure 2. Standard Jerusalem artichoke (experimental line NC10-69) type used in this research. Plants are in the postbloom, exponential tuber-filling stage of growth.

potassium fertilizers were side-band applied at the respective rates of 100, 60 and 75 kg/ha when the plants were 15 cm high. Weeds were controlled by cultivating twice during the growing season, until the canopy shaded the interrow area.

Year effects from July to October in both years were generally above normal for temperature and precipitation. Degree-day temperature accumulation above 5°C for this period each year were 1167 and 1296, respectively. The amount of rainfall received was 317 mm and 268 mm for the respective years.

Figure 3.   Jerusalem artichoke nursery showing differences in variability
among genotypes.

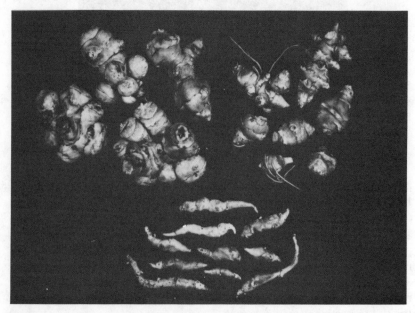

Figure 4.   Variation in Jerusalem artichoke tuber morphology.

Plant sampling was initiated at early flower bud development. This time was established by monitoring plants as well as using previous photoperiodic response observations. Two adjacent plants were randomly sampled within each plot at weekly intervals after the bud stage sampling date until flowering ceased. Following this relatively constant eight-week period, samples were taken at biweekly intervals until the end of the test.

Measurements made on the forage component were length of longest stem, number of stems/plant, forage DM yield and content, and number of flowers in the bud, flowering or senesced (postbloom) stages. Tuber development and growth was characterized by determining number of tuber initials (numbers of stolons and initially thickening stolons), number of tubers (swollen stolons), total length of stolons and tubers, and fresh weight of stolons and tubers obtained from each plant of the two-plant sample.

Syncrony of plant development was very similar for the two years of study. The data presented are combined over years. Standard deviations were calculated for each character at each sampling date.

The objectives were to establish tuber development characteristics and yield by determining the number of potential tuberizing sites, the average number of developing tubers and their growth rate, and to determine the influence which top growth and flower development have on tuber initiation and development.

Experiment 2: Variability in Forage Yield and Composition

Experiment 2A

Utilizing the forage material harvested in Experiment 1, the forage was chopped and a subsample was taken. The subsample was oven-dried at 55°C for 72 hr. Analysis for total nitrogen, acid-detergent fiber, lignin and ash content were all determined using official Association of Official Agricultural Chemists (AOAC) methods [21]. Crude protein was estimated by multiplying N content by 6.25.

Experiment 2B

In an earlier study comparing Jerusalem artichoke silage quality at different stages of maturity, plant tops were harvested at weekly intervals as in Experiment 2A. Plots were grown and maintained in a similar manner. By comparing crude protein, acid-detergent fiber and lignin contents with those values in Experiment 2A, and using date of harvest, morphological development and composition of the forage, the data were aligned. Since cellulose and ash contents varied little, the possibility of misrepresenting forage quality is minimized. Cellulose determinations were made by the method of Goering and Van Soest [22].

Experiment 2C

Variability in yield and quality among 67 experimental Jerusalem artichoke lines was studied. Single-row yield plots were grown at the Agriculture Canada Research Station, Morden, Manitoba, Canada. Plots were established from whole seed tubers on May 5, 1978, as single, 7.5-m row plots. Treatments were replicated once. Plots were planted and maintained as described in Experiment 1. The forage component was harvested from the total plot when the first flower appeared on a majority of the plants in the plot. Late-maturing lines that did not flower were harvested on September 6, 1978, which was prior to the first frost.

Total forage plot weights were made for yield estimates. Whole-plant subsamples were chopped and sub-subsamples were taken for DM and chemical analyses. The sub-subsample was oven-dried at 55°C for 72 hr. Crude-protein, acid-detergent fiber and lignin contents were determined using the official AOAC methods [21].

The experimental accession data presented in this paper were selected from the total number in the test to demonstrate the range of variability found in the species. The objectives of these studies were to identify the yield and quality of Jerusalem artichoke forage as it matured, and to establish the amount of variability which exists in both of these parameters among the available germplasm.

Experiment 3: Variability in Tuber Yield and Composition

Experiment 3A

Total reducing sugar (TRS), fructose (Fru) and glucose (Glu) content and variability were analyzed in experimental Jerusalem artichoke lines, and in the standard accession NC 10-69. The procedure used is the method of Chubey and Dorrell [9].

Experiment 3B

Storage studies were conducted using the standard Jerusalem artichoke line (NC 10-69) grown at the Agriculture Canada Research Station, Morden, Manitoba, as previously described. Following harvest on October 26, 1976, tubers were sorted to remove those which were decayed or visibly injured. Samples (20-kg) were stored at varying temperatures and relative humidities (RH) in controlled-environment cabinets. The treatments imposed were 3 ± 1°C,95% RH; 3 ± 1°C, 75% RH; and −40°C, ambient RH. Samples were taken at weeks zero and eight after storage. Dry matter and TRS contents were determined using the procedures of Chubey and Dorrell [9].

Experiment 3C

Analysis of carbohydrate-extracted Jerusalem artichoke pulp was made on tuber samples harvested in late October 1973 from plots grown at Morden, Manitoba, Canada. Plants were grown as reported for Experiment 1. After harvest, tubers were stored at 2 ± 1°C for several months. The pulp was prepared by thinly slicing tubers and extracting the carbohydrates by refluxing a 50-g fresh weight sample in 200 ml distilled water. The extraction was repeated five times. The extracted pulp was oven-dried and sealed in plastic bags.

Samples were analyzed for crude protein by nitrogen determination (N × 6.25), neutral- and acid-detergent fiber, acid-detergent lignin and ether extract. Digestible energy (DE) content (Mcal/kg) was derived from percent digestible dry matter (DDM) data. The latter was determined using in vitro digestion of the pulp. A regression equation based on actual digestion trials with sheep fed green forages and silages were used to convert percent DDM to DE content [23]. The other fractions were analyzed using official AOAC methods [21].

Experiment 3D

Amino acid analysis of Jerusalem artichoke carbohydrate-extracted pulp was determined on dried samples prepared in a manner similar to those reported in Experiment 3C. Amino acid determinations were made using standard procedures at the Agriculture Canada Grain Research Institute [24]. No basic hydrolysis was performed to determine tryptophan. Performic acid oxidation was not done. Correction factors used to determine certain amino acid contents were threonine, 1.04; serine, 1.08; valine, 1.05; and isoleucine, 1.04. The hydrolysates were incubated at 110°C for 24 hr.

Experiment 3E

The TRS, fructose and crude protein content variability resulting from crosses and selections were determined on individual plants selected from our Jerusalem artichoke nursery. Individual selections were maintained as reported in Experiment 2C. At anthesis, controlled crosses were made between the branching and nonbranching standard lines with selected lines carrying unique characteristics. Seeds were harvested, preconditioned at cold temperatures, germinated, and the seedlings were potted and grown in the glass house. Young plants were transplanted to the field after attaining sufficient growth. Plants from these crosses were reselected for desirable characteristics or new crosses were made using selected desirable parents.

Analyses for total reducing sugar and fructose were made as noted by Chubey and Dorrell [9]. Crude protein was determined according to official AOAC methods [21].

The objectives of these studies were to determine variability in tuber yield and their compositional characteristics, especially TRS, fructose and protein contents. To a limited extent, the inheritability of these traits was determined.

RESULTS AND DISCUSSION

Experiment 1. Growth Functions

The growth of plant tops and tubers are influenced by the flowering process. Considerable variation in time of flowering occurs within the species, the earliest lines flowering in early July, and the latest at the end of September. The experimental line used in this study flowers in early August. When flower buds appeared, the rate of dry matter accumulation in the aerial parts decreased (Figure 5). Maximum DM yields of plant tops were 12.2 metric ton/ha and occurred after flowering. The subsequent loss of dry matter was, in part, the result of leaf senescence and abscission. Tubers develop from stolons which enlarge with the onset of flowering (Figure 5). The number of tubers increased continuously until 50% flowering occurred and subsequently declined to approximately 25 tubers/plant. Throughout the flowering period, individual tuber growth was exponential. It is hypothesized that translocation of material from some tubers, as well as leaf senescence, contributed to increasing the rate of individual tuber dry matter accumulation. Tuber yields were 283.5 g DM/plant, which is equivalent to 9450 kg DM/ha (42 metric ton/ha fresh weight). Plant growth was terminated by killing frosts.

Experiment 2. Variability in Forage Yield and Composition

Forage composition changed with advancing maturity (Table I). Protein content decreased continually, with a significant reduction between

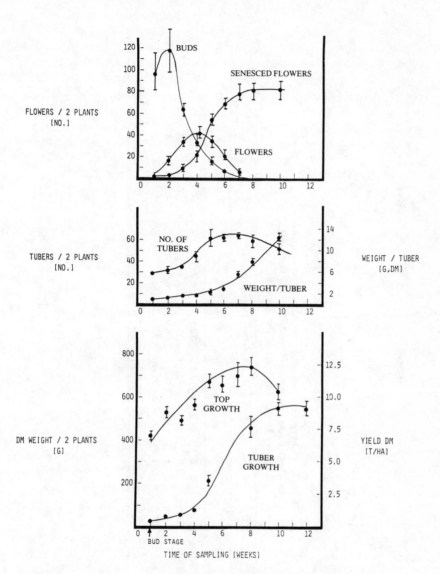

Figure 5. Flowering pattern and growth rates of tubers and plant tops sampled at weekly intervals beginning at the bud stage.

weeks six and seven. Conversely, ADF and lignin fractions increased between weeks seven and eight. These changes were associated with cessation of flowering. Cellulose and ash contents remained relatively constant throughout the sampling period.

Considerable variation in forage dry matter yield and composition exists among accessions (Table II). High forage-yielding lines generally

Table I.   Change in Composition and Content of Jerusalem Artichoke Forage Sampled at Weekly Intervals Following the Flower Bud Stage of Development

Plant Fraction	Sampling Period (weeks)											
	1	2	3	4	5	6	7	8	9	10	11	12
Protein (% DM)	18.0	17.0	15.0	14.3	13.5	12.5	9.7	9.7		7.9		6.2
ADF (% DM)	35.3	31.8	33.9	31.1	32.1	35.5	35.4	40.9		45.6		46.6
Lignin (% DM)	6.21	6.0	7.2	6.1	6.3	6.9	6.7	7.8		9.4		8.9
Cellulose (% DM)	23.0	23.2	22.3	22.3	23.6	23.9	24.2					
Ash (% DM)	2.1	2.0	2.2	2.3	2.3	2.4	2.3					

Table II.   Forage Content of Crude Protein (CP),
Acid-Detergent Fiber (ADF) and Lignin (Lig) Content of
Selected Jerusalem Artichoke Accessions Sampled at
Early Flowering or Prior to Frost

Accession	DM Yield (metric ton/ha)	DM Content (%)	% of Total DM		
			CP	ADF	Lig
NC 10- 5	9.81	32	11.96	52.21	11.40
8	31.78	28	13.85	45.76	8.48
9	8.35	35	10.72	50.69	12.02
13	14.67	38	15.95	42.16	12.77
18	7.26	40	11.92	44.14	7.69
44	22.97	22	9.54	34.37	7.10
50	2.30	23	17.28	25.21	3.78
60	26.04	31	13.21	40.25	7.21

were those which flowered late. All accessions were harvested at first flower. Variability in DM content therefore, was not influenced by stage of maturity. Rather, DM content appeared to be associated with prevailing climatic conditions at the time of harvest. The low-yielding line (NC 10-50) is a leafy, relatively fine-stemmed type having the highest protein and lowest ADF and lignin contents among all accessions evaluated. The greatest amount of protein, ADF and lignin in the respective highest value accession was 1.8, 2.1 and 3.4 times that of the lowest value accession. Based on the magnitude of variability, it appears that forage composition could be readily improved through plant breeding.

Experiment 3. Variability in Tuber Yield and Composition

Total reducing sugar content and fructose:glucose (F/G) ratios were used to estimate inulin content and its molecular size. Actual inulin content and molecular size found in Jerusalem artichoke is not clearly understood, but research is currently underway at a western Canadian university to identify degree of polymerization and the changes which occur during tuberization and later during storage.

Ontogenic changes in carbohydrate content and composition were recorded in two Jerusalem artichoke accessions (Table III). The native Manitoba accession had a higher average TRS content and a wider range of values over the sampling period than the higher yielding Russian accession. A trend toward lower reducing sugar content and percent fructose was evident in the native strain, whereas TRS content remained relatively constant in the Russian line.

Table III.   The Effect of Harvest Dates on the Content and
Composition of Reducing Sugars in Fresh Tubers of Two
Strains of Jerusalem Artichoke

Strain	Harvest Date	Reducing Sugar (%)	Frustose (%)	Glucose (%)	F/G Ratio
Manitoba	Sep 28	21.5	87.6	9.1	9.6
	Oct 12	23.3	80.9	12.6	6.5
	20	21.8	81.8	12.0	6.8
	26	19.9	78.7	13.2	6.0
	Nov 2	22.7	74.3	16.1	4.6
	14	18.5	78.9	12.6	6.3
	23	16.5	76.0	13.6	5.6
Mean		20.6	79.7	12.7	6.5
Russian	Sep 28	14.5	82.4	11.2	7.4
	Oct 12	18.7	78.7	12.9	6.1
	20	16.8	76.8	14.4	5.3
	26	16.2	74.9	14.8	5.1
	Nov 2	19.9	74.1	14.6	5.1
	14	13.8	79.2	11.8	6.7
	23	18.4	71.8	14.9	4.8
Mean		16.9	76.8	13.5	5.8

Duration and storage conditions altered carbohydrate and DM content of Jerusalem artichoke tubers (Table IV). Holding the tubers at 3°C and 75% relative humidity for eight weeks allowed some dehydration of the tubers and the greatest reduction in TRS content. Freezing the tubers caused a small increase in DM content, and the least reduction in TRS of those storage treatments imposed. These data suggest that at northern latitudes where prevailing winter temperatures are below 0°C, Jerusalem artichoke tubers can be maintained in storage piles with minimal loss of TRS. However, occurrence of an eight-week period between harvesting and freezing would significantly reduce the carbohydrate content of the tubers.

The variability in carbohydrate content among several selected accessions ranged from 13.2 to 27.7% when harvested late in the season (Table V). Fructose:glucose ratios also differed, but no relationship existed between high TRS yields and high F/G ratios. In another study to determine composition of the carbohydrate-extracted pulp, significant variability was identified for NDF, ADF, acid-detergent lignin (ADL) and ether extract (Table VI). Neither protein nor digestible energy varied greatly. The amount found in the pulp suggests that it has good feeding value. Amino acid analysis of the pulp revealed high lysine and methione contents (Table VII). Protein quality is considered to be very good.

Table IV.   Dry Matter and Total Reducing Sugar (TRS)
Content in Jerusalem Artichoke Tubers Stored at Different
Temperature and Relative Humidity Levels

	0 Weeks	8 Weeks Storage and		
		3°C/95% RH	3°C/75% RH	−40°C/Ambient RH
Dry Matter (%)	25.8	25.5	28.2	26.5
TRS (%) DM)	78.4	66.7	62.9	72.5

Table V.   Content and Composition of Reducing Sugars in
Fresh Tubers of Six Strains of Jerusalem Artichoke Harvested
on October 23, 1972

Strain	Reducing Sugar (%)	Fructose (%)	Glucose (%)	F/G) ratio
MS #1	27.7	75.3	14.9	5.1
MS #5	20.7	74.7	15.1	4.9
HMR #1	17.1	78.2	13.4	5.8
HMR #2	18.6	71.0	16.1	4.4
HMR #3	17.3	75.2	16.0	4.7
Commercial	13.2	80.6	12.3	6.6

Table VI.   Composition of Several Experimental Jerusalem
Artichoke Accessions

Accession	Digestible Energy (Mcal/kg)	% of DM				
		CP	NDF	ADF	AD Lig	Ether Extract
Morder #5	3.509	26.9	47.8	34.4	1.43	3.9
Perron	3.551	26.7	45.1	42.8	2.62	2.9
Branching	3.603	25.6	47.6	38.1	0.92	3.3
Nonbranching	3.563	25.4	50.4	38.1	0.60	3.2

It is evident that considerable variability in yield and carbohydrate content exists within the species, although accessions high in carbohydrate often produce low yields. A small breeding effort showed interesting results. As summarized in Table VIII, high TRS lines crossed with intermediate carbohydrate–high yielding branching and nonbranching lines (standards) produced significant ranges in percent TRS, fructose

Table VII.   Amino Acid Analysis of Inulin-Extracted
Jerusalem Artichoke Pulp

Amino Acid	Content (g/100 g Sample N)	Amino Acid	Content (g/100 g Sample N)	Amino Acid	Content (g/100 g Sample N)
Lys.	48.98	Ser.	30.42	Meth	11.79
His.	13.08	Glut.	71.43	Iso	30.80
NH₃	12.67	Pro.	21.83	Leu	46.45
Arg.	31.70	Gly	32.01	Tyr.	22.37
Asp.	59.52	Ala	35.44	Phe	28.63
Thr	33.36	Vil	38.31		
Moisture	5.9%				
Protein (DB N × 5.7)	16.16%				
Recovery	86.9%				

Table VIII.   Variability in Percent Total Reducing Sugar
(TRS), Fructose (F) and Protein Among Jerusalem Artichoke
Advanced Selections and New Crosses Compared to Standard
Branching and Nonbranching Types

	No. of Comparisons	Range (%)		
		TRS	F	Prot.
Adv. Selections	21	17.4–22.4	68.0–80.4	23.1–27.0
New Crosses	41	10.66–22.8	65.9–80.1	21.0–30.3
Branching (Standard)		21.2	75.3	25.6
Nonbranching (Standard)		20.3	75.5	25.4

and protein. This suggests that rapid progress could be made when plant utilization is determined and plant breeding objectives are established.

Production Requirements

The highest Jerusalem artichoke yields [75 metric ton/ha (34 short ton/ac)] occurred in a year when temperatures were below normal and precipitation was above normal. The importance of soil water availability during the vegetative stage is apparent for establishing vigorous plants. Since tubers contain approximately 77.5% water, tuber yield is highly dependent on soil water potential during tuberization. Production requirements for Jerusalem artichoke are similar to potatoes. Generally, the requirements are:

- soils: sandy loam, sandy clay loam
- fertilizers (kg/ha): N – 90, P_2O – 56, K_2O – 50
  (or according to regional potato recommendations)
- weed control; interrow cultivation
- fungicide: (possibly)
- pesticides: (none required to date)

Estimated energy requirements for producing the crop are derived using existing data [25]. The data are presented in terms of energy resource depletion (ERD).

Distinction is made between fossil fuel and total energy resources consumed in producing the crop (Table IX). ERD is based on calorific values of fuel used (converted from Btu) times the supply system as losses or expenditures.

Stating Jerusalem artichoke tuber and forage yields in terms of alcohol production, the respective components would yield 4580 and 1920 kg/ha (Table X). The energy output of each is 1181×10^6 kcal/ha from the tubers and 495×10^6 kcal/ha from the forage component.

Potentially, the energy yield/per hectare from Jerusalem artichoke is 1676×10^6 kcal. This exceeds the energy requirement for producing the

Table IX. Energy Requirements in Terms of Energy
Resource Depletion (ERD) of Fossil Fuel (FF) and Total (T) to
Produce One Hectare of Jerusalem Artichoke[a]

	ERD (FF) (10^3 kcal)	ERD (T) (10^3 kcal)
Cultivation	81	83
Planting	171	174
Fertilizer[b]	2327	2402
Interrow Cultivation (2)	145	146
Fungicide	184	193
Sprout Inhibition	148	160
Harvesting Tops[c]	311	317
Tubers	767	780
Hauling to Storage	878	878
Total	5012	5133

[a] Potato production energy requirements by category according to Southwell and Rothwell [25].
[b] Fertilizer rates (kg/ha) in Manitoba: N, 90; P_2O_5, 56; K_2O, 50.
[c] Assumes values similar to corn silage.

Table X.   Theoretical Yields of Ethyl Alcohol from
Several Crops (Manitoba Yield Basis)

Crop	Plant Part	Yield (kg/ha) Fresh Wt.	Carbohydrate	EtOH (kg/ha)
J. Artichoke	Tuber	42,000	7,088	4,580
	Forage[a]	40,000		1,920
Sugar Beet	Root	33,600	4,930	3,185
Corn	Grain	6,700	4,150	2,680
Wheat	Grain	3,360	2,240	1,447

[a]DM equivalent is 12,000 kg/ha.

crop. The energy requirement for alcohol production has not been derived.

REFERENCES

1.   Breen, J. J. *De Landbouwkundige en industriel Betekenis van de Aardpeer* (Helianthus tuberosus L.) (Delft, The Netherlands: Van Markens Drukkerij Vennootschop, 1964).
2.   Corduroux, J. C. "Mecanisme Physiologique de la Tuberiasation du Topinambour," *Bull. Soc. Franc. Physiol. Veget.* 12:213 (1966).
3.   Pilnik, W., and G. J. Vervelde. "Jerusalem Artichoke (*Helianthus tuberosus* L.) as a Source of Fructose, a Natural Alternative Sweetener," *Z. Acker- und Pflanzenbau* 142:153 (1976).
4.   Shoemaker, D. M. "The Jerusalem Artichoke as a Crop Plant," USDA Technical Bulletin No. 33 (1927).
5.   Schoth, H. A. "The Jerusalem Artichoke," Oregon Agricultural Experiment Station Circular No. 89 (1929).
6.   Boswell, J. R., C. E. Steinbauer, M. F. Babb, W. L. Burlison, W. H. Alderman and H. A. Schoth. "Studies of the Culture and Certain Varieties of the Jerusalem Artichoke," USDA Technical Bulletin No. 514 (1936).
7.   Underkoffler, L. A., W. K. McPherson and E. I. Fulmer. "Alcoholic Fermentation of Jerusalem Artichokes," *Ind. Eng. Chem.* 29:1160 (1937).
8.   Yamazaki, J. "Manufacture of Levulose from Jerusalem Artichoke," *Chem. Soc. Japan Bull.* 27:375 (1954).
9.   Chubey, B. B., and D. G. Dorrell. "Jerusalem Artichoke, a Potential Fructose Crop for the Prairies," *Can. Inst. Food Sci. Technol. J.* 7:98 (1974).
10.   Stauffer, M. D., B. B. Chubey and D. G. Dorrell. "Jerusalem Artichoke," Agriculture Canada Canadex 164 (1975).
11.   Haber, E. S., W. G. Gaessler and R. M. Hixon. "Levulose from Chicory, Dahlias and Artichokes," *Iowa State College J. Sci.* 16:291 (1941).
12.   Hoehn, E. Personal communication.

13.   Bacon, J. S. D., and R. Loxley. "Seasonal Changes in the Carbohydrates of the Jerusalem Artichoke Tuber," *Biochem. J.* 51:208 (1952).
14.   Rutherford, P. P., and E. W. Weston. "Carbohydrate Changes During Cold Storage of Some Inulin-Containing Roots and Tubers," *Phytochemistry* 7:175 (1968).
15.   Edelman, J., and T. G. Jefford. "The Metabolism of Fructose Polymers in Plants," *Biochem. J.* 93:148 (1964).
16.   Rutherford, P. P., and A. E. Flood. "Seasonal Changes in the Invertase and Hydrolase Activities of Jerusalem Artichoke Tubers," Phytochemistry 10:953 (1971).
17.   Edelman, J., and M. A. Hall. "Enzyme Formation in Higher Plant Tissues," *Biochem. J.* 95:403 (1965).
18.   Küppers-Sonnenberg, G. A. "Reichaltiges Topinambursortiment," *Saatgutwirtschaft* 17:381 (1965).
19.   Nash, M. J. *Crop Conservation and Storage in Cool Temperature Climates* (Oxford: Pergamon Press, 1978), pp. 165–167.
20.   Huang, H. C., and M. D. Stauffer. "Scerlotinia Wilt in Jerusalem Artichoke," Agriculture Canada Canadex 164–630 (1979).
21.   *Official Methods of Analysis,* 11th ed. (Washington, DC: Association of Official Agricultural Chemists, 1970).
22.   Goering, H. K., and P. J. Van Soest. "Forage Fiber Analysis (Apparatus, Reagents, Procedures and Some Applications)," USDA Agric. Handbook No. 379 (1970).
23.   Waldern, D. Personal communication.
24.   Tkatchuk, R. Personal communication.
25.   Southwell P. H., and T. M. Rothwell. "Report on Analysis of Output/Input Energy Ratios of Food Production in Ontario," Guelph School of Engineering, University of Guelph, Ontario, Canada (1977).

CHAPTER 6

CARBON AND LIGHT LIMITATION
IN MASS ALGAL CULTURE

D. E. Brune

Department of Agricultural Engineering
University of California
Davis, California

J. T. Novak

Department of Civil Engineering
University of Missouri
Columbia, Missouri

A variety of potential applications for mass algal culture is often proposed. The suggested uses for algal biomass include use as fertilizers, raw material for extracted commercial chemicals, animal or human protein supplements either directly or indirectly through incorporation into aquaculture systems, as well as energy via conversion to methane gas [1]. Despite the apparent potential usefulness of algal biomass and years of research centered around understanding algal growth in both laboratory and field culture, mass algal culture has not yet been commercially realized to any large extent.

There are a number of reasons for this lack of success. Difficulty in economically harvesting microalgae biomass and algal culture instability have been two of the major obstacles. The complex problem of culture stability has received much attention particularly with regard to algal species dominance. Attempts at modeling the species-specific responses of algal cultures have been made with varying degrees of success [2–7]. The continued development of such models will likely be the only successful means of answering the many important questions concerning optimum design and operation of large-scale algal culture.

Most mathematical models of algal growth revolve around three central submodels. These submodels attempt to define the growth of a particular alga as a function of the important environmental parameters, nutrient concentration, light levels and temperature. These models attempt to simulate the response of algal cell production as a function of these three factors. Although simple in concept, because of the many possible limiting nutrients, multitudes of possible dominating algal species, combined with interaction between the three central variables, the resultant models become extremely detailed and complex. However, in the situation of high-density algal culture, it is often the case that two factors, in particular, become most important in controlling production. The supply of inorganic carbon at a sufficient rate and concentration to meet algal carbon uptake rates and the availability of sufficient light intensity to supply the energy needs of the growing culture, are often suggested as controlling net cell productivities as well as playing an important role in controlling the dominance of certain algal species [2,8,9].

For these reasons, attention has been directed at a better understanding of light and inorganic carbon limitation of algal growth. It is the purpose of this chapter to examine the nature of the carbon-limited growth response of algae and to combine a quantitative model of this behavior with models describing the carbonate equilibrium chemistry and flow-through algal culture. These relationships will be examined under both nonlight- and light-limiting conditions.

PREDICTING THE RESPONSE OF A CARBON-LIMITED CONTINUOUS ALGAL CULTURE

Algal Response to Dissolved Carbon Dioxide

Over the years, considerable controversy has developed over the interpretation of data concerning the uptake of the various forms of inorganic carbon by unicellular algae [4,10–14]. Early investigators felt that most algae were capable of using either dissolved carbon dioxide (CO_2) or bicarbonate (HCO_3^-) as a carbon source. The basis for the belief in HCO_3^- uptake was centered on observations of algal culture growth to pH values as high as 11.0. Since in vitro studies on the K_{SCO_2} of the enzyme ribulose diphosphate carboxydismutase yielded values as high as 10^{-4} mol/l, it was felt that the CO_2 concentrations in high-pH cultures (10^{-6}–10^{-8} mol/l) was simply too low to supply the carbon needs of the growing culture [12]. However, Goldman et al. [14] suggested that the interconversion of one carbon form to another was much more rapid than the carbon uptake rate of a growing algal culture. This led them to conclude that carbon-limited response of an algal culture may best be represented as a Monod model of the specific growth rate (μ) vs the total carbon concentration (C_T). They further suggested that this relationship must be modified by effects of culture pH.

Work by Brune [5] has, however, led to the continued development of yet another model, first proposed by King [8]. The basis for this model exists in an array of experiments in which the batch growth of laboratory cultures of various freshwater algae was studied. Typical behavior of these cultures is illustrated in Figures 1 to 4. It was found that for cultures grown over a wide range of pH values (7-11), the carbon-limited growth response could best be modeled as a Monod fit of specific growth rate vs dissolved carbon dioxide (Figure 1). In contrast to this, fits of μ to HCO_3^- (Figure 2) or μ to C_T (Figure 3) yielded plots atypical of what is considered normal microbial response to limiting nutrient levels.

In an attempt to quantify any effects from varying culture pH on this relationship, several cultures were grown in which the initial culture alkalinity was varied. The net result of this modification was the observation of μ at similar dissolved CO_2 concentrations but at different pH values. A sample of the data (Figure 4) indicated an "alkalinity effect," which was first interpreted as a suppression of growth rate by increased culture pH. However, attempts to relate μ to pH did not prove particularly successful. For example, μ vs pH is plotted in Figure 5 from data

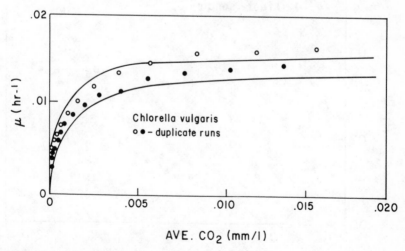

Figure 1. Monod fits of μ vs average CO_2 concentration from batch cultures of *Chlorella vulgaris*.

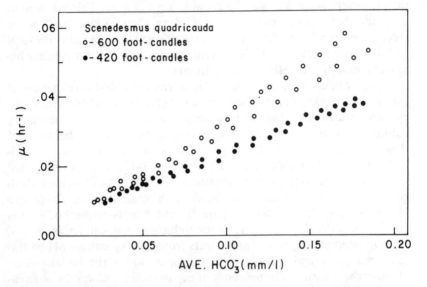

Figure 2.   Plots of μ vs average HCO_3^- concentration from batch cultures of *Scenedesmus quadricauda*.

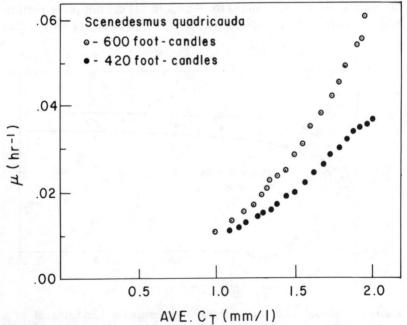

Figure 3.   Plots of μ vs average C_T concentration from batch cultures of *Scenedesmus quadricauda*.

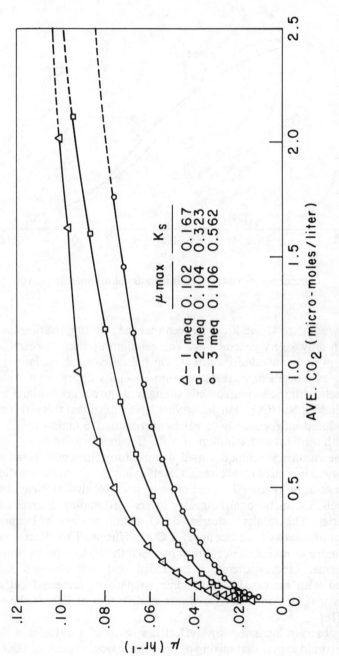

Figure 4. Comparison of μ vs CO_2 concentration from batch cultures of *Chlorella vulgaris* grown at varying initial alkalinities.

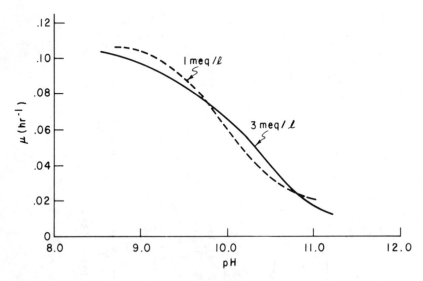

Figure 5.   Comparison of culture pH at various culture-specific growth rates.

taken from Figure 4, and it can be seen that the resulting relationship is similar in only a very general way. This relationship becomes even less uniform in higher-alkalinity cultures. On the other hand, variations in the K_{SCO_2} as a result of increased culture alkalinity appear to correlate very closely with the increased ionic strength of the culture medium as a result of added $NaHCO_3$. Further studies have shown that this effect can be reproduced independently of pH by increasing the ionic strength of the growth medium with addition of NaCl (Figure 6).

For the culturing techniques used in this study, increased alkalinity cultures yield increased algal biomass levels at any dissolved carbon dioxide concentration (Figure 7). Lindsey suggested [16] that at higher biomass levels, the diffusional boundary layers surrounding discrete cells may overlap. This results in decreased CO_2 uptake per unit of biomass, as a result of decreased surface area for CO_2 diffusion. This effect would be even more pronounced in cultures in which the algae grow in clumps or filaments. Lindsey therefore concluded that the increased K_{SCO_2} associated with increased culture ionic strength or increased culture alkalinity may be due to decreased rate of diffusion of CO_2 to the cell surface [16].

If the observed increased K_{SCO_2} effect is a result of a diffusional barrier, one would expect that mixing of the culture would break up the cell clusters and reverse this effect by decreasing the film thickness. Both of

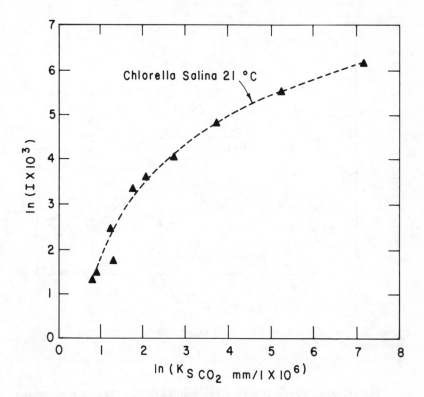

Figure 6. Relationship between K_{SCO_2} and increasing culture ionic strength (as NaCl).

these actions should result in an increased rate of diffusion of CO_2 to the cell surface. Mixing experiments (Figures 8 and 9) indeed confirmed this expectation. An initial alkalinity of 1 meq/l was observed to be less affected by mixing as compared to a 3-meq/l culture, as would be expected, since the 1-meq/l culture contained less biomass. In general, mixing a 3-meq/l culture had the effect of decreasing the K_{SCO_2} to values comparable to those obtained in 1-meq/l cultures.

The influence of different salts and varying salt concentrations on the diffusion coefficient of CO_2 in water has been characterized [17]. Since different salts affect CO_2 diffusivity differently, one would expect that the effect of additions of different salts to algal cultures would likewise produce a differential effect on the observed K_{SCO_2} of the algal culture. This effect has also been confirmed by experimental data (Figure 9). Qualitatively, the effects of mixing, salt concentration and salt composition on the apparent carbon uptake kinetics are identical to those sug-

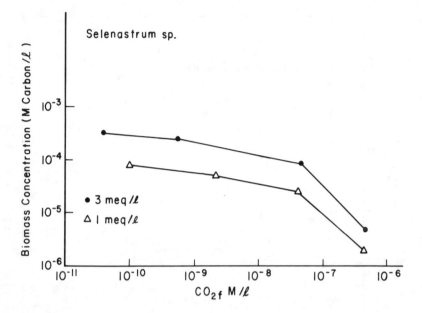

Figure 7.   Algal biomass levels in batch cultures of differing initial alkalinities.

gested by the model proposed by Gavis and Ferguson [18]. In their model (Figure 10), they show how substrate uptake velocity as a function of substrate concentration is affected by reductions in the diffusivity of the surrounding medium. As a measure of the diffusivity of the medium, they introduce a factor F, which corresponds to a ratio of the microbial uptake rate compared to the substrate diffusion rate. When F is large, it signifies a slow uptake rate relative to the rate at which nutrients can diffuse to the cell. This corresponds to the lowest ionic strength curve in Figure 11. As F gets small (or ionic strength increases), the overall uptake rate is reduced.

Thus, to date, the simplest model capable of simulating the carbon-limited algal growth response over a wide range of environmental conditions appears to be a Monod fit of μ vs dissolved CO_2, with K_{SCO_2} varying over a range of values depending on the ionic strength of the culture medium and the biomass concentration. Ultimately, as more data become available, it should become possible to express the carbon-limited growth response as a pure response to CO_2, with the CO_2 levels modified by the diffusivity of the culture medium and the diffusional boundary

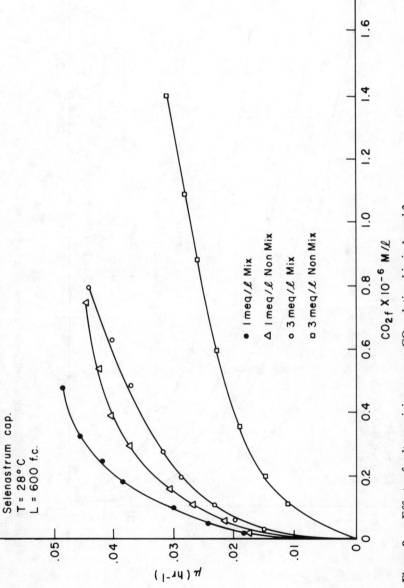

Figure 8.   Effects of culture mixing on μ vs CO_2 relationship in 1- and 3-meq/l initial alkalinity cultures.

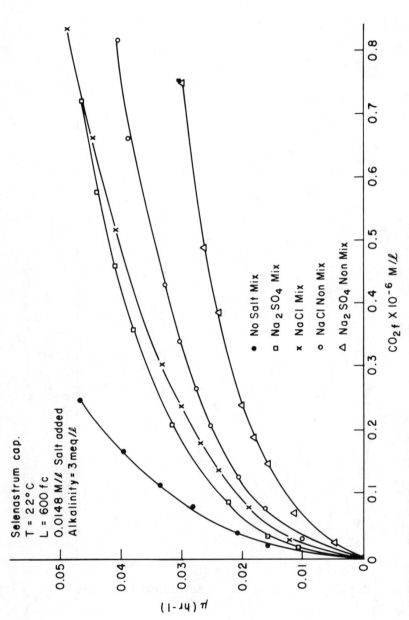

Figure 9.   Effects of mixing and different anions on μ vs CO_2 relationship.

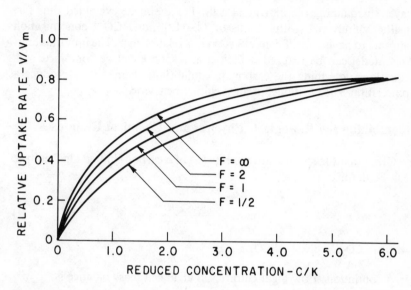

Figure 10. Relative uptake rate as a function of reduced concentration [18].

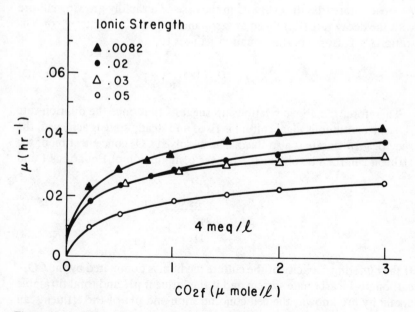

Figure 11. Effect of ionic strength (as NaCl) on μ vs CO_2 relationship at constant alkalinity.

layer surrounding the individual cells. For the algae examined thus far, neither culture pH (within limits of 7.0–11.0) nor HCO_3^- concentration appears to have a significant effect on this relationship. The importance of this model can be realized if biological response can be combined with equations describing the carbonate equilibrium chemistry to produce a powerful predictive model of algal culture behavior.

Combining the Biological, Physical and Chemical Responses

Given that the specific growth rate (μ) of a carbon-limited algal culture can be defined as:

$$\mu = \frac{\mu_{max}(CO_{2_f})}{K_S + (CO_{2_f})} \tag{1}$$

where CO_{2_f} = dissolved CO_2.

In a continuous flow algal culture, an algal cell mass balance is:

$$\frac{dx}{dt} = DX_1 - DX_2 + \mu X_2 - K_d X_2 \tag{2}$$

At steady state ($dx/dt = 0$) and in the case of a rapidly growing culture with the decay rate (K_d) taken as zero, and with the influent cell concentration (X_1) also zero, this equation reduces to:

$$\mu = D = 1/\theta \tag{3}$$

Therefore, the above relationship suggests that once the dilution rate (D) of a continuous algal culture is fixed and steady state is achieved, the specific growth rate is also fixed. The dissolved CO_2 concentration of the effluent can then be obtained from the combination of Equations 1 and 3:

$$[CO_2]_2 = \frac{K_{SCO_2}}{\dfrac{\mu_{max}}{D} - 1} \tag{4}$$

If the buffering capacity of the culture medium is dominated by the CO_2–carbonate–bicarbonate system and if the influent pH and total titratable alkalinity are known, the cell concentration and pH of the effluent can be obtained by combining Equation 4 with the carbonate equilibrium equations [19] where:

$$\text{algal biomass} = X_A = C_{T_1} - C_{T_2} \qquad (5)$$

and

$$C_{T_1} = \frac{\text{AlK} + \text{H}^+ - \text{OH}^-}{\alpha_1 - 2\alpha_2} \qquad (6)$$

The effluent C_T concentration is determined by μ_{max}, D and K_{SCO_2}, and is given by:

$$C_{T_2} = \frac{CO_2}{\alpha_0} = \left[\frac{K_{SCO_2}}{\dfrac{\mu_{max}}{D} - 1} \right] \frac{1}{\alpha_0} \qquad (7)$$

where:

$$[CO_2] = \left(\frac{\text{AlK} + \text{H}^+ - \text{OH}^-}{\alpha_1 - 2\alpha_2} \right)(\alpha_0) \qquad (8)$$

Total titratable alkalinity is affected to a small degree by algal growth [20], and this effect may be accounted for. Since culture alkalinity at any time can be determined, and since $[CO_2]_2$ is determined by D, μ_{max} and K_{SCO_2}, which are also known, the effluent culture pH can be obtained. The solution is in the form of a fourth-order equation [21]. The predicted culture pH for a hypothetical alga (a composite of pooled data) with a μ_{max} of 0.10 hr^{-1} and K_{SCO_2} ranging from 0.17×10^{-6} to 8.1×10^{-6} mol/L (depending on culture alkalinity) is given in Figure 12. As can be seen, the tendency toward higher culture pH as a result of increasing alkalinity is eventually overpowered by the decreasing ability of the algae to extract CO_2 to low levels at the increased ionic strength due to higher alkalinity levels. The net result is that at high alkalinity, continuous cultures will stabilize at lower pH values at a given detention time. Support for the theoretical model has been obtained in the form of data from actual continuous cultures of the alga *Scenedesmus quadricauda*. As can be seen (Figure 13) the general form of the pH response as a function of detention time is as predicted.

The important implications of this model are summarized in Figures 14 and 15. If culture pH is allowed to drift without control, a large percentage of total carbon in the influent medium will not be utilized. At the detention time giving optimum production (Figure 14), the carbon

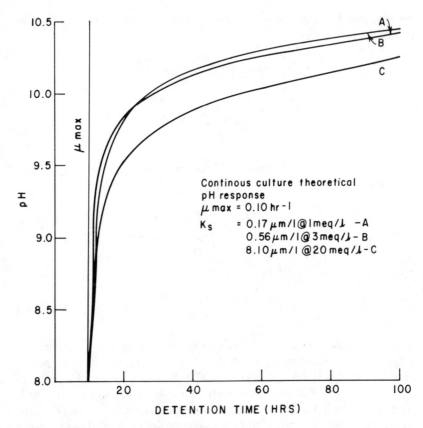

Figure 12. Effect of continuous culture detention time on culture pH as predicted by limiting CO_2 model.

utilization will range from only 10 to 30% depending on culture alkalinity (Figure 15). Therefore, attempts to increase carbon supply by alkalinity addition alone (as $NaHCO_3$) will not provide for efficient utilization of inorganic carbon. pH control through acid addition would markedly improve the situation; however, the costs of continuous acid addition combined with dangers of instability produced by destroying the culture buffering capacity do not favor this technique.

As previously shown, mixing can also be used to increase the availability of inorganic carbon. Unfortunately, the decrease in K_{SCO_2}, at any given ionic strength, as a result of mixing, represents a rather small increase in carbon availability as a fraction of the total carbon concentration. Mixing represents an expensive and energy-intensive method which produces only a small increase in carbon utilization.

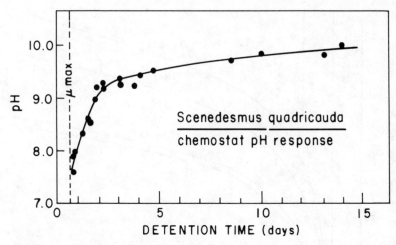

Figure 13.   Actual continuous culture pH as a function of detention time.

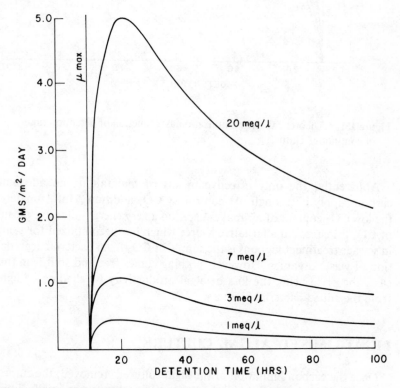

Figure 14.   Theoretical culture productivity as a function of detention given at varying influent alkalinities.

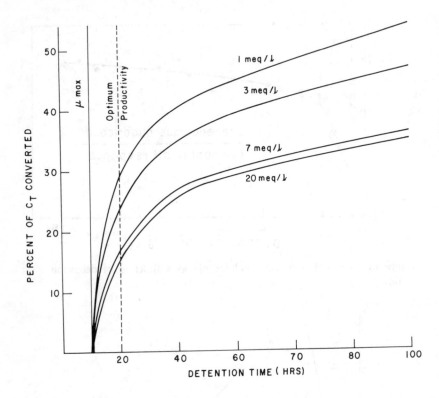

Figure 15. Theoretical carbon utilization as a function of detention time in continuous culture.

Apparently, the only effective means of maintaining an adequate carbon supply is through pH control by CO_2 addition. Unfortunately, the low CO_2 content of air makes aeration a very energy-intensive means of CO_2 transfer. An alternative source which has been utilized for years in sewage treatment lagoons is the supply of CO_2 from bacterial degradation of waste organics. However, nothing comes free and so it is in this case, the price being the loss of algal productivity by shading of light from the added bacterial biomass.

LIGHT-LIMITED ALGAL CULTURE

Once the carbon limitation of the algal culture is removed, the culture will respond by increasing cell density until another factor finally limits cell production. In many cases, this factor will be the availability of light.

It has been demonstrated that net algal cell production in a light-limited culture is independent of culture detention time [22]. Thus, the algal cell density (X) is a linear function of detention time (θ):

$$X \cong \frac{K\theta}{V} \tag{9}$$

It has been shown [23] that the overall productivity (P) can be related to the biological response of the alga to limiting light and the incipient light levels by the equation [24]:

$$P = \frac{\alpha I}{\left(1 + \left(\dfrac{I}{P_m}\right)^2\right)^{1/2}} \tag{10}$$

Using the integrated form of this equation with values of the extinction coefficient of algal biomass of 1.2×10^{-7} liter/cell-mol, the response of the cell density of a shallow light-limited culture of an alga with $P_{max} = 0.10 \text{ hr}^{-1}$, $I_0 = 5000$ fc, to increasing detention time is given in Figure 16 (computer generated solutions to Equation 10). As seen in this figure, cell density responds linearly to hydraulic detention time as predicted by the earlier equation [22]. The ideal behavior illustrated in this figure will be modified by many factors; of prime importance will be the additional light shading by the added bacterial biomass and the effects of bacterial CO_2 production. The bacterial biomass present may be obtained from equations describing the decay of influent BOD where:

$$BOD_E = BOD_I \, 10^{-kt} \tag{11}$$

The bacteria biomass may then be predicted from [25]:

$$X_B = Y_B \frac{(BOD_I - BOD_E)}{1 + K_b \, \theta} \tag{12}$$

and the rate of supply of CO_2 from the bacterial decomposition of the incoming BOD by:

$$\frac{d\,CO_2}{dt} = -C_1 \frac{d\,BOD}{dt} \tag{13}$$

Using these equations, Figure 16 is modified to account for added bacterial biomass and CO_2 production and the resultant modifications are presented in Figures 17 and 18. Assuming a strong waste influent with a BOD of 500 mg/l, a bacterial yield coefficient Y = 0.55, decay rate b =

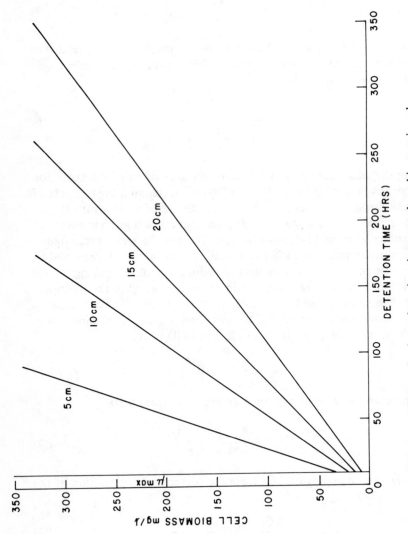

Figure 16. Effects of culture detention time on culture biomass in cultures of varying depth.

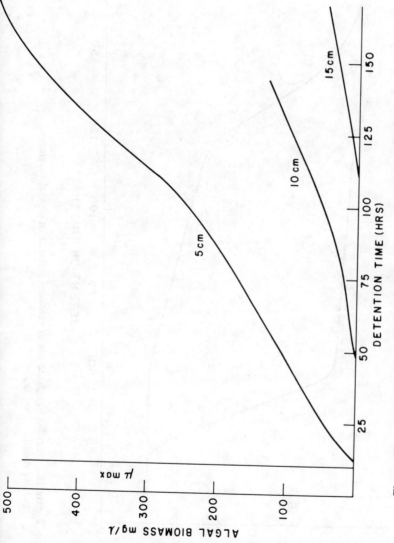

Figure 17. Effect of bacterial biomass production on algal cell density with constant influent BOD.

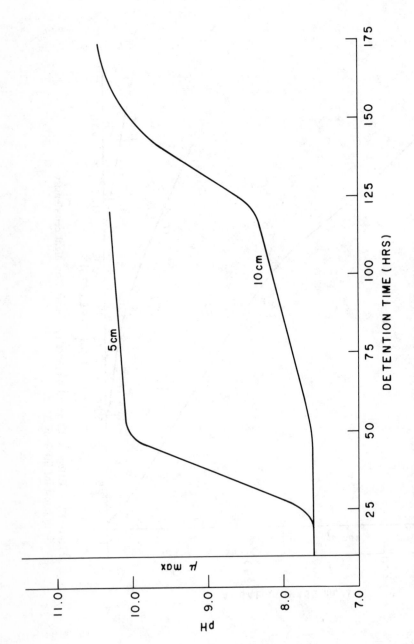

Figure 18. Predicted culture pH at constant influent BOD and varying culture depth.

0.05, C_1 = 0.018 mmol CO_2/mg BOD oxidized, and a coefficient of extinction of light from the bacterial biomasss the same as for the algal biomass, the effects of the added bacterial biomass are illustrated in Figure 17. Perhaps the most important result of this effect can be seen as the requirement for longer and longer detention time at increasing depth to achieve a stable algal cell population. This effect is due solely to the slower algal growth rate as a result of lowered average light levels per unit of algal biomass.

With increasing detention time, the final upper limit on cell growth will again come through carbon limitation and this factor has the effect of producing a decline in biomass increases at longer detention time (Figure 19). These figures illustrate these combined effects on culture pH and algal cell biomass. When bacterial CO_2 production exceeds algal CO_2 fixation, the effect will be to drive the pH below the atmospheric equilibrium pH. The lowest level that pH will fall to will depend on the rate at which CO_2 transports across the water surface and exits from the culture. At steady-state pH:

$$\frac{d\ CO_2}{dt} = \begin{matrix} \text{net} \\ CO_2 \\ \text{production} \end{matrix} = K_{La}(CO_{2_S} - CO_2) \tag{14}$$

On the other hand, if algal CO_2 uptake exceeds bacterial CO_2 production, the culture pH will rise according to the carbonate equilibrium chemistry and carbon uptake behavior of the algae as detailed in Equation 8. Unfortunately, because of low atmospheric CO_2 levels, CO_2 input (unless aggressively supplied) from surface transport will not usually create a significant pH stabilizing effect as will CO_2 transport out of the solution. The total algal cell biomass will respond by increasing in density with increasing detention time until, as a result of the pH rise, the dissolved CO_2 concentration again limits cell production. The effect of either increasing culture depth or increased influent BOD levels will both delay the onset of carbon limitation and increase the detention time for a stable algal biomass population.

Figures 18 and 19 summarize these effects. In Figure 18, the culture pH is seen to remain at low levels at a detention time of less than 25 hr (corresponding to bacterial culture only). With the establishment of a stable algal culture, pH rises rapidly with increases in detention in shallow culture (5 cm). On the other hand, if the culture is deeper (10 cm), the pH rise is delayed as CO_2 production from the increased BOD loading per unit of surface area helps to meet the carbon demand of the algal culture. Ultimately with increasing detention time, the carbon limitation of the growing algal culture will dominate as the uncontrolled culture pH

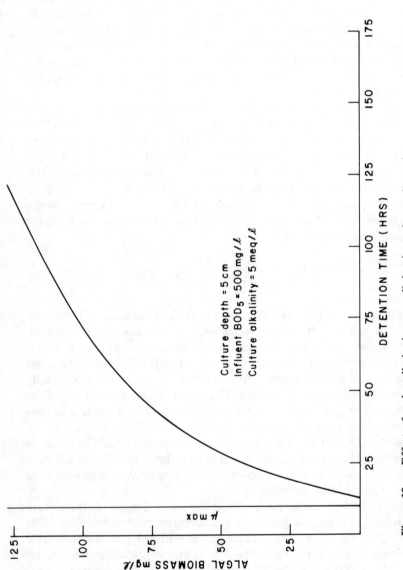

Figure 19.   Effect of carbon limitation on cell density at increasing culture detention time.

levels off at some elevated value. This ultimate limitation is seen in the biomass concentration as a declining biomass level with increasing detention time (Figure 19).

It should be noted that the solutions presented here are not tied to the culture depths in any absolute sense. It would be quite possible to use much deeper cultures, and in some cases, this may be desirable, particularly if concentrated levels of toxic substances (i.e., NH_3) in the influent are at excessive levels. The effect of increasing culture depth while applying the same total BOD loading/unit area will be to dilute total algal cell density.

LIGHT AND CO_2 MODELS AS A PREDICTIVE TOOL

The model presented here considers only the case of carbon- and light-limited growth of algal culture. It is, of course, an oversimplification of complex algal bacterial culture. This model is viewed as a starting point for a more comprehensive model which will be developed to include the important modifiers of the relationships presented here. Of particular importance will be additions to the chemical model to account for various noncarbonate buffers such as ammonia, phosphates and borates. Field determination of the many empirical constants must be made, as well as an assessment of the validity of applying the laboratory-derived kinetic data to field situations. In addition, there are a large number of modifiers to the biological model, such as toxicity effects due to various substances; of particular importance will be toxic effects due to high ammonia levels [25]. Furthermore, interaction among the key variables such as between K_{SCO_2} and light or temperature which has recently been demonstrated [15,26] must be accounted for.

Although the model may be simplistic in nature, the power of a simple carbon and light limitation model in predicting, in general, responses of field algal culture should not be dismissed. Observations of algal cell production from a recent pilot study [17] indicate that the theoretical behavior describes in general the actual culture responses (Figure 20). Although complicated by the changing influent BOD loading rates used in this study, the observations of cell density compare well with the predicted light- and carbon-limited values. The culture pH, which was observed to rise to 10 in shallow cultures, and level off at 8.0–9.0 in deeper cultures, while dropping to 7.6 in the bacterial cultures, behaves as predicted by the dissolved CO_2 limitation model. A simple yet often unappreciated corollary of the carbon model suggests that whenever culture pH rises above the atmospheric equilibrium value, external carbon is

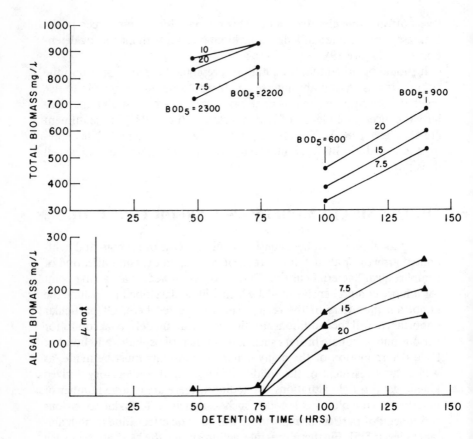

Figure 20.   Data taken from pilot-scale algal culture.

not being supplied at a rate fast enough to meet the algal carbon fixation rate. Thus the culture obtains the needed carbon by extracting it from the carbonate system. Unless this situation is carefully controlled, the pH may stabilize at values which yield dissolved CO_2 concentrations that will limit algal growth rates.

EXTRACTING ENERGY FROM ALGAL CULTURE; RECYCLING CARBON

An attempt has been made to show quantitatively the importance of dissolved CO_2 concentration in controlling algal cell production. One promising method of dissolved CO_2-pH control is through careful selection of detention time, influent BOD, and depth of a combined algal-bacterial culture. Even though bacterial decomposition of organics to CO_2 followed by CO_2 fixation by algae does not represent a net organic carbon fixation, it can be used to obtain a net energy fixation. This may be particularly applicable to the situation in which algal biomass is converted to methane gas via anaerobic digestion. The resulting CO_2 from gas combustion and the organics in the digester effluent represent a recyclable carbon supply to be returned to the algal ponds. In this situation, the importance of proper balancing of algal CO_2 uptake against bacterial CO_2 production cannot be overemphasized. An imbalance in either direction will result in a loss of efficiency in carbon utilization. Proper selection of the control parameters will likely come through continued development and refinement of models such as presented here.

SUMMARY

The carbon-limited kinetic responses of various fast growing algal species have been summarized. These results suggest that the growth responses of many algae used in mass culture may best be represented as a Monod fit of the specific growth rate (μ) and the free carbon dioxide concentration (CO_2). The environmental modifiers of primary importance appear to be light levels, temperature and the diffusivity of CO_2 in the culture medium. The various mathematical models describing algal biological response to limiting CO_{2_f} concentration, carbonate equilibrium chemistry, and the physical configuration of a flow-through microbial culture are combined to predict the general form of equations which predict the pH, total carbon concentration (C_T) and algal cell concentration of a continuous algal culture, given μ_{max} and K_{SCO_2} for the alga of interest. This model is further used to illustrate the underutilization of inorganic carbon in mass algal cultures in which the pH is uncontrolled.

One method of pH control in such cultures involves the utilization of CO_2 supply from bacterial degradation of waste organics in the influent culture medium. In such a situation, both the culture pH and algal cell production will often be governed by either carbon or light limitation depending primarily on the influent BOD loading, detention time, and culture depth. An example is given in which the light-dependent response of a particular alga is combined with equations describing the bacterial cell and CO_2 production as a function of influent BOD. The resultant calculations are used to explain why algal populations in combined algal-bacterial culture are often observed to be unstable at detention times considerably longer than theoretical minimum detention times based on laboratory culture data. The effect of increasing culture depth at constant BOD influent concentration is shown to amplify this effect.

Despite the obvious oversimplification of considering only light and carbon limits in describing the behavior of mass algal culture, examination of actual field data suggest that these two parameters will be of paramount importance in controlling net algal cell production rates.

NOMENCLATURE

ALK = total titratable alkalinity
BOD_I = influent BOD_5
BOD_E = effluent BOD_5
CO_{2_f} = free carbon dioxide concentration
$[CO_2]_2$ = effluent CO_{2_f}
C_{T_1} = influent total carbon concentration
C_{T_2} = effluent total carbon concentration
CO_{2_S} = atmospheric equilibrium CO_{2_f} concentration
C_1 = moles CO_2 produced per mg BOD oxidized
D = dilution rate
I = effective light level
k = BOD decay coefficient
K_b = bacterial decay coefficient
K_d = algal decay coefficient
K = overall algal productivity (from Pipes)
K_{La} = CO_2 transfer coefficient
K_{SCO_2} = CO_{2_f} concentration at which $\mu = \frac{1}{2}\mu_{max}$
P_m = light saturated photosynthetic rate
P = average photosynthetic rate
t = time
V = reactor volume
X_1 = influent algal cell concentration
X_2 = effluent algal cell concentration
X_A = algal cell concentration
X_B = bacterial cell concentration
Y_B = bacterial yield coefficient
α = slope of photosynthetic rate vs. light intensity curve
α_0 = CO_{2_f} fraction of C_T
α_1 = HCO_3 fraction of C_T
α_2 = CO_3 fraction of C_T
μ = specific growth rate
μ_{max} = maximum specific growth rate
θ = hydraulic detention time = $1/D$

ACKNOWLEDGMENT

This research was funded, in part, by a grant from the National Science Foundation.

REFERENCES

1. Goldman, J. C. "Outdoor Algal Mass Cultures. I. Application," *Water Res.* 13:1–19 (1979).
2. Lehman, J. T., D. B. Botkin and G. E. Likens. "Lake Eutrophication and the Limiting CO_2 Concept: A Simulation Study," *Verh. Int. Verein. Limnol.* 19:300–307 (1975).
3. Lehman, J. T., D. B. Botkin and G. E. Likens. "The Assumption and Rationales of a Computer Model of Phytoplankton Population Dynamics," *Limnol. Oceanog.* 29:343–364 (1975).
4. Lehman, J. T. "Enhanced Transport of Inorganic Carbon into Algal Cells and Its Implications for the Biological Fixation of Carbon," *J. Physiol.* 14:33–42 (1978).
5. Toerien, D. G., and C. H. Huang. "Algal Growth Predictions Using Growth Kinetic Constrants," *Water Res.* 17:1673–1681 (1973).
6. *Modeling the Eutrophication Process,* Proceedings of a Workshop, St. Petersburg, FL, National Technical Information Service No. PB–217–383.
7. Bierman, V. J., F. H. Verhoff, T. L. Poulson and M. W. Tenney. "Multinutrient Dynamic Models of Algal Growth and Species Completion in Eutrophic Lakes," in *Modeling the Eutrophication Process,* J. M. Middlebrooks, Ed. (Utah State University Water Research Laboratory, 1973), pp. 89–109.
8. King, D. L. "The Role of Carbon in Eutrophication," *J. Water Poll. Control Fed.* 42:2035–2051 (1970).
9. King, D. L. "Carbon Limitation in Sewage Lagoons," in *Nutrients and Eutrophication,* *Vol. 1* Special Symposium (American Society of Limnology and Oceanography, 1972), pp. 98–105.
10. Shelef, G., M. Schwarz and H. Schecter. "Predication of Photosynthetic Biomass Production in Accelerated Algal-Bacterial Waste Water Treatment Systems," Proceedings of the Sixth International Conference on Water Pollution Research, Jerusalem, Israel (1973), A/5/9/1–A/5/9/10.
11. King, D. L., and J. T. Novak. "The Kinetics of Inorganic Carbon Limited Growth," *J. Water Poll. Control Fed.* 48:1812–18154 (1974).
12. Steemann, N. E., and P. K. Jensen. "Concentration of Carbon Dioxide and Rate of Photosynthesis in *Chlorella pyrenoidosa,*" *Physiol. Plant* 11:170–180 (1958).
13. Osterlind, S. "Inorganic Carbon Sources of Green Algae. III. Measurements of Photosynthesis in *Scenedesmus quardicauda* and *Chlorella pyrenoidosa,*" *Physiol. Plant* 4:242–254 (1951).
14. Goldman, J. C., W. J. Oswald and D. Jenkins. "Kinetics of Inorganic Carbon Limited Growth for Green Algae in Continuous Culture, Its Relationship to Eutrophication," Sanitary Engineering Laboratory Report No. 72–11, University of California, Berkeley, CA (1972).

15.   Brune, D. E. "The Growth Kinetics of Freshwater Algae," PhD Dissertation, University of Missouri–Columbia (1978).

16.   Lindsey, W. B. "Media Effects on Algal Growth Kinetics," MS Thesis, Department of Civil Engineering, University of Missouri – Columbia (1980).

17.   Boersma, L. L., E. Gasper, J. R. Miner, J. E. Oldfield, H. K. Phinney and P. R. Cheeke. "Managment of Swine Manure for the Recovery of Protein and Biomass," final report to the National Science Foundation (1978).

18.   Gavis, J., and J. F. Ferguson. "Kinetics of Carbon Dioxide Uptake by Phytoplankton at High pH," *Limnol. Oceanog.* 20:211 (1975).

19.   Stumm, W., and J. J. Morgan. *Aquatic Chemistry* (New York: John Wiley & Sons, Inc., 1970).

20.   Brewer, P. G., and J. C. Goldman. "Alkalinity Changes Generated by Phytoplankton Growth," *Limnol. Ocean.* 21:108 (1976).

21.   Ricci, J. E. *Hydrogen Ion Concentration* (Princteon, NJ: Princeton University Press, 1952).

22.   Pipes, W. O., "Light Limited Growth of *Chlorella* in Continuous Culture," *Appl. Microbiol.* 10:1–5 (1962).

23.   Smith, E. L. "Photosynthesis in Relation to Light and Carbon Dioxide," *Proc. Nat. Acad. Sci. U.S.* 22:504–511 (1936).

24.   Groden, T. W. "Modeling Temperature and Light Adaptation of Phytoplankton," Center for Ecological Modeling, Renssalaer Polytechnic Institute, Troy, NY (1977).

25.   Abeliovich, A., and Y. Azov. "Toxicity of Ammonia to Algae in Sewage Oxidation Ponds," *Appl. Environ. Microbiol.* Vol. 31 (1976).

26.   Garrett, M. K., S. T. C. Weatherup and M. O. B. Allen. "Algal Culture in the Liquid Phase of Animal Slurry," *Environ. Poll.* Vol. 15 (1978).

SECTION 2

BIOLOGICAL AND THERMAL GASIFICATION

CHAPTER 7

METHANE PRODUCTION BY ANAEROBIC DIGESTION OF WATER HYACINTH (EICHHORNIA CRASSIPES)

D. L. Klass and S. Ghosh

Institute of Gas Technology
Chicago, Illinois

Water hyacinth *(Eichhornia crassipes)* is an aquatic biomass species that exhibits prolific growth in many parts of the world [1]. It has been suggested as a strong candidate for production of methane because of high biomass yield potential [2]. Several studies have been carried out which establish that methane can be produced from water hyacinth under anaerobic digestion conditions [3–6]. Both batch and semicontinuous digestion experiments were performed. The highest apparent gas yields reported were obtained in the batch mode of operation over long detention times [4], but the yields were based on wet hyacinth containing unspecified amounts of water and ash. Some of the data in the literature on gas yield and production rate are difficult to interpret because they are experimentally observed values and are not reduced to standard conditions. The energy recovery efficiencies in the product gas are also not available because the energy contents of the feed were not determined. The work described in this study was initiated to develop more quantitative data in terms of the physical and chemical characteristics of water hyacinth.

MATERIALS AND METHODS

Digesters

The digestion runs were carried out in the semicontinuous mode in cylindrical, complete-mix, 7-liter digesters [7]. The culture volume for all experiments was 5 liters, and the internal diameter of the digesters was 19 cm. Continuous mixing at 130 rpm was provided with two 7.6-cm propeller-type impellers located 7.6 and 15.2 cm from the digester bottom on a central shaft.

Analytical Techniques

Most analyses were performed in duplicate; several were performed in triplicate or higher multiples. The procedures were either ASTM, *Standard Methods,* special techniques as reported previously [8] or other techniques as indicated by footnotes in the tables.

Data Reduction

Gas yield, methane yield, volatile solids reduction and energy recovery efficiency were calculated by methods described previously [8]. All gas data reported are converted to normal cubic meters at 0°C and 101.325 kPa on a dry basis, or to standard cubic feet (scf) at 60°F and 30 in. Hg on a dry basis.

Digester Feeds

Water hyacinth samples (0.5–1 ton) were harvested from an experimental sewage-treatment lagoon of the NASA National Space Technology Laboratory in Bay St. Louis, MS. Whole adult and young plants were collected and fed directly to an agricultural chopper that provided particles 7.6 cm (3 in.) or smaller in size. The chopped hyacinth was placed in polyethylene-lined fiber drums, frozen, shipped by refrigerated truck to IGT, stored at -23°C (-10°F), ground in a laboratory grinder, mixed in a double-ribbon blender to ensure homogeneity, placed in 0.5-gal cartons, and stored at about -29°C (-20°F).

During one of the harvests (June 3, 1977), small samples of whole plants were also collected, in addition to the chopped plants, and shipped separately to IGT overnight in sealed bottles without freezing, for moisture, volatile matter and ash analyses. The results are shown in Table I along with the corresponding analyses for the hyacinth treated as described above.

Whole water hyacinth plants were collected from a 0.1-ha (0.25-ac) freshwater pond in the Lee County Hyacinth Control District, Fort Myers, FL. This pond is located northeast of Fort Meyers in an unincorporated area known as Buckingham and receives both surface runoff and groundwater. The pond is stagnant, has no outlet, is about 3 m deep and has a mucky bottom. The whole plants were shipped by unrefrigerated truck to IGT in polyethylene-lined fiber drums. After arrival at IGT, the water hyacinth was treated in the same manner as the Mississippi shipments.

Grinding of the water hyacinth in the laboratory was achieved with an Urschel Laboratory Grinder (Comitrol 3600) equipped with a 0.076-cm (0.030-in.) cutting head. A typical particle size analysis is shown in Table II, and the effects of storage time on the moisture, volatile matter and ash contents are shown in Table III.

The characteristics of the particular lots of hyacinth used to make the feed slurries for the digestion runs reported in this paper are summarized in Table IV. Feed slurries were prepared fresh daily by blending the required amounts of ground hyacinth and demineralized water. The properties of the slurries are compared in Table V. The pH of the digester contents was maintained in the desired range by adding a predetermined amount of caustic solution to the feed slurry before dilution to the required amount of water. When added nutrient solutions were used, the compositions of which are shown in Table VI, preselected amounts were

Table I.   Moisture, Volatile Matter and Ash Content (wt %)
of Mississippi Water Hyacinth Plant Parts
Harvested June 3, 1977

Plant Part	Moisture	Volatile Matter	Ash
Roots	91.2	63.6	36.4
Stem, Stolon	90.4	80.5	19.5
Stem, Subfloat	90.9	81.2	18.8
Stem, Float	91.1	80.5	19.5
Leaf	87.5	82.6	17.4
Average	90.2	77.7	22.3
Whole (Chopped, Frozen, Thawed, Ground)[a]	95.3	77.7	22.4

[a]After shipment to laboratory, thawing and grinding.

Table II.   Typical Particle Size Analysis of
Ground Water Hyacinth

U.S. Sieve Size (mm)	Retained on Sieve (wt %)
1.180	0
0.600	12.7
0.297	34.5
0.250	72.7
0.212	78.2
0.180	85.5
0.149	89.1
0.105	94.8
0.063	98.2

Table III.   Effect of Source, Harvest Time, and Storage
on Moisture, Volatile Matter, and Ash Content
of Water Hyacinth[a]

Source	Harvest Date	Storage Time (mo)	Moisture (wt %)	Volatile Matter (wt %)	Ash (wt %)
Bay St. Louis, Mississippi	June 3, 1977	2.5	95.3	77.5	22.5
		2.8	95.3	77.9	22.1
		7.8	95.0	76.9	23.1
	June 21, 1978	0.2	94.3	76.5	23.5
		2.2	94.3	75.2	24.8
	July 19, 1978	2.8	94.5	78.8	21.2
Fort Myers, Florida	March 13, 1978	0.5	94.7	79.9	20.1
		5.0	94.3	80.9	19.1

[a]All samples ground with 0.076-cm (0.030-in.) cutting head of Urschel grinder, homogenized, stored at -29°C (-20°F) and thawed before analysis in triplicate.

also blended with the feed slurries before dilution to the final feed volume.

Inoculum, Startup and Operation

The inoculum for the initial replicate digestion runs (runs 1M-B and 2M-B) was developed by accumulating daily effluents from existing laboratory digesters operating on giant brown kelp and primary-

Table IV.   Physical and Chemical Characteristics of
Whole Water Hyacinth Plants after Grinding

Characteristic	Mississippi June 3, 1977		Florida March 13, 1978	
Ultimate Analysis, wt %				
C	41.1		40.3	
H	5.29		4.60	
N	1.96		1.51	
S	0.41		0.49	
P	0.46		0.39	
Ca	2.15		5.80	
Na	1.85		0.47	
K	1.48		1.00	
Mg	0.35		1.40	
Proximate Analysis, wt %				
Moisture	95.3		94.5	
Volatile Matter	77.7[a]	(77.5)[b]	80.4[a]	
Ash	22.4[a]	(22.5)[b]	19.6[a]	
Organic Components, wt % of TS				
Crude Protein[c]	12.3		9.4	
Cellulose[a]	16.2			
Hemicellulose[a]	55.5			
Lignin	6.1[a]	(5.4)[b]		
High Heating Value,				
MJ/dry kg (Btu/dry lb)	16.02	(6,886)	14.86	(6,389)
MJ/kg (MAF) (Btu/lb (MAF))	20.61	(8,862)	18.48	(7,947)
MJ/kg C (Btu/lb C)	38.97	(16,750)	36.87	(15,850)
C/N Weight Ratio	21.0		26.7	
C/P Weight Ratio	89.3		103	
Theoretical Methane Yield,[d] normal m³/kg VS reacted (scf/lb VS reacted)	0.554 (9.36)		0.486 (8.20)	
Theoretical Gas Composition,				
mol % CH_4	56.1		51.8	
mol % CO_2	43.9		48.2	
Theoretical Heat of Reaction, MJ/kg VS reacted (Btu/lb VS reacted)	+1.41	(+606)	+0.809	(+348)

[a] USDA Agricultural Handbook methods.
[b] *Tappi* method.
[c] Kjeldahl N × 6.25.
[d] Based on empirical formulas; yields are not corrected for cellular biomass production.

Table V. Comparison of Feed Slurries[a]

	Mississippi	Florida
Density (g/ml) at 25°	1.0249	1.0170
Total Solids (wt % of slurry)	2.47	2.41
Volatile Matter (wt % of slurry)	1.92	1.92
Total Alkalinity (mg/L) as $CaCO_3$	425	1,443
pH	5.01	6.10
Bicarbonate Alkalinity		
(mg/L) as $CaCO_3$	302	556
Conductivity (μmho/cm)	3,500	2,100
Volatile Acids (mg/L)		
Acetic	50	747
Propionic	102	323
Butyric	47	63
Isobutyric	19	2
Valeric	0	9
Isovaleric	0	11
Total as Acetic	173	1,065
Chemical Oxygen Demand (mg/L)	15,860	17,479
Ammonia N (mg/L) as N	28.0	9.4

[a]Formulated for loading of 1.6 g VS/liter-day (0.1 lb VS/ft³-day), 12-day detention time, 5-liter culture volume.

Table VI.   Composition of Nutrient Solution

Component	Mixed Nutrient Formulation (g/L)	Ammonium Chloride Solution (g/L)
NH_4Cl	30.0	120.0
NaH_2PO_4	20.0	
KI	2.0	
$FeCl_3$	2.0	
$MgCl_2$	2.0	
$CoCl_2$	0.25	
$CaCl_2$	0.25	
$NaMoO_2$	0.10	
$CuCl_2$	0.10	
$MnCl_2$	0.10	
N Concentration (mg/ml)	7.85	31.42
P Concentration (mg/ml)	0.26	

activated sewage sludge as described previously [7]. These digesters were then operated in the semicontinuous mode with initial mixed inoculum volumes of 2.5 liters and a daily feeding and wasting schedule aimed at increasing the working volume to 5 liters over an 8-day period, after which a transition period was begun to change the feed to 100% hyacinth [7]. The total time required from startup to conversion to hyacinth feeds was 42 days. A second transition period was then used to adjust the operating conditions to a loading of 1.6 g volatile solids (VS) /liter-day (0.1 lb VS/ft^3-day) and a detention time of 12 days; this required 21 days [7]. Digestion was then continued at the target operating conditions with hyacinth feed only.

The experimental results obtained at steady state with runs 1M-B, 2M-B, and subsequent runs are shown in Table VII. Steady-state digestion was defined in this work as operation without significant change in gas production rate, gas composition and effluent characteristics. Usually, operation for two or three detention times established steady-state digestion.

Mesophilic runs 1M-4, 1M-7, 1M-8, and 1M-9 were each successively derived starting from the initial run 1M-B. Run 1M-4 shows the effects of added nitrogen as an ammonium chloride solution. Run 1M-7 shows the effects of terminating caustic additions to maintain pH. Run 1M-8 was developed by replacing the Mississippi hyacinth in the feed slurry with Florida hyacinth. Run 1M-9 is a continuation of run 1M-8 except caustic additions were made to control pH. Run 2M-3 was derived from run 2M-B and was carried out with additions of the mixed nutrient solution.

Thermophilic run 1T-5 was developed from the effluents of mesophilic runs 1M-B and 2M-B. Successively, the effluents were collected and used as inoculum (16 days); the digester was operated at the conditions of runs 1M-B and 2M-B to stabilize the new digester (16 days); the temperature was increased to 55°C and the digester was kept in the batch mode (14 days); the semicontinuous mode of operation was started with gradual change of the detention time from 106 to 16.7 days and of the loading from 0.16 to 2.4 g VS/liter-day (0.01 to 0.15 lb VS/ft^3-day) (27 days); and run 1T-5 was continued. Runs 1T-8, 1T-10 and 1T-11 were each successively derived starting from run 1T-5. Runs 1T-8 and 1T-10 were operated at higher loading rates and lower detention times than run 1T-5; ammonium chloride solution was added to each of these runs. Run 1T-11 is identical to run 1T-10 except that nitrogen additions were terminated.

Table VII.   Summary of Selected Steady-State
Digestion Data

Operating Conditions	1M-B Miss.	2M-B Miss.	1M-4 Miss.	2M-3 Miss.
Temperature (°C)	35	35	35	35
pH[a]	7.05	7.05	7.02	6.99
Caustic Dosage (meq/l feed)	49	45	47	50
Loading Rate (kg VS/liter-day)	1.6	1.6	1.6	1.6
Detention Time (days)	12	12	12	12
TS in Feed Slurry (wt %)	2.47	2.47	2.47	2.47
VS in Feed Slurry (wt %)	1.92	1.92	1.92	1.92
Nutrients Added[b]	0	0	N	MN
C/N Ratio in Feed Slurry	21.0	21.0	8.2	8.2
C/P Ratio in Feed Slurry	89.3	89.3	89.3	73.2
Detention Times Operated	5.1	5.1	2.8	2.8
Gas Production[c]				
Rate (vol/vol-day)	0.480 (13)	0.497 (10)	0.477 (6)	0.483 (7)
Yield (normal m³/kg VS added)	0.285 (13)	0.295 (10)	0.282 (6)	0.285 (8)
CH₄ Concentration (mol %)	64.0	62.8	62.3	60.6
CH₄ Yield (normal m³/kg VS added)	0.182	0.185	0.176	0.173
Specific CH₄ Production Rate (normal m³/kg VS added-day)	0.015	0.015	0.015	0.014
Efficiencies				
VS Reduction (%)	28.8	29.8	28.5	28.9
Feed Energy Recovered as CH₄ (%)	35.2	35.7	33.9	33.3
Effluent Volatile Acids (mg/l as HOAc)	27	26	26	51

[a] pH was maintained as indicated by addition of sodium hydroxide solution, except for run 1M-9, where lime was used. No caustic additions were made for runs 1M-7 and 1M-8.
[b] 0 denotes no nutrients added to feed slurry; MN denotes mixed nutrient solution added to feed slurry; N denotes ammonium chloride solution added to feed slurry.
[c] Mean values; the values in parentheses are coefficients of variation.

Dewatering Tests

Gravity sedimentation tests were conducted by a modification of the AEEP Method [9] in which a 400-ml sample of the effluent was examined in a 1-liter graduated cyclinder giving a fluid depth of 140 mm [7]. Vacuum filtration tests were conducted by a modification of the AEEP Method [10] in which a 417-ml sample of effluent was filtered through a monofilament filter cloth [7].

DISCUSSION

Feed Properties

The roots of water hyacinth had higher ash and lower volatile matter contents than other parts of the plant, as shown by the data in Table I.

1M-7 Miss.	1M-8 Fla.	1M-9 Fla.	1T-5 Miss.	1T-8 Miss.	1T-10 Miss.	1%-11 Miss.
35	35	35	55	55	55	55
6.72	6.57	6.87	7.08	7.00	6.82	6.80
0	0	31	21	17	4	5
1.6	1.6	1.6	2.4	3.36	4.8	4.8
12	12	12	16.7	12	6	6
2.47	2.41	2.41	3.70	5.19	7.41	7.41
1.92	1.92	1.92	2.87	4.03	5.76	5.76
0	0	0	0	N	N	0
21.0	26.7	26.7	21.0	11.8	15.1	21.0
89.3	103	103	89.3	89.3	89.3	89.3
2.7	6.6	3.5	1.0	1.4	3.0	1.0
0.488 (15)	0.268 (13)	0.179 (21)	0.688 (10)	0.865 (11)	1.062 (6)	1.026 (6)
0.289 (15)	0.159 (12)	0.106 (21)	0.271 (8)	0.243 (10)	0.210 (5)	0.202 (8)
57.4	61.8	66.2	57.5	58.7	57.9	57.3
0.166	0.098	0.070	0.156	0.143	0.122	0.115
0.014	0.0082	0.0058	0.0093	0.012	0.020	0.019
29.2	17.0	11.3	27.4	24.6	21.3	20.4
32.0	21.1	15.2	30.0	27.6	23.5	22.3
9	5	63	7	10	21	16

Harvesting and storage times as well as the source of the plant seemed to have little effect on the moisture, volatile matter and ash contents of the plants, as illustrated by the data in Table III. Samples harvested many months apart in Mississippi had essentially the same volatile matter and ash contents. The sample harvested in Florida had slightly higher volatile matter and slightly lower ash contents than the Mississippi samples, but this might expected in view of the different growth media from which the hyacinth harvests were taken. The Mississippi hyacinth was grown in a sewage-fed lagoon, and hyacinth is known to take up heavy metals from such media [1].

The data on the chemical and physical properties of the Mississippi and Florida hyacinths used in this work (Table IV) indicate some interesting differences. The C/N and C/P weight ratios are each lower for the Mississippi hyacinth than the Florida hyacinth, but both sets of ratios appear to be somewhat high when compared with the corresponding ratios supplied by suitable feeds for anaerobic digestion such as giant brown kelp and sewage sludge [7]. Although analytical data for the organic components in Florida hyacinth were not obtained, the relatively high hemicellulose content of the Mississippi hyacinth indicates potentially good digestibility [7]. Interestingly, the theoretical methane yield derived from the empirical formula and stoichiometric conversion [7] of the Mississippi hyacinth has a maximum value about 14% higher than that of the Florida hyacinth.

Comparison of the feed slurries (Table V) also reveals some interesting differences. The slurry made with the Mississippi hyacinth had a lower pH and buffering capacity than the Florida hyacinth slurry and therefore needed more caustic for pH control. However, the ammonia nitrogen concentrations in each slurry appeared too low for good digestion when compared to the beneficial range for sewage digestion [11]. Concentrations of calcium, potassium, sodium and magnesium calculated from the data in Table IV for the feed slurries, assuming each element is totally dissolved, were either in the stimulatory range or less than the inhibitory range [11]. Addition of sodium hydroxide for pH control, although increasing the sodium ion concentration several times, was still estimated to be insufficient to raise the sodium ion concentration to the inhibitory range. Also, addition of lime for pH control (run 1M-9) at the level required raised the calcium ion concentration in the feed slurry but not enough to inhibit digestion based on sewage digestion and inhibition by metallic cations [11].

Mesophilic Digestion

Operation of replicate runs 1M-B and 2M-B on Mississippi hyacinth without added nutrients showed good reproducibility and balanced digestion. Typical operating performance over a period of several detention times is shown in Figure 1. It was found that to maintain pH in the desired range, about 45–50 meq of sodium hydroxide per liter of feed had to be added.

To attempt to increase methane yields, pure and mixed nutrient solution additions were made in runs 1M-4 and 2M-3, respectively, while controlling pH with added caustic. Little change was observed in digester performance; the gas production rates and yields were about the same as those observed without nutrient additions.

Elimination of both pH control and nutrient additions in run 1M-7 resulted in small decreases in pH, methane yield and methane concentration in the product gas, but overall performance in terms of volatile solids reduction and energy recovery efficiency as methane were about

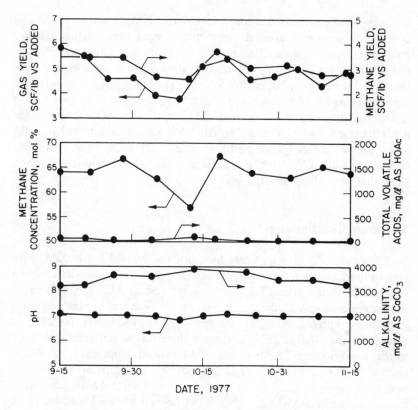

Figure 1. Typical performance, run 1MB.

the same as those of the runs with pH control and with or without nutrient additions.

Conversion from Mississippi hyacinth to Florida hyacinth in run 1M-8, which did not incorporate pH control or nutrient additions and which was identical to run 1M-7 except for the feed source, showed significant reduction in most of the gas production parameters. Gas production rate and yield and methane yield decreased, but digester performance was still balanced as shown by low volatile acids in the digester effluent and the methane concentration in the product gas. From the elemental analyses and the theoretical methane yields (Table IV), the methane yield for run 1M-8 would be expected to be about 14% less than that of run 1M-7; it decreased by about 41%. Prolonged operation of run 1M-8 for more than six detention times did not result in any improvement; the run exhibited steady-state performance with no change in methane yield or gas production rate. Use of pH control (run 1M-9) and continued operation reduced the methane yield even further. It was concluded from these experiments that the Florida hyacinth sample contained unknown inhibitors or that the Mississippi water hyacinth contained unknown stimulatory components. The latter possibility was considered more likely because the Mississippi hyacinth was grown in a sewage-fed lagoon, and it is well established that normal sewage has good digestion characteristics [11]. Also, it is known that water hyacinth, when grown in laboratory media enriched with nickel and cadmium, components often found in sewage, incorporates these metals and shows good digestion characteristics [4].

Thermophilic Digestion

Digestion of Mississippi water hyacinth was carried out at 55°C with and without nitrogen supplementation. Balanced digestion was achieved with all four runs, runs 1T-5, 1T-8, 1T-10, and 1T-11. The gas production rate increased with decreases in detention time and increases in loading rate as expected. Also, as expected, the gas production rate at 55°C was higher than that at 35°C, and again there was no apparent benefit of nitrogen additions. The methane yield ranged from 0.115 to 0.156 normal m³/kg VS added (1.95 to 2.63 scf/lb VS added) over the detention time range studied (6–16.7 days). At the same 12-day detention times, the methane yield at 55°C, 0.143 normal m³/kg VS added (2.42 scf/lb VS added) (run 1T-8) was lower than those observed for all of the mesophilic runs at 35°C with Mississippi hyacinth. However, comparison of the specific methane production rates [methane production rate/(loading × detention time)] in Table VII shows that at the highest loading and shortest detention time studies in this work (runs 1T-10 and 1T-11), the rate of methane production per unit mass of volatile solids added is higher at 55°C than at 35°C, although the methane yields are lower.

Carbon and Energy Balances

The difficulty of calculating carbon and energy balances for digestion experiments in which additions of alkali and nutrients are made has been discussed before [7]. These additives contribute to ash weights. The two methods used to circumvent this problem in previous work [7] were also used in this work. They are described in the footnotes to Table VIII, which presents sample calculations by each method for runs 1M-B, 2M-B

Table VIII. Summary of Carbon and Energy Balances

Run	Accounted For			
	Feed Carbon (%)		Feed Energy (%)	
1M-B	99.5,[a]	102[b]	105,[a]	107[b]
2M-B	98.3,[a]	100[b]	104,[a]	106[b]
1M-9	80.8,[a]	87.0[b]	85.7,[a]	91.9[b]

[a]Calculated from experimental determinations for moisture, volatile solids, ash, carbon and heating values of feed and digested solids, and yield and composition of product gas. Volatile solids in digested solids calculated from percent volatile solids reduction.
[b]Calculated from parameters in footnote a except that ash in digested solids estimated by assuming original ash in feed is in digested solids, NaOH used for pH control is converted to $NaHCO_3$ on ashing at 550°C and remains in ash, and that NH_4Cl, if added, is volatilized on ashing.

and 1M-9. Run 1M-9 exhibited the largest deviation from the theoretical carbon and energy balances: both balances were quite low and only accounted for 81–87% of the feed carbon and 86–92% of the feed energy. The major reason for this is probably the deviation in the experimental gas production measurements. Run 1M-9 had the largest coefficients of variation of all the runs for both gas production rate and yield (Table VII).

Properties of Effluent and Digested Solids

A comparison of fresh feed slurries and effluents from runs 1M-B, 1M-4 and 1M-8 is presented in Table IX. The addition of sodium hydroxide for pH control in run 1M-B had the expected effects on total and bicarbonate alkalinities, pH and conductivity. The effluent from run 1M-4, which was subjected to both caustic and nitrogen additions, showed the same trends except that the ammonia nitrogen concentration also increased. Run 1M-8, which had neither caustic nor nitrogen additions, showed a significant increase in alkalinity and a major reduction in volatile acids. The volatile acids present in the fresh feed slurry were expected to undergo a large decrease on balanced digestion. However, the conversion of nonammonia nitrogen in the feed to ammonia nitrogen in the effluent is not apparent in these runs in contrast to the usual increase observed on digestion [7]. Also, because of the moderate to low volatile solids reductions in these experiments, the chemical oxygen demands of the digester effluents are relatively high.

Table 7.9.   Comparison of Feed and
Digester Effluent Slurries

Reactor	Mississippi Hyacinth Slurry	Run 1M-B	Run 1M-4	Florida Hyacinth Slurry	Run 1M-8
Total Alkalinity (mg/L as $CaCO_3$)	425	3,400	3,460	1,443	2,3000
pH	5.01	7.05	7.02	6.10	6.57
Bicarbonate Alkalinity (mg/L as $CACO_3$)	302	3,390	3,430	556	2,290
Conductivity (ωmho/cm)	3,500	5,620	9,870	2,100	2,680
Volatile Acids (mg/L as HOAc)	173	27	26	1,065	5
Chemical Oxygen Demand (mg/L)	15,860	12,020		17,479	14,630
Ammonia N (mg/L) as N	28	27	640	9	2

A few experiments were carried out to examine the gravity sedimentation and filtration characteristics of digester effluent from run 1M-B. The sedimentation results for unconditioned and conditioned effluent are shown in Figure 2. The settling characteristics were poor and the con-

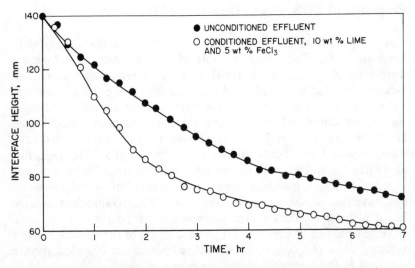

Figure 2.   Interface height vs time for gravity settling of unconditioned effluent from run 1MB.

ditioning treatment improved settling only slightly. A more detailed study is necessary to optimize the conditioning method. Similarly, the filtration characteristics of the conditioned and unconditioned effluent shown in Table X were poor.

The properties of the dry feeds and digested solids from runs 1M-B, 2M-B and 1M-9 are listed in Table XI. Carbon content, volatile matter and heating value of the total digested solids decreased on digestion as expected while ash content increased. The heating value per mass unit of contained carbon remained reasonably constant from dry feed to dry digested solids, but there appeared to be a significant reduction in the heating value of the volatile matter in the Florida hyacinth residual solids, while the heating value of the Mississippi hyacinth residual solids remained about the same as that of the feed. As indicated in previous work [7], this may be due to the difference in degradabilities of different organic components.

Thermodynamic Estimates

The maximum theoretical methane yields uncorrected for cellular biomass production for the Mississippi and Florida water hyacinth samples used for the digestion runs were estimated to be 0.554 and 0.486 normal m^3/kg VS reacted (9.36 and 8.20 SCF/lb VS) (Table IV). Assuming that 7% of the protein and 20% of the carbohydrate is converted to cells on one pass through the digester, the maximum theoretical yield of methane for Mississippi hyacinth is given by [7]:

$$(1 \text{ lb VS added} - 0.195 \text{ lb VS to cells}) \left(9.36 \; \frac{\text{scf } CH_4}{\text{lb VS reacted}} \right)$$

$$= 7.53 \; \frac{\text{scf } CH_4}{\text{lb VS-pass}} = \frac{0.446 \text{ normal } m^3 \; CH_4}{\text{kg VS-pass}}$$

If the same conversion factor is assumed to be valid for the Florida hyacinth sample, the corresponding yield is 0.391 normal m^3/kg VS-pass (6.60 scf CH_4/lb VS-pass). The highest experimental methane yields observed for the Mississippi and Florida hyacinth samples used in this work are 0.185 and 0.0983 normal m^3/kg VS added (3.13 and 1.66 scf/lb VS added), or about 42% an 25% of these theoretical values.

Table X.   Vacuum Filtration Characteristics of Digestion Effluent
(Run 1M-B)

Effluent		Cake			Yield[a]		
TS (wt %)	VS (wt % of TS)	TS (wt %)	VS (wt % of TS)	Dry Cake (Vg/m²-hr)	Filtrate (kg/kg dry cake)		Conditioned[b]
1.63	60.7	11.5	82.1	8.54	136		No
1.60	61.3	14.4	73.5	2.17	420		Yes

[a]30-sec cycle time, 6-sec form time, 12-sec drying time, 12-sec removal time, 67.5 kPa (20 in. Hg).
[b]Flocculent doses were FeCl₃, 5 wt % TS; Ca(OH)₂, 10 wt % TS.

Table XI.   Comparison of Dry Feed and Digested Solids[a]

	Mississippi Hyacinth	Run 1M-B	Run 2M-B	Florida Hyacinth	Run 1M-9
Ultimate Analysis (wt %)					
C	41.1	31.7	31.3	40.3	27.3
H	5.29	3.82	3.78	4.60	3.30
N	1.96	1.98	1.98	1.51	
Proximate Analysis (wt %)					
Moisture	95.3			94.5	
Volatile Matter	77.7	60.7	60.7	80.4	69.4
Ash	22.4	39.3	39.3	19.6	30.6
Heating Value					
MJ/dry kg (Btu/dry lb)	16.02 (6,886)	12.28 (5,280)	12.21 (5,249)	14.86 (6,389)	10.21 (4,391)
MJ/kg (MAF) [Btu/lb (MAF)]	20.61 (8,862)	20.23 (8,698)	20.11 (8,647)	18.48 (7,947)	14.72 (6,327)
MJ/kg C (Btu/lb C)	38.96 (16,750)	38.75 (16,660)	39.01 (16,770)	36.87 (15,850)	37.40 (16,080)

[a]The dry digested solids were prepared by evaporation of the total effluent to dryness on a steam bath, pulverization, and drying in an evacuated desiccator to a constant weight.

Comparison With Other Substrates

The methane yields, volatile solids reductions and energy recovery efficiencies as methane in the product gas from experiments carried out under similar high-rate conditions with other substrates are summarized in Table XII [7] along with the results from run 2M-B. The relatively narrow span of the yields and efficiencies when considered together suggests that standard high-rate conditions in the conventional range tend to afford about the same digestion performance with degradable substrates. The basic organic component groups in these substrates are similar. Usually, the largest fraction consists of mono- and polysaccharides and the smallest fraction is lignin, if present at all. The protein content is usually intermediate in concentration. Experimental data indicate that the hemicelluloses are generally more degradable than the cellulosics on digestion [7], and that the cellulosics and protein fraction are lower in degradability than the monosaccharides [12]. Thus, feeds high in hemicelluloses and monosaccharides should exhibit high gasification rates, but the actual concentrations of these components in the feeds might be expected to govern gas yields. Further improvements in yields and energy recovery efficiencies are therefore more likely through post- or pretreating procedures that increase the degradabilities of the resistant organic components in biomass, or through longer residence times. For example, about 90% of the monosaccharide glucose was converted to product gas on anaerobic digestion at an overall residence time of about 4.5 days in a two-phase system [13], while long-term digestion of cellulose indicates an ultimate anaerobic biodegradability of about 75% [14]. A mixed biomass-waste feed containing water hyacinth has been estimated to have an ultimate biodegradability of 66% [15].

CONCLUSIONS

Water hyacinth under conventional high-rate digestion conditions exhibited higher methane yields and energy recovery efficiencies when grown in sewage-fed lagoons as compared to the corresponding values obtained with water hyacinth grown in a freshwater pond. Mesophilic digestion provided the highest feed energy recovered in the product gas as methane while thermophilic digestion, when operated at sufficiently high loading rates and reduced detention times, gave the highest specific methane production rates. Methane yields, volatile solids reduction and energy recovery as methane for the sewage-grown water hyacinth were in the same range as those observed for other biomass substrates when digested under similar conditions.

Table XII.   Comparison of Steady-State Methane Yields and Efficiencies under Standard High-Rate Conditions[a]

	Coastal Bermuda Grass[b]	Kentucky Bluegrass	Giant Brown Kelp	Mississippi Water Hyacinth[c]	Primary Sludge
CH_4 Yield, normal m^3/kg	0.208	0.150	0.229	0.185	0.31
VS added (scf/lb)	(3.51)	(2.54)	(3.87)	(3.13)	(5.3)
VS Reduction, %	37.5	25.1	43.7	29.8	41.5
Energy Recovered as CH_4, %	41.2	27.6	49.1	35.7	46.2

[a] Loadings of about 1.6 g VS/m^3-day (0.1 lb VS/ft^3-day), detention time of 12 days, 35°C.
[b] Supplemented with added nitrogen.
[c] Run 2M-B.

ACKNOWLEDGMENT

This research was supported by United Gas Pipe Line Company, Houston, Texas. The project was performed under the management of UGPL and is currently managed by the Gas Research Institute. The assistance of Dr. B. C. Wolverton of NASA and Mr. E. S. Del Fosse of the Lee County Hyacinth Control District, Fort Myers, Florida, in obtaining the hyacinth samples is gratefully appreciated. The authors also wish to thank Dr. Victor Edwards and Mr. Robert Christopher of United for their many valuable discussions and suggestions. We also acknowledge the efforts of Michael Henry, Janet Vorres, Alvin Iverson, Mona Singh, Frank Sedzielarz, Phek Hwee Yen and Ramanurti Ravichandran in performing the experimental work.

REFERENCES

1. Del Fosse, E. S. "Water Hyacinth Biomass Yield Potentials," Symposium Papers, Clean Fuels from Biomass and Wastes, sponsored by IGT, Orlando, FL, January 25–28, pp. 73–99.
2. Klass, D. L. "A Perpetual Methane Economy—Is It Possible?" *Chemtech* 4:161–168 (1974).
3. "Technology for the Conversion of Solar Energy to Fuel Gas," University of Pennsylvania, Annual Report NSF/RANN/SE/GI34991/PR/7314 (1974).
4. Wolverton, B. C., R. C. McDonald and J. Gorden. "Bio-conversion of Water Hyacinths into Methane Gas: Part 1," National Space Technology Laboratories, Bay St. Louis, MS, NASA Technical Memorandum TM-X2725 (1975).
5. Chin, K. K., and T. N. Goh. "Bioconversion of Solar Energy: Methane Production Through Water Hyacinth," Symposium Papers, Clean Fuels from Biomass and Wastes, sponsored by IGT, Washington, DC, August 14–18, 1978, pp. 215–228.
6. Ryther, J. H. et al. "Biomass Production by Marine and Freshwater Plants," Proceedings of the Third Annual Biomass Energy Systems Conference Proceedings, sponsored by DOE, Golden, CO, June 5–7, 1979, pp. 13–23.
7. Klass, D. L., and S. Ghosh. "Methane Production by Anaerobic Digestion of Bermuda Grass," ACS Div. Pet. Chem. Preprint. 24(2):414–428 (1979).
8. Klass, D. L., and S. Ghosh. "The Anaerobic Digestion of *Macrocystis pyrifera* under Mesophilic Conditions," Symposium Papers, Clean Fuels from Biomass and Wastes, sponsored by IGT, Orlando, FL, January 25–28, 1977, pp. 323–351.
9. O'Conner, J. T., Ed. *Environmental Engineering Unit Operations and Unit Processes Laboratory Manual, Vol. 3* (Knoxville, TN: Association of Environmental Engineering Professors, 1972).

10. O'Conner, J. T., Ed. *Environmental Engineering Unit Operations and Unit Processes Laboratory Manual, Vol. 2* (Knoxville, TN: Association of Environmental Engineering Professors, 1972).
11. McCarty, P. L. "Anaerobic Waste Treatment Fundamentals, Part III," *Public Works* (November 1964), pp. 91-94.
12. Klass, D. L., S. Ghosh and D. P. Chynoweth. "Methane Production from Aquatic Biomass by Anaerobic Digestion of Giant Brown Kelp," *Process Biochem.* 14(4):18-22 (1979).
13. Cohen, A. et al. "Anaerobic Digestion of Glucose with Separated Acid Production and Methane Formation," *Water Res.* 13:571-580 (1979).
14. Ghosh, S. et al. "A Comprehensive Gasification Process for Energy Recovery from Cellulose Wastes," paper presented at the Symposium on Bioconversion of Cellulosic Substances into Energy, Chemicals and Protein, New Delhi, India, February 1977.
15. Ghosh, S., M. P. Henry and D. L. Klass. "Bioconversion of Water Hyacinth-Coastal Bermuda Grass-MSW-Sludge Blends to Methane," *Biotechnol. Bioeng.* (10):163-187 (1980).

CHAPTER 8

ENZYMATIC ENHANCEMENT OF THE BIOCONVERSION OF CELLULOSE TO METHANE

G. M. Higgins, L. D. Bullock and J. T. Swartzbaugh
Systems Technology Corporation
Xenia, Ohio

Much of the current interest in bioconversion technologies has been focused on the conversion of cellulosic materials to readily usable fuel products such as ethanol or methane. These technologies have traditionally been two-step processes in which the cellulosic material is first hydrolyzed to glucose monomers and is then biologically converted to the final fuel product. Since the efficiency of the biological conversion processes is highly dependent on the feedstock presented to the organisms, great interest has been focused on improving the yield of readily metabolized simple sugars from various hydrolysis techniques.

This chapter reports on a three-year program to examine the potential of enzymatic hydrolysis as a means to enhance the bioconversion of cellulosic wastes to methane. The major portion of this work was focused on the application of enzymatic hydrolysis to the anaerobic digestion of municipal solid waste (MSW). Preliminary studies on the feasibility of this process for bioconversion of other cellulosic wastes have also been conducted. Four major areas of work are described in this chapter:

1. operation of a 60-liter laboratory-scale anaerobic digester on MSW feedstock;
2. operation of a laboratory-scale process line incorporating a 60-liter MSW digester, an enzyme reactor for hydrolysis of the digester effluent, and a second anaerobic digester for biogasification of the enzyme reactor effluent;

3. optimization of the enzymatic hydrolysis process to maximize the conversion of cellulose to simple sugars; and
4. investigation of the utility of enzymatic hydrolysis as a means to enhance the bioconversion of food processing wastes and paper mill sludges.

BACKGROUND

The difficulties encountered in the hydrolysis of cellulosic materials can be traced directly to their chemical and physical structure. Cellulosic material is composed primarily of linear cellulose chains of β (1 → 4) glucopyranose arranged into long bundles referred to as microfibrils. These bundles are arranged into larger structural units with varying levels of organization, portions of which have been operationally identified as either crystalline regions (highly structured) or amorphous or paracrystalline regions (less highly structured) [1]. Organization of these structures is maintained through the operation of interchain hydrogen bonding and van der Waals forces. The molecular weight of cellulose ranges from about 50,000 to 2,500,000 in various species, with 300–15,000 glucose monomers per molecule. Associated with the cellulosic material are a number of other chemical compounds, including primarily hemicelluloses, pectins and lignins, which form the matrix within which the cellulose fibers are arranged. Table I shows the variability of the relative composition of these associated compounds between different cellulosic materials [1–4].

The combined effect of these properties of cellulosic materials ultimately controls their accessibility to hydrolysis. In general, the relatively high crystallinity of cellulose and the high level of substitution with associated compounds of limited biodegradability make the hydrolysis of cellulosic materials a difficult proposition. A great number of techniques have been employed to enhance the rate of hydrolysis to simple sugars

Table I.   Composition of Various Cellulosic Materials

Composition	Cellulose (wt %)	Lignin (wt %)	Noncellulosic Polysaccharides (wt %)
Softwoods	40–50	20–30	25–30
Hardwoods	40–55	18–25	24–40
Cotton	~90	~5	~0
Grasses	25–40	10–30	25–50
Newsprint	~60	~20	~15

[2,5], the majority of which fall into three categories of hydrolytic methods: chemical, thermal and biological.

Chemical treatments for the enhancement of cellulose hydrolysis have included a comprehensive list of hydrolytic agents: acids, bases, cadmium oxide, dimethyl sulfoxide, ethylenediamine tartrate, cadoxen, hydrazine, ferric tartrate, cuprammonium and others [6,7]. The most widely employed of these methods have been acidic and basic hydrolysis.

Various methods of acidic hydrolysis are described in the literature. Converse and Grethlein [8] reported that the application of 1% acid hydrolysis at 240°C in a high-pressure, steam-injection, plug-flow reactor system produced a 50–57% yield of glucose from 5–13.5% cellulose slurries. This method has been applied as a pretreatment method prior to enzymatic hydrolysis to glucose, and the overall conversion of cellulose to glucose has been found to increase from 20 to 90% for oak, from 50 to 100% for corn stovers and from 60 to 93% for newsprint. O'Neil [9] reported that with recycle of delignified cellulose through this type of plug-flow reactor, the overall glucose yield can be expected to approach 80%.

Rogers [10] and Brenner et al, [11] reported on a high-temperature extrusion reactor process in which the acidic hydrolysis of 20–30% cellulose slurries and of solid sawdust at 240°C produced up to a 60% yield of glucose. Tsao [12] reported on an acidic hydrolysis process that converts the highly crystalline cellulose in lignocellulosics into the amorphous state, so that they can subsequently be converted rapidly to glucose by enzymatic hydrolysis. This process has been applied successfully to cornstalks, bagasse and other nonwoody lignocellulosics. Hydrolysis is accomplished by the application of 85% aqueous H_2SO_4, alkaline tartrate and cadoxen (5% CaO in 28% aqueous ethylenediamine).

Alkaline hydrolysis has also received wide attention as an effective method of cellulose hydrolysis. Perhaps the most widely employed form of alkaline hydrolysis can be found in the kraft or alkaline processes used in manufacturing paper. These processes normally involve cooking the wood for 2–3 hr at approximately 175°C and 110 psig in an alkaline solution of NaOH and Na_2S [13]. Ishida et al. [14] reported on the use of alkaline hydrolysis as a pretreatment method for the anaerobic digestion of municipal waste. With a 3-hr pretreatment at 60°C and pH 9.8, the cellulose-rich waste was converted at high yield into volatile acids and subsequently to methane. Millett et al. [2] studied the digestibility of various cellulosic materials in rumen fluid following alkaline pretreatment. The pretreatment of wood with 1% NaOH for 1 hr at room temperature and pressure was shown to cause up to a tenfold increase in the digestibility of cellulose in rumen fluid. Subsequent analysis of these results indicated that the digestibility of the material was highly related to its lignin content, and that at 20–35% lignin, digestibility fell to near zero. As in the various kraft processes, one of the major effects of the alkaline pretreatment process was to partially delignify the material thereby making it more susceptible to hydrolysis.

Thermal processes for the conversion of cellulosic materials have been directed primarily to pyrolysis for gaseous fuels, char and pyrolytic oil residues. Kosstrin [15] and Diebold and Smith [16] recently reported on two representative forms of the pyrolysis process. Thermal processes for the hydrolysis of cellulosic materials to simple sugars have appeared much less frequently in the literature. One such scheme is the Iotech Process®, which involves high-temperature steam treatment of the material followed by explosive depressurization [5,17]. The effect of this process is a significant increase in the surface area of the material, a reduction in the molecular weight of the hemicellulose fraction and partial liberation of the lignin from the cellulosic matrix. As a result, the extent of cellulose hydrolysis is significantly increased, as evidenced by rumen fluid digestibility studies. McCarty and co-workers [18,19] reported on a similar process termed autohydrolysis in which slurried lignocellulosic materials are heated at temperatures from 150 to 225°C in the absence of oxygen, resulting in significant increases in hydrolysis and subsequent anaerobic digestibility to methane. The autohydrolysis process is thought to occur as a result of the low autogenic pH of about 2.5 produced by organic acids released during high-temperature treatment. Up to 85% of the cellulose content of fir wood has been hydrolyzed by this process into soluble products suitable for anaerobic fermentation.

A number of biological processes have been employed to enzymatically hydrolyze cellulosic materials to glucose. Thayer [20] examined the ability of various bacterial species to hydrolyze cellulose. *Pseudomonas* sp. was found to possess both cell-bound and extracellular cellulase enzymes capable of hydrolyzing mesquite wood to 17% of its initial weight in 10 days at 30°C. The same organism was also moderately active against newsprint and lignin. A number of *Bacillus* sp. and *Brevibacterium* sp. also exhibited limited ability to hydrolyze cellulose. Ben-Bassat et al. [21] examined the hydrolysis of cellulose by three thermophilic bacteria: *Thermoanaerobium brockii, Methanobacterium thermoautotrophicum* and *Clostridium thermocellum*. All three species were capable of converting various hexoses, pentoses and, in particular, cellobiose, an intermediate product of complete cellulose hydrolysis, to the end products ethanol, lactate, acetate and H_2/CO_2. Only *C. thermocellum*, however, was capable of hydrolyzing cellulose. In co-culture with *C. thermohydrosulfuricum, C. thermocellum* degraded 90% of a commercial cellulose substrate (after 48–96 hr), 50% of a delignified wood substrate (after 168 hr) and < 1% (after 168 hr) of an aspen wood sample.

The most studied source of cellulolytic enzymes is the mesophilic fungi. Research into the enzymatic digestion of cellulose by fungal enzymes has been conducted at the U.S. Army Natick Laboratories [6,22,23]. One particular source of this enzyme, that secreted by the fungus *Trichoderma reesei,* has been extensively characterized and found highly suitable for the digestion of insoluble cellulose to readily metabolized sugars [22]. Conversions to soluble sugars approaching 100% using various pure cellulosic substrates are reported for this enzyme complex [24]. In practice, the application of this enzyme to substrates of less than pure cellulose content, and particularly to those containing the phenolic compound lignin, has produced a considerably lower conversion. This results from a number of factors, perhaps the more significant of which are product inhibition of the conversion process and the tendency for lignin to complicate the process by strongly cementing the cellulosic polymers together and rendering them less susceptible to hydrolysis [6]. The inhibition of cellulose hydrolysis by the end products cellobiose and glucose has been well characterized in the literature [26,27]. Bullock et al. [27,28] examined the effect of glucose inhibition of the enzymatic hydrolysis process and found that the extent of hydrolysis of pure cellulose decreased significantly at concentrations of glucose well below 1%.

A number of studies have been reported in the literature in which enzymatic hydrolysis of cellulose has been coupled with the ethanol fermentation [6,7,29]. Wilke and Mitra [29] reported that fermentation by the yeast *Saccharomyces cerevisiae* proceeds to 83% of the theoretical conversion rate in a coupled hydrolysis/fermentation process. Mandels [6] noted that one of the principal advantages of this coupling is the removal of the end products inhibitory to the hydrolysis process by ethanol fermentation. Bullock et al. [27,28] coupled the hydrolysis of cellulosic materials to anaerobic digestion and methanogenesis, with similar success in optimizing the overall destruction of cellulosic materials.

This chapter discusses the application of enzymatic hydrolysis to MSW that had already been degraded partially by anaerobic digestion. This process provides the advantage of rendering the cellulosic fraction of MSW more amenable to hydrolysis because it has already been subjected to mechanical and thermal stresses in the digester, and much of the rapidly biodegradable associated material has already been metabolized. After enzymatic hydrolysis, the MSW is introduced to a second anaerobic digester to further degrade the material to biogas.

METHODS

Figure 1 is a diagrammatic presentation of the laboratory-scale enzyme treatment and anaerobic digester process used for MSW. The first digester vessel in the process is a 60-liter laboratory digester operated with a controlled artificial MSW (Table II) feedstock, and with efficient mixing provided by a reciprocating agitator. The artificial MSW was employed to permit standardization of the process so that changes in digester performance will not be the result of random variations in feedstock. Feeding and sampling were accomplished through a 3-in.-diameter pipe attached to the side of the vessel, located below the level of the liquid. The gastight lid was broached by a gas exit line, a thermocouple and two leads for the immersion heater, all of which were sealed and gasketed. Gas production was monitored by a water displacement system, and gas composition was determined by gas chromatography (GC). Gas volumes were converted to standard conditions (15°C, 1 atm, dry). Vessel temperature was maintained at 37°C in the mesophilic range appropriate for the anaerobes.

The 60-liter digester was operated continuously for more than one year during this study. Digester performance was evaluated at hydraulic retention times (HRT) of 60, 30 and 15 days and at influent MSW total solids (TS) concentrations from 3 to 5 wt %. The digester was also subjected to extensive testing while incorporated into the process line. During this period, the digester was operated on a 5% MSW TS feed and a 15-day HRT.

Enzymes used in this work were obtained from NOVO Laboratories Incorporated. The cellulase complex was Celluclast® 100L, NOVO's liquid preparation from the fungus *Trichoderma reesei* containing 100 C_1U/g as determined on Avicel® microcrystalline cellulose at 50°C in 20 minutes. The cellobiase complex, obtained from *Aspergillus niger*, was Cellobiase 250L with 250 CBU/g. It was used in a 1:10 ratio by weight of Cellobiase 250L to Celluclast® 100L as recommended by NOVO.

To assess the energy recovery process in this 5% study, a portion of the digester effluent was passed into the enzyme reactor. This consisted of a 2-liter stirred Erlenmeyer flask that was incubated at a constant temperature of 40°C. A gas trap was placed on top of the flask to allow the evolution of gases, while keeping the reactor oxygen-free. Enzyme load and retention time were chosen as 60 C_1U/g MSW TS and 15 CBU/g MSW TS and 3 days, levels selected from enzyme treatment studies discussed within this chapter.

The reactor effluent was passed through a polypropylene filter to remove undigested MSW as a filter cake. The filtrate was then passed

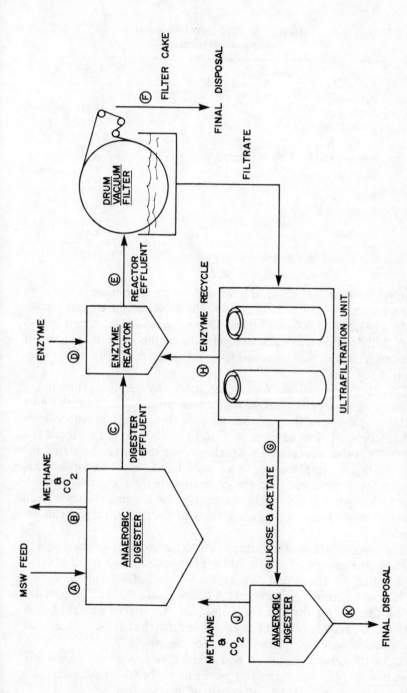

Figure 1. Laboratory-scale process flow diagram.

Table II.   Artificial MSW as Prepared
from Standard Materials[a]

Item	Percent
Cellulosics	
Newspaper	73
Wood (Sawdust)	8
Cloth	7
Proteinaceous Food Waste (Puppy Chow)	4
Oil	3
Sugar	2
Starch	2
Detergent	1

[a] Composition is based on several solid waste studies performed by SYSTECH.

through an Amicon® ultrafiltration unit with a membrane designed to have a capture cutoff of 5000 mol wt. The unit traps the enzyme (on the order of 45,000–70,000 mol wt) and allows glucose and other nutrients to pass. Although enzyme was not recycled during this study, the amount of enzyme recovered was measured to examine the feasibility of this concept.

Solutions were analyzed for enzyme activity by measuring the amount of reducing sugars produced during the incubation of enzyme with a pure cellulosic substrate. The solutions were incubated for 1 hr with 1.0g of Whatman No. 1 filter paper at 40°C in 13-× 100-mm screw-capped test tubes. The tubes were placed in a preheated water vessel and agitated at 2 Hz with a 2-cm displacement in a heated Lab-Line orbital shaker. Reducing sugars produced from the substrate were then determined colorimetrically by the ferricyanide reaction, and the amount of enzyme read from a standard curve of reducing sugar production vs enzyme concentration.

The glucose stream was then passed into a second digester vessel to measure the amount of additional methane obtained as a result of cellulose hydrolysis. The second digester was a 6-liter polyethylene vessel kept at 37°C in a circulating water bath. A feed and sample tube that extended below the level of the liquid was installed through the gastight lid. A sealed and gasketed gas exit line also penetrated the lid. Gas volume and composition were measured as stated for digester I.

Table III shows the analyses performed at each point during the process line studies. Those analyses pertinent to the 60-liter digester were performed throughout the 1-yr course of its operation.

Table III.   Analyses Performed at Each Test Point

Type of Test	A	B	C	D	E	F	G	H	J	K
Volume	X[a]		X	X	X		X	X		X
Total Solids	O[b]		X		X	X	X	X		X
Volatile Solids	O		X		X	X	X	X		X
Glucose	O		X		X		X	X		X
Suspended Solids	O		X		X					
Biogas Composition		X							X	
Volatile Acids			X		X					X
Alkalinity			X		X					X
Enzyme Activity	O		X		X					
Weight				X		X				
Filter Yield						O				
Solids Captured						O				
pH			X		X					X
Gas Volume		X							X	
Heat Content	O		O			O				
Nutrients	O		c				c			c

[a] X = daily.
[b] O = once.
[c] Analysis was performed three times.

Additional studies were conducted in this program to identify the optimum conditions for enzymatic hydrolysis of MSW and to investigate the hydrolysis of other waste cellulosic substrates. In some of this work, Whatman No. 1 filter paper was employed as a pure cellulose substrate; anaerobically digested MSW, paper mill sludge or apple processing wastes were used for the remainder of the studies. Enzyme loadings, incubation temperatures and retention times are indicated in the results presented. In some of these experiments, where noted, a variant form of NOVO's cellulase complex was used.

RESULTS AND DISCUSSION

Anaerobic Digester

The data obtained over the course of 1 year of operation of the 60-liter MSW digester have been summarized in terms of a kinetic model for the steady-state operation of the digester. This approach has been employed as a means of presenting the data in a manner that is readily comparable to other studies of anaerobic digestion processes [30–32].

It has been assumed for this analysis that the 60-liter digester represents a continuous-flow, well-mixed digester. In addition, the mass of the digester's microbial population (M) has been approximated by the mass of biogas produced in the digestion process. Substrate concentration (S) has been measured from the total volatile solids (TVS) concentration in the digester, and the rate of substrate utilization has been measured from the total volatile solids destruction (TVS_d).

The operation of the digester can then be described by the following equations:

$$\frac{dM}{dt} = a\,\mu - b\,M$$

$$\mu = \frac{\mu_{max}\,S\,M}{K_2 + S}$$

where

M = microorganism concentration, mass/vol (g biogas/ml digester vol-day)
t = time (day)
a = growth yield coefficient, time^{-1} (day^{-1})
b = microorganism decay coefficient, time^{-1} (day^{-1})
μ = rate of waste utilization, mass/vol-time (g TVS_d/ml digester vol-day)
μ_{max} = maximum rate of waste utilization per unit weight of microorganisms, time^{-1} (day^{-1})
S = waste concentration mass/vol (g TVS/ml digester vol)
K_s = waste concentration when $\mu = \frac{1}{2}\mu_{max}$, mass/vol (g TVS/ml digester vol)

These equations can be combined to yield a kinetic model for the steady-state operation of the digester which is valid over a range of HRT and substrate concentrations:

$$\frac{dM/dt}{M} = \frac{a\,\mu_{max}\,S}{K_s + S} - b$$

Figure 2 is a plot of the solids retention time (SRT), as approximated by the HRT, vs the mean values of the mass of TVS_d/day per mass of biogas produced/day (g TVS_d/g biogas) observed over a 1-yr period. The bars represent the range of one standard deviation about the mean values. It is apparent that a definite functional relationship exists between these variables (linear regression analysis of mean values yields and r = 0.81). Table IV indicates the numerical values obtained for the parameters a and b by this graphical method.

Figure 3 is a modified Lineweaver-Burk plot of the mean values of the mass of biogas produced/day per mass of total volatile solids de-

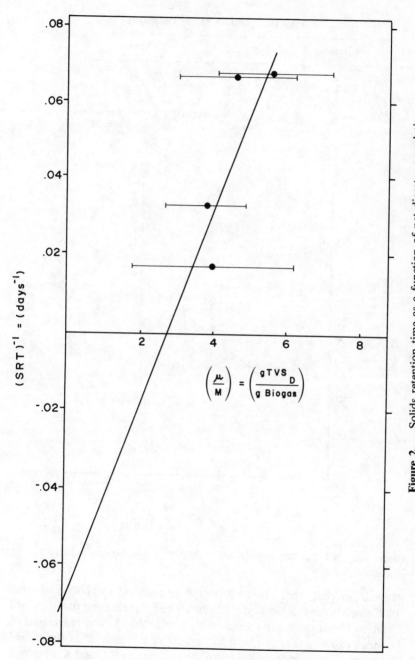

Figure 2. Solids retention time as a function of net digester population growth.

Table IV.   Growth Kinetic Parameters

Parameter	Value
a	0.0258 day^{-1}
b	0.0709 day^{-1}
μ_{max}	17.15 day^{-1}
K_s	0.0653 g VS/mL digester vol

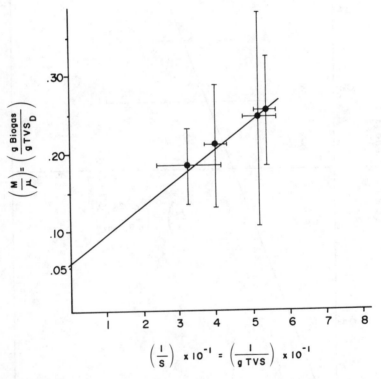

Figure 3.   Net digester population growth as a function of substrate concentration.

stroyed/day that were observed over a 1-yr period vs 1/mass of total volatile solids per ml of digester volume. The bars represent one standard deviation about the mean values of both variables. Linear regression of the mean values yields an r = 0.97, indicating a high level of functional relationship. the numerical values of the parameters μ_{max} and K_s obtained by this graphical method are shown in Table IV.

As an example of the utility of this kinetic approach to characterizing digester performance, the data obtained during the period of operation at a 15-day HRT on 5% MSW TS feed has been examined more closely. Solution of the kinetic model for this period of operation yields a value for dM/dt of 2.03 g biogas/60 liter-day, within 1% of the experimentally observed rate. Given a value for either MSW substrate concentration or HRT, the performance of the digester can be predicted with considerable accuracy.

Process Line Study

The entire process line shown in Figure 1 was operated during one phase of this study using a 5 wt % TS by volume artificial MSW feed. Figure 4 illustrates the general operating conditions observed in digester I during one 15-day HRT period. It is apparent the digester I was generally healthy during this period, maintaining a relatively stable pH and ratio of volatile acids to alkalinity. Although this ratio was generally higher than the 0.5 value recommended for the operation of sludge digesters [33,34], considerable buffering capacity is indicated by the limited variability of this ratio. The general operating conditions of digester II are shown in Figure 5. The pH remained relatively stable throughout this period, indicating that the buffering capacity of the digester was sufficient to accommodate the high levels of volatile acids observed.

The biogas production from each digester during the 5% study is shown in Figure 6. Since only a portion of the effluent from digester I (following enzymatic hydrolysis) was actually added to digester II, a correction factor of 6.8 was applied to determine the amount of biogas which would have been obtained if all of the effluent had been used. The average biogas production from digester I was 21.4 ± 3.19 liter/day and from Digester II was 24.7 ± 5.4 liter day. An average of 115% more biogas was obtained by using the enzyme reactor and second digester. Figure 7 shows the quantity of methane in the biogas. Digester I produced an average of 9.5 ± 1.5 liter/day, and digester II produced an average of 14.5 ± 3.6 liter/day. This represents a total increase in methane production of 153% over that observed for digester I alone.

The overall impact of the enzymatic hydrolysis component on the MSW digestion process can be seen most clearly from the system mass balance over the entire 15-day HRT, shown in Table V. The overall mass closure for total solids was 96%, with 4% of the input mass not accounted for. Mass inputs were digested MSW, enzyme and buffer. Mass outputs consisted of measured biogas from the digesters, CO_2 absorbed

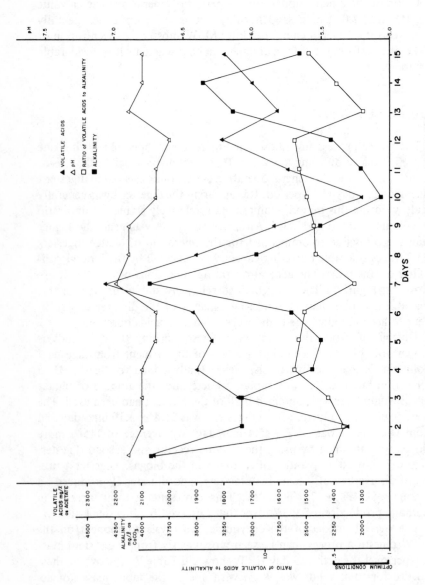

Figure 4.   Digester I—general operating conditions, 5% solids.

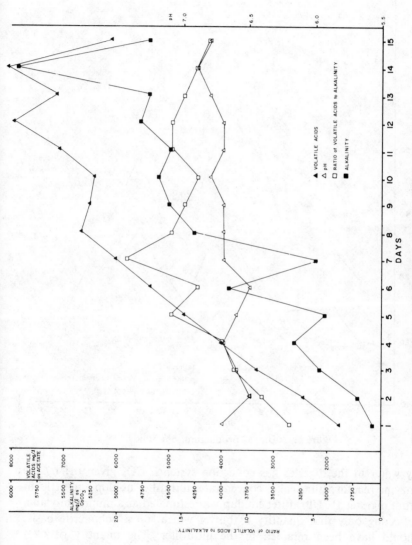

Figure 5. Digester II — general operating conditions, 5% solids.

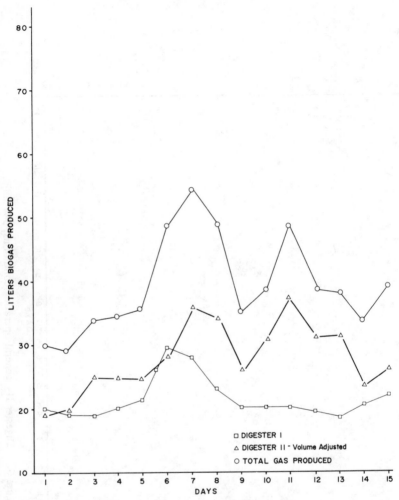

Figure 6. Biogas production, 5% solids.

by water in the digester gas collection systems, CO_2 given off by the enzyme reactor, filter cake, concentrate, ultrafilter holdup and effluent from digester II. Ultrafilter holdup was estimated at a minimum value, reflecting only that quantity of buffer which, on a volumetric basis, should have been retained in the ultrafilter. The quantity of CO_2 absorbed in the gas collection system water was calculated by Henry's law from the water temperature and partial pressure of CO_2 in the gas. Because this quantity of CO_2 is not derived from direct experimental measurements, it has not been included in any calculations regarding

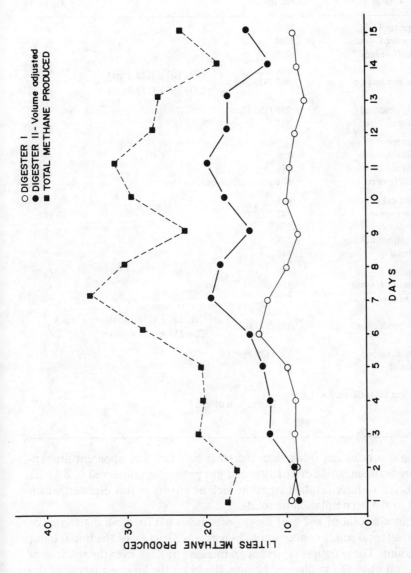

Figure 7. Methane production, 5% solids.

Table V.   15-Day Mass Balance—5% MSW
Total Solids Feed

Item	Mass Total Solids (g)		Reference Point on Figure 1
Digester I—in	2820		A
Digester I—out	2084		C
Difference	736		
Gas production	467 (321 liters)	CH₄ 102 g (143 liters) CO₂ 365 g (178 liters)	B
CO₂ absorbed	298 (152 liters)		
Total	765		
Enzyme			
Reactor—in	4150		D
Reactor—out	3576		E
Difference	574 given off as CO₂		
Filter cake	1648		F
Concentrate	342		H
Permeate	1370		G
Ultrafilter holdup	~188		
Total	3548		
Digester II—in	1370		G
Digester II—out	524		K
Difference	846		
Gas production	395 (370 liters)	CH₄ 156 g (218 liters) CO₂ 239 g (152 liters)	J
CO₂ absorbed	256 (130 liters)		
Total	651		

$$\text{Percent Mass Closure} = 100 - \frac{\text{inputs} - \text{outputs}}{\text{inputs} \times 0.01}$$

$$= 96\%$$

biogas production other than the mass balance. The apparent discrepancy between solids destruction and gas production observed in digester II is felt to have resulted from the lack of mixing in this digester, which led to the accumulation of solids.

Since a total of 862 g of biogas was measured for both digesters, the overall total solids conversion to biogas was 31 wt % of the initial input amount. These figures represent an increase of 82% over the solids conversion observed in digester I alone. Based on the kinetic analysis of the 60-liter digester, described earlier, it can be estimated that if the MSW had simply remained in digester I for a time period effectively equal to that of this process line (33 days), a total of 42.8 g biogas would have been produced per day. The process line system generated 57.5 g biogas per day, or 34% more than could have been achieved without the use of enzymatic hydrolysis.

If the solids conversion observed in this study is examined in terms of the composition of the MSW feedstock (see Table II), the increase in cellulose destruction in comparison to total solids conversion is considerably more dramatic. It can be assumed that the majority of the detergent, oil and protein in the feedstock will not be metabolized with subsequent biogas production by the microbial population of the digester. Therefore, only 92 wt % of the actual input solids can be expected to be metabolized with the production of biogas. Using this adjusted input solids concentration, the extent of conversion of metabolizable solids to biogas becomes 18 wt % for digester I and 33 wt % for the process line system. Again, if it is assumed that all of the sugar and starch in the digester feed was rapidly converted to biogas, digester I converted 16 wt % of the cellulosic material in the feed and the process line system converted 33 wt %. These values represent an increase in cellulose conversion of more than 100% by the application of the process line system.

The results of recovery assays conducted during the 5% solids study indicated that no activity was detectable in either digester I feed, digester I effluent or enzyme reactor ultrafilter permeate. The mean recovery of enzyme from ultrafilter concentrates corrected for volume loss in the filter cake was 55 wt % of the initial enzyme load. To maintain the enzyme reactor at a constant loading level, this recovery data indicate that the addition of 45 wt % of the initial enzyme load would have been required to make up for enzyme lost in the recovery procedures.

Enzyme Hydrolysis Optimization

During the course of this program, a number of studies were conducted to identify the optimum conditions for enzymatic hydrolysis of the cellulosic materials in MSW. The effects of pH, temperature, time, enzyme concentration and product inhibition on the hydrolysis process were examined. In addition, studies were conducted to identify the most effective method of recovering enzyme from enzyme-treated MSW.

Figure 8 illustrates the activity of the enzyme with a pure cellulose substrate as a function of pH. It is apparent that activity decreases with increasing pH, and that the activity measured at pH 5 is nearly three times greater than that observed a pH 7. This relationship was the major rea-

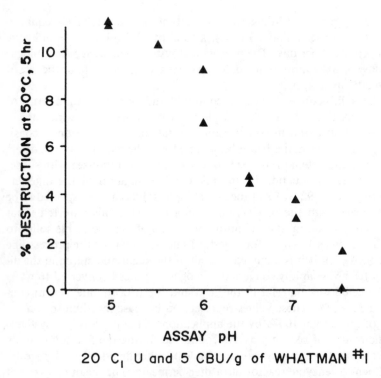

ASSAY pH

20 C_I U and 5 CBU/g of WHATMAN #1

Figure 8.   Enzyme activity as a function of pH.

son for operating the enzyme reactor in the process line at pH 5, rather than pH 7, the average pH of the digester effluent.

The effect of temperature on the enzymatic hydrolysis of the effluent of the 60-liter MSW digester is shown in Figure 9. Digested MSW was incubated with 50 C_1U and 12.5 CBU/g MSW TS. Although the optimum temperature for enzymatic hydrolysis of pure cellulose has been identified as 50°C [35], it is apparent that the digestion of MSW proceeds to a greater overall extent at 40°C. Since 40°C represents the optimum temperature of the *Acetobacter* sp. population of the digested MSW, it seems likely that the optimization of hydrolysis is strongly related to maximizing the growth of those microorganisms which remove the soluble sugar end products of hydrolysis. This phenomenon is a result of product inhibition of hydrolysis.

The effect of time on the hydrolysis of effluent from the 60-liter MSW digester is illustrated in Figure 10. Digested MSW was incubated with 75 C_1U and 18.8 CBU/g MSW TS, and the destruction of total suspended solids (TSS) was measured as a function of incubation time. The extent of hydrolysis is a linear function of time for nearly 18 hr, after which it

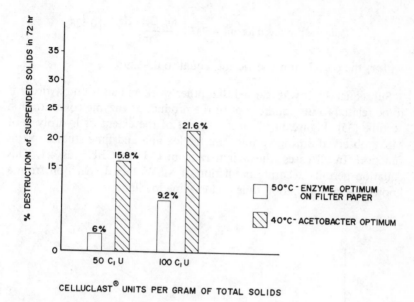

Figure 9. Temperature comparison of enzyme destruction of digested MSW.

gradually decreases, approaching a limit near 30 wt % TSS destruction. Since 72-hr incubation represented a near maximum level of TSS destruction, it was chosen as the HRT for the enzyme reactor in the process line.

The extent of hydrolysis as a function of enzyme concentration is shown in Figures 11 and 12. Figure 11 represents the destruction of TSS as a function of enzyme concentration using Whatman No. 1 as a pure cellulosic substrate. The ratio of C_1U to CBU was 4:1 on a weight basis. This figure illustrates that a loading of 70 C_1U and 17.5 CBU/g Whatman No. 1 resulted in the maximum level of destruction observed, 75% of the initial TSS input. Figure 12 shows the destruction of TSS as a function of enzyme concentration for digested MSW as substrate. Again, the ratio of C_1U to CBU was 4:1 on a weight basis. It is apparent that maximum destruction occurs at approximately the same level of enzyme concentration as that observed on a pure cellulosic substrate. On the basis of this curve a near optimal loading level of 60 C_1U and 15 CBU/g of MSW TS was chosen for the 5% process line study.

Figure 13 shows the enzyme assay standard curve which illustrates the production of glucose from 1.0 g of Whatman No. 1 filter paper. A least-squares linear regression fit of this data to the logarithmic form of a power curve yielded the following equation:

$$\text{glucose produced } \mu\text{g/ml} = 747 \left[\frac{\text{mg Celluclast}}{\text{ml}} \right] 0.538$$

with r, the correlation coefficient, equal to 0.9629

Subsequently, it was learned that other workers had found hydrolysis to be related to the square root of the product of enzyme concentration × time [35]. Figure 14 illustrates the fit of the extent of hydrolysis of MSW observed during various enzyme loading and time studies to this function. In all cases, the weight ratio of C_1U to CBU was 4:1. This equation permits accurate prediction of MSW TSS destruction from a knowledge of incubation time and enzyme loading.

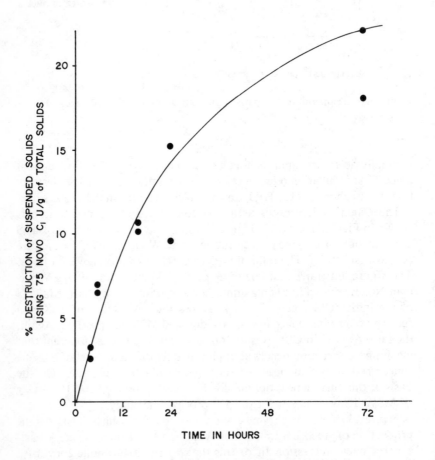

Figure 10.   Enzyme time study on digested MSW.

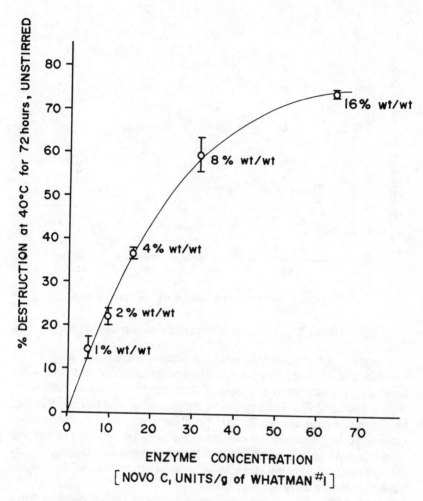

Figure 11.   Destruction as a function of enzyme loading.

The effect of glucose inhibition on the extent of hydrolysis of pure cellulose is illustrated in Figure 15. It is apparent that a significant decrease in TSS destruction is observed at levels of glucose well below 2 wt %. At 4 wt % glucose, the destruction of suspended solids has been reduced by 50%. Figure 16 indicates the inhibition of TSS destruction observed at extremely low levels of glucose. Since the mean concentration of glucose found in the enzyme reactors during the 5% process line study was 0.6 wt %, it seems likely that considerably more of the cellulose could have been hydrolyzed had the glucose concentration not reached an inhibitory level.

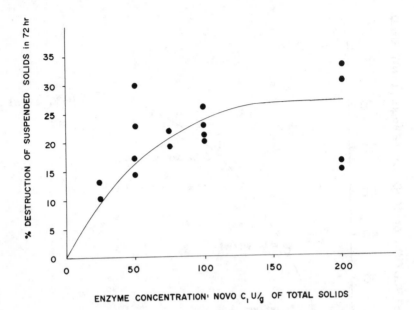

Figure 12.   Enzyme loading study on digested MSW.

A number of experiments were conducted during this program to optimize the recovery of enzyme. The results obtained are considered promising for ultimately maximizing the amount of enzyme retrievable from digested MSW. The amount of enzyme recovered can be increased by continued examination of two problem areas: the development of more effective filtration and extraction methods for recovery of enzyme from digested MSW, and the maximization of cellulose destruction by glucose removal, thereby eliminating product inhibition and any potential adherence of enzyme to substrate.

It is likely that enzyme recovery optimization by extraction procedures is ultimately limited by the fact that enzyme is preferentially adsorbed to its substrate. This concept has been proposed by a number of workers in cellulose chemistry and offers an explanation for the low recoveries observed after 72 hr incubation if hydrolysis were not complete. For example, Huang [22] and Wilke and Mitra [29] have found that only 35–50% of the initial enzyme load can be found in suspension above newsprint that has been hydrolyzed to 50–100% of completion [22,29]. Since work conducted in this program to date has achieved at best only a 30% destruction of suspended solids in 72 hr, it is possible that the remaining unhydrolyzed cellulose in the enzyme reactors prevents achieving complete recovery of enzyme.

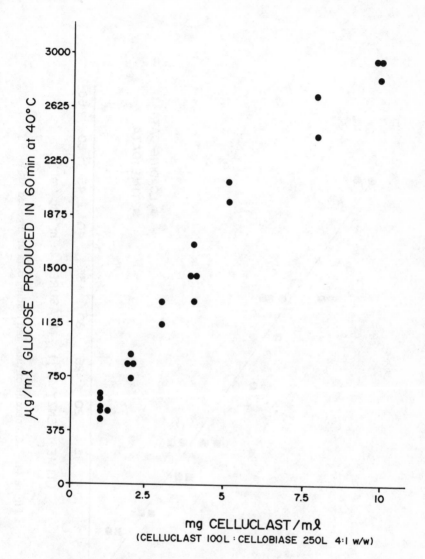

Figure 13. Enzyme assay standard curve—low concentration.

To test this hypothesis, a series of experiments was run in which the concentration of enzyme was increased with a constant amount of digested MSW substrate. Incubation was continued for 72 hr at 40°C, and activities were run on the filtrates of the reactors. Figure 17 shows the results of these experiments as percent recovery of enzyme vs enzyme concentration. It is apparent that at high enzyme loading levels, where

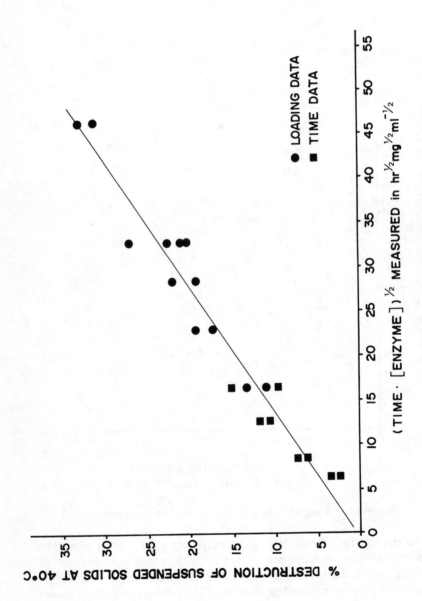

Figure 14. Composite of time and loading data on digested MSW.

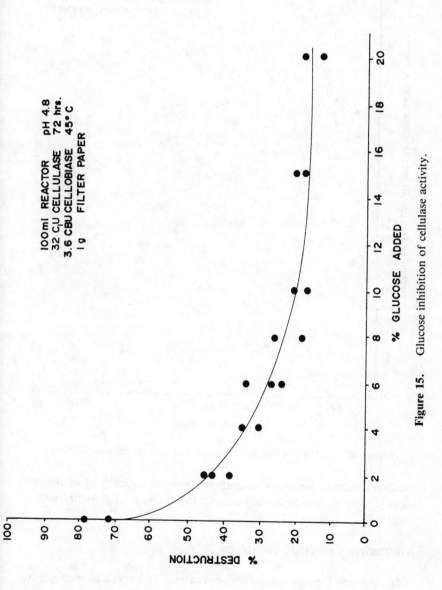

Figure 15. Glucose inhibition of cellulase activity.

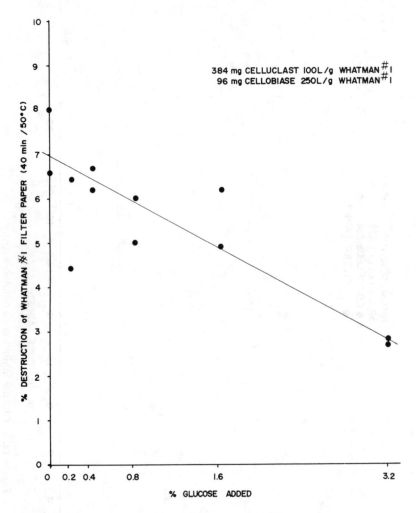

Figure 16. Effect of low-level glucose addition on cellulase activity.

product inhibition becomes insignificant and theoretically all of the cellulose present should be destroyed, enzyme recovery approaches 100%.

Alternative Cellulosic Substrates

To examine the applicability of enzymatic hydrolysis to waste cellulosic substrates other than MSW, preliminary investigations were conducted with apple pomace, a waste produce of the manufacture of apple

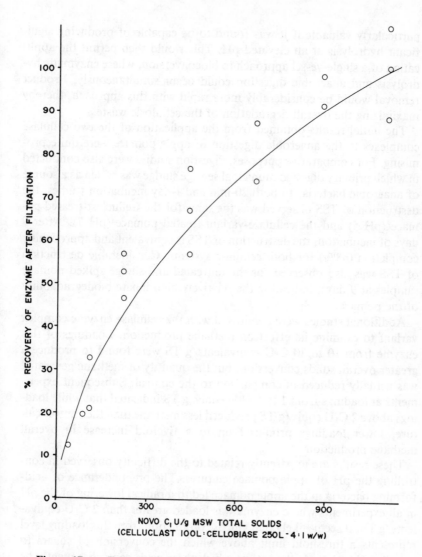

Figure 17..   Enzyme recovery at high enzyme:substrate loading levels.

juices, and paper mill wastes. The objective of this work was to increase the rate and extent of bioconversion and biomethanation normally observed in the degradation of these wastes.

Studies on the bioconversion of apple pomace were conducted with the Celluclast complex and with a variant cellulast enzyme complex, similar in activity to the Celluclast but with a somewhat greater stability toward pH excursions. It was felt that information on this complex would be particularly valuable if it was found to be capable of producing significant hydrolysis at an elevated pH. This would then permit the application of a single-vessel approach to bioconversion, where enzymatic hydrolysis and anaerobic digestion could occur simultaneously. Product removal would be considerably more rapid with this approach, thereby maximizing the overall degradation of the cellulosic waste.

The initial results obtained from the application of the two cellulase complexes to the anaerobic digestion of apple pomace were quite promising. For comparative purposes, digestion studies were also conducted in which primary digested municipal sewage sludge was added as a source of anaerobic bacteria. In both 20-min and 1-day incubation studies, the destruction of TSS observed was the same for the Celluclast-treated pomace (pH 5) and the cellulase-variant-treated pomace (pH 7). After 9 days of incubation, the destruction of TSS is equivalent and approaching completion (68%) for both cellulase systems. Considerable destruction of TSS was also observed for the untreated and sludge spiked pomace samples at 9 days, indicating the relatively high innate biodegradability of the pomace.

Additional studies were conducted with the cellulase enzyme complex variant to examine its effect on methane production. Loadings of this enzyme from 10 to 30 C_1U equivalents/g TS were found to produce a greater overall solids conversion, but the quantity of methane generated was actually reduced in comparison to the control. Subsequent experiments at loadings from 1 to 5 C_1U equiv/g TS indicated that while loadings above 2 C_1U equiv/g TS produced less methane than the control culture, lower loadings produced up to a fivefold increase in overall methane production.

These results are apparently related to the difficulty observed in controlling the pH of apple pomace cultures. The preponderance of acid-forming bacteria in the apple pomace led to a radical lowering of the pH in all experiments where enzyme was loaded greater than 2 C_1U equivalents/g TS. It seems likely that glucose production from this loading level represents a threshold limit above which the conversion of sugars to volatile acids exceeds the rate at which these acids can subsequently be converted to methane. As the pH is lowered, the viability of the methane-producing bacteria is then severely reduced.

Initial studies on the applicability of enzyme hydrolysis to paper mill waste anaerobic digestion have produced similar optimism regarding this substrate. Loading studies have been conducted on a 10% TS aqueous slurry of this material, with Celluclast enzyme added at levels from 10 to 1000 C_1U/g TS at pH 5.8. It was found that overall increases of 10, 48 and 74% in CO_2 production in comparison to the control were observed at loadings of 100, 500 and 1000 C_1U/g TS, respectively.

Subsequent work on this substrate has examined the effect of enzyme application on methane production. As in the studies on apple pomace, it has been found that maximum production of methane occurs at relatively low enzyme loadings. With enzyme concentration at 10 C_1U/g TS, a 40–60% increase in methane production over the control has been observed.

CONCLUSIONS

The work presented in this chapter indicates that the 60-liter MSW digester can be operated efficiently at steady-state conditions over a wide range of substrate concentrations and HRT. A kinetic model has been presented which permits the prediction of biogas production from a knowledge of substrate concentration and HRT. During the 5% MSW TS study, this digester converted 17 wt % of the total solids and 16 wt % of the cellulosic total solids to biogas. On the average, 0.11 liter (0.17 g) of biogas was produced per gram of input solids, 45 vol % of which was methane.

The incorporation of this digester into the process line system illustrates the gain in overall conversion of solids to biogas resulting from enzymatic hydrolysis. An additional 0.13 liter (0.14 g) of biogas was produced per gram of input solids, 58 vol % of which was methane. As a result of the enzymatic hydrolysis process, 115% more biogas, 153% more methane and 82% more total solids conversion were observed than with digester I alone. In addition, approximately 100% more of the total cellulosic solids were converted to methane.

The optimization studies have indicated conditions of pH, temperature, reaction time and enzyme concentration which are most appropriate for maximizing the conversion of cellulose-rich MSW to biogas. The effects of glucose inhibition and the relationship between enzyme recovery and extent of cellulose hydrolysis both indicate the need for further study of the enzymatic hydrolysis process. If the rate of removal of glucose could be enhanced, hydrolysis should proceed to completion, accompanied by an increase in the overall recovery of enzyme. This would result in a significant increase in the efficiency of the process, and ultimately to a much greater destruction of MSW solids.

Preliminary studies on apple pomace and paper mill wastes have indicated that the application of enzymatic hydrolysis to these substrates is quite promising for the overall increase in bioconvertibility. With appropriate levels of enzyme applications, both solids destruction and methane production should be greatly enhanced.

This program has demonstrated the feasibility of enzymatic hydrolysis as a means of optimizing the anaerobic digestion of various cellulosic wastes. Given the rapidly increasing costs of conventional methods of disposal, this process deserves further attention as a feasible alternative.

ACKNOWLEDGEMENTS

Much of the work on the digestion of MSW described in this paper was conducted under Department of Energy Contract No. DEACO1-78 CS 20451.

REFERENCES

1. Cowling, E. B., and T. K. Kirk."Properties of Cellulose and Lignocellulosic Materials as Substrates for Enzymatic Conversion Processes," in *Enzymatic Conversion of Cellulosic Materials: Technology and Applications,* Biotechnol. Bioeng. Symp. No. 6, (New York: John Wiley and Sons, Inc., 1976), pp. 95–123.
2. Millett, M. A., A. T. Baker and D. Satter. "Physical and Chemical Pretreatments for Enhancing Cellulose Saccharification," in *Enzymatic Conversion of Cellulosic Materials: Technology and Application,* Biotechnol. Bioeng. Symp. No. 6 (New York: John Wiley and Sons, Inc., 1976), pp. 125–153.
3. Wilke, C. R., R. D. Yang and U. Von Stockar. "Preliminary Cost Analyses for Enzymatic Hydrolysis of Newsprint," in *Enzymatic Conversion of Cellulosic Materials: Technology and Application,* Biotechnol. Bioeng. Symp. No. 6 (New York: John Wiley and Sons, Inc., 1976), pp. 155–176.
4. Andren, R. K., R. T. Erickson and J. E. Madeiros. "Cellulose Substrates from Enzymatic Saccharification," in *Enzymatic Conversion of Cellulosic Materials: Technology and Application,* Biotechnol. Bioeng. Symp. No. 6 (New York: John Wiley and Sons, Inc., 1976), pp. 177–204.
5. Lipinski, E. S., D. A. Scantland and I. A. McClure. "Systems Study of the Potential Integration of U. S. Corn Production and Cattle Feeding with Manufacture of Fuels via Fermentation," U. S. DOE Report BMI 2033, Vol. 1 (1979), pp. 1–147.
6. Mandels, M. "Enzymatic Saccharification of Waste Cellulose," in *Proceedings of the 3rd Annual Biomass Energy Systems Conference* (Golden, CO: Solar Energy Research Institute, 1979), pp. 281–289.
7. Emert, G. H., and R. Katzen. "Gulf's Cellulose-to-Ethanol Process," *Chemtech* 10:610–615 (1980).
8. Converse, A., and H. Grethlein. "Acid Hydrolysis of Cellulosic Biomass," in *Proceedings of the 3rd Annual Biomass Energy Systems Conference* (Golden, CO: Solar Energy Research Insitute, 1979), pp. 91–95.

9. O'Neil, P. T. "Design, Fabrication, and Operation of a Biomass Fermentation Facility," Georgia Institute of Technology Technical Progress Report No. 2, GIT/EES Project No. A 2256-000 (1979).
10. Rogers, C. "Emerging Technology for Maximum Conversion of Waste Cellulose to Ethanol Fuel," In: *Proceedings of the Waste to Energy Technology Update, 1980,* U.S. EPA/IERL, Cincinnati, OH (1980), pp. 47–48.
11. Brenner, W., B. Rugg and C. Rogers. "Utilization of Waste Cellulose for Production of Chemical Feedstocks via Acid Hydrolysis," in *Second Symposium on Biotechnology in Energy Production and Conservation,* Biotechnol. Bioeng. Symp. No. 10 (New York: John Wiley and Sons, Inc., 1979), pp. 201–212.
12. Tsao, G. E. "Fermentable Sugars from Cellulosic Wastes as a Natural Resource," in *Proceedings of the Second Annual Symposium on Fuels and Biomass, Vol. I,* (Troy, NY: Rensselaer Polytechnic Institute, 1978).
13. Manahan, S. *Environmental Chemistry* (Boston, MA: Willard Grant Press, 1977), p. 532.
14. Ishida, M., Y. Adauara, T. Gejo and H. Okumura. "Biogasification of Municipal Waste," in *Recycling Berlin, Vol. 2* (Berlin: Springer-Verlag, 1979), p. 797.
15. Kosstrin, H. M. "Pilot Scale Pyrolytic Conversion of Mixed Wastes to Fuel," in *Proceedings of the Waste to Energy Technology Update, 1980,* U.S. EPA/IERL, Cincinnati, OH (1980), pp. 1–10.
16. Diebold, T. P., and G. D. Smith. "Thermochemical Conversion of Biomass to Gasoline," in *Proceedings of the 3rd Annual Biomass Energy Systems Conference* (Golden, CO: Solar Energy Research Institute, 1979), pp. 139–147.
17. Wayman, M., and T. H. Lora. *Tappi* 61(6) (1978).
18. McCarty, P., L. Young, W. Owen, D. Stuckey and P. Colberg. "Heat Treatment of Biomass for Increasing Biodegradability," in *Proceedings of the 3rd Annual Biomass Energy Systems Conference* (Golden, CO: Solar Energy Research Institute, 1979), pp. 411–418.
19. McCarty, P. L., L. T. Young, J. B. Healey, W. F. Owen and D. C. Stuckey. "Thermochemical Treatment of Lignocellulosics and Nitrogenous Residuals for Increasing Anaerobic Biodegradability," in *Proceedings of the Second Annual Symposium on Fuels and Biomass, Vol. I.* (Troy, NY: Rensselaer Polytechnic Institute, 1978).
20. Thayer, D. W. "Celluloytic and Physiological Activities of Bacteria During Production of Single Cell Protein From Wood," *Food Pharmaceutical and Bioengineering,* Am. Inst. Chem. Eng. Symp. Series No. 172 (1978), pp. 126–135.
21. Ben-Bassat, A., R. Lamed, T. K. Ng and J. G. Zeikus. "Metabolic Control for Microbial Fuel Production," Proceedings of the IGT Symposium on Energy from Biomass and Wastes IV, Lake Buena Vista, FL (1980), pp. 275–294.
22. Huang, A. A. "Enzymatic Hydrolysis of Cellulose to Sugar," Biotechnol. Bioeng. Symp. No. 5 (New York: John Wiley and Sons, Inc., 1975), pp. 245–252.
23. Steinberg, D. "Production of Cellulase by Trichoderma," in *Enzymatic Conversion of Cellulosic Materials: Technology and Application,* Biotechnol. Bioeng. Symp. No. 6 (New York: John Wiley and Sons, Inc., 1976), pp. 35–54.

24. Mandels, M., R. Andreotti and C. Roche. "Measurement of Saccharifying Cellulose," in *Enzymatic Conversion of Cellulosic Materials: Technology and Application,* Biotechnol. Bioeng. Symp. No. 6 (New York: John Wiley and Sons, Inc., 1976), pp. 21–34.

25. Mangat, M. N., and J. A. Howell. "Product Inhibition of Trichoderma Viride Cellulase," in *Food Pharmaceutical and Bioengineering,* Am. Inst. Chem. Eng. Symp. Series No. 172 (1978), pp. 77–81.

26. Ghose, T., and K. Das. *Adv. Biochem. Eng.* 1(55) (1971).

27. Bullock, L., G. Higgins, R. B. Smith and J. T. Swartzbaugh. "Enzymatic Enhancement of Solid Waste Bioconversion," Proceedings of the IGT Symposium on Energy From Biomass and Wastes IV, Lake Buena Vista, FL, (1980), pp. 319–331.

28. Bullock, L., G. Higgins, R. B. Smith and J. T. Swartzbaugh. "Energy Recovery From the Effluent of Plants Anaerobically Digesting Cellulosic Urban Solid Waste," draft final technical report on U.S. DOE Contract No. DEACO1-78 CS 20451 (1980), pp. 1–112.

29. Wilke, C. R., and G. Mitra. "Process Development Studies on the Enzymatic Hydrolysis of Cellulose," Biotechnol. Bioeng. Symp. No. 5. (New York: John Wiley and Sons, Inc., 1975), pp. 253–274.

30. Vesilind, P. Aarne. *Treatment and Disposal of Wastewater Sludges* (Ann Arbor, MI: Ann Arbor Science Publishers, Inc., 1974).

31. Ghosh, S., and D. L. Klass. "Two-Phase Anaerobic Digestion," Proceedings of the Symposium on Clean Fuels from Biomass and Wastes, Orlando, FL (1977), pp. 373–415.

32. Lawrence, A. W., and P. L. McCarty. "Kinetics of Methane Fermentation in Anaerobic Treatment," *J. Water Poll. Control. Fed.* 41:2 (1969).

33. "Process Design Manual for Sludge Treatment and Disposal," U.S. EPA Technology Transfer (1974).

34. McCarty, P. L. "Anaerobic Waste Treatment Fundamentals, Part Three, Toxic Materials and Their Control," *Public Works* 95(11):94 (1964).

35. Posorske, L., NOVO Laboratories, Inc. Wilson, CT. Personal communications (1979–1980).

CHAPTER 9

BIODEGRADABLE POTENTIAL OF PRETREATED MUNICIPAL SOLID WASTES

P. D. Chase and J. H. Singletary

Public Service Electric and Gas Company
Newark, New Jersey

One of the more innovative methods for producing energy is to convert biologically a mixture of municipal solid waste (MSW) and sewage sludge into a methane-containing gas. This process, commonly referred to as anaerobic digestion, is not a new concept. It has been practiced traditionally for nuisance prevention (putrescibility), health considerations (reduction of pathogens) and economic waste disposal (reduction of solids requiring final disposal). The methane-rich gas can be upgraded as a supplement to natural gas by removing the carbon dioxide and trace gases. Simply stated, the process requires the separation of organics from MSW, digestion of the mixture and gas refining.

Currently, there are commercial facilities that separate organics from MSW. The resulting material, known as refuse-derived fuel (RDF), is used for burning with oil or coal. RDF is a good substrate to mix with sewage sludge for anaerobic digestion; however, it is believed that pretreatment procedures in the preparation of MSW can influence the overall gas yield. Some RDF plants consist of processing schemes which may be considered as pretreatment steps in processing MSW as a supplementary feedstock for biological gasification. Thus, PSE&G Research Corporation investigated the biodegradable potential of substrates from four RDF facilities. Results from this project will be considered in selecting a front-end MSW process for a future commercial digestion plant. This chapter describes the results of these investigations. Factors considered in selecting the existing RDF plants were: (1) physical treatment of the

MSW; (2) chemical treatment of the organics; and (3) final particle size. The RDF samples selected included:

1. feedstock, without secondary shredding, from the U.S. Department of Energy (DOE) facility in Pompano Beach, FL;
2. shredded MSW from the Madison, WI, plant;
3. Eco-Fuel from the Bridgeport, CT, RDF plant; and
4. shredded MSW, prior to pelletizing, from the Baltimore County RDF plant.

The samples, although processed as RDF, had the following variations:

1. substrate from the Pompano Beach facility was not processed through a secondary shredder;
2. material from the Madison plant passed through a patented operation before RDF collection;
3. Eco-Fuel was produced by treating the light fraction of shredded MSW with acid to embrittle the cellulosic materials and then ball milling to a fine powder; and
4. material, prior to pelletizing, from the Baltimore County RDF plant represented substrate from a typical RDF operation.

METHODS

A mobile trailer was modified for use as the test facility. The inside dimensions are $8.5 \times 2.4 \times 2.4$ m ($28 \times 8 \times 8$ ft). The trailer is partitioned into two compartments. The smaller front portion is an office and the rear section serves as a laboratory.

Two separate batch digestion systems were assembled in the laboratory and each consisted of a digester, pH controller, sampling tube and gas meter. Figure 1 shows a laboratory digester assembly. Each digester has a 14-liter capacity vessel, Model M-19-1400, manufactured by the New Brunswick Scientific (NBS) Company, Edison, NJ. The heat transfer medium (tapwater) can be raised between 5 and $70 \pm 0.25°C$ above its entering temperature. Agitation can be adjusted in a range from 80 to 800 rpm. Each pH controller is a Model pH-21 manufactured by NBS, and will control pH to within 0.1 pH units. The sampling tubes are 250-ml glass cylinders with a stopcock at each end. The tubes were purchased from Ace Scientific Supply Company, Linden, NJ. The gas meters were standard wet test meters manufactured by the American Meter Company, Inc., Albany, NY. Gas volumes can be read to the nearest 2.68×10^{-5} m^3 (0.001 ft^3).

Figure 1. Laboratory digester assembly.

The main objective of the experiments was to determine and compare the gas volumes from the readily digestible fraction (initial production stage) of each pretreated substrate against a control. The Baltimore substrate (typical RDF without pretreatment) was the control for these experiments. It is believed that the biconversion potential of MSW will influence the total gas production from a mixture of MSW and sewage sludge. In addition, gas production volumes from the initial production stage may be a more appropriate basis for comparing the various substrates.

Each substrate was digested in a similar manner, using the batch digesters, until a substantial reduction in the gas production rate was observed. One run was permitted to extend beyond the initial gas production stage. These data were used to confirm that more than one gas production stage occurred during the fermentation of MSW. In addition, the data provided an estimate of potential total gas yield and a profile of the gas production phases during the process.

A standard inoculum was prepared for the experiments from 440 g of cellulose, 600 g of wet sewage sludge and 9000 g of water to form a 4.5% solids mixture. The mixture was digested until maximum growth of microorganisms was achieved. The growth of microbes was monitored by the net volume of gas produced daily. The developed culture was harvested and frozen in 500-ml aliquots and used to inoculate each run, except the Pompano Beach substrate run, which was inoculated with a freshly prepared (unfrozen) culture.

In each run, 500 g of inoculum was mixed with approximately 445 g of substrate and water to produce a mixture which was approximately 4% solids. Each mixture was supplemented with NH_4Cl and KH_2PO_4 to adjust the C/N ratio to about 30:1 and the potassium concentration to about 200 mg/l. Each mixture was digested at pH 7.0 and 35°C [1-3]. Agitation was maintained at approximately 200 rpm for each run. Each run, except the Pompano Beach substrate, was terminated after exhaustion of the readily biodegradable materials, which were monitored by measurement of gas production rates. Cumulative gas volumes were taken at the end of each run to use as a basis for comparing the pretreated substrates against the Baltimore substrate. Gas volumes measured at 65°F (trailer laboratory ambient temperature controlled thermostatically except during minor power failures) and atmospheric pressure were corrected to normal conditions (0°C, 1 atm, dry). Average atmospheric pressure over a given run was used in the pressure correction calculations. These readings were computed from the daily average pressure readings. All gas volumes reported in this chapter were corrected to normal conditions unless otherwise noted. Samples were taken periodically and analyzed during each run by gas chromatography to determine methane content. Each substrate was analyzed for total solids, volatile solids and ash content. Total solids were determined by oven drying at 103-105°C for 24 hr. Volatile and fixed solids were determined by igniting the dried material in a muffle furnace at 550°C for 1 hr.

RESULTS AND DISCUSSION

Mixing the shredded substrates presented problems. The digesters were equipped with magnetically coupled agitators and baffle inserts. Excluding Eco-Fuel, a powdered material, the relatively large particle size of the shredded MSW limited the amount of material that could initially be slurried and mixed. A solids concentration of 4% for the shredded materials was established as the upper limit that could effectively be mixed in these digesters. Even with this relatively low solids limitation, problems were experienced with agitator decoupling. However, mixing improved as digestion progressed. An additional mixing problem, relating again to the shredded substrates, was caused by fabric fragments and their tendency to wrap around the agitator. Severe buildup of this material caused the agitator to jam and decouple. To alleviate these problems, a reversal switch was installed which allowed periodic reverse rotation of the agitator. Although 4% suspensions of Eco-Fuel were easily agitated, it was determined that this material contains some ash and grit which acted as an abrasive. When Eco-Fuel was tested, the Teflon® seats for the agitator were subject to abnormal wear, and poor mixing resulted.

Power failures caused brief interruptions during the experimental operation of three runs. These minor interruptions, however, caused no serious problems and probably resulted in only minor temperature changes within the digesters.

When adding either acid or base to a digester for pH control, it is necessary to have adequate and fairly rapid mixing. On several occasions, pH was improperly controlled because of inadequate mixing. Once the mixing problems were overcome, pH control was no longer a problem.

The pH of the Eco-Fuel slurry tended to rise slightly as digestion progressed and had to be adjusted periodically by the addition of acid. Elevation in pH is unexpected during the early stages of a fermentation cycle. It usually occurs during later stages when the culture is starved for carbon and the microorganisms begin to deaminate proteins spilled into the culture from lysing cells. This causes a release of NH_3 and a subsequent rise in pH. It seems unlikely that this occurred during early stages of the Eco-Fuel runs. Similar behavior was observed in the Pompano Beach run, but only for a short period of time. As a result, it is difficult to attribute this behavior solely to Eco-Fuel, even though the problem was much more pronounced and chronic for that substrate.

Gas volumes were not corrected for fluctuation of temperature in the trailer laboratory during the minor power failures. Diurnal temperature changes were estimated to cause up to a 7% variation in measured gas volumes during these periods. However, extremely small gas volumes were generated, thus, the variations were insignificant when figured in the total gas yield of a run.

As previously mentioned, failures in the agitation systems caused several problems. In the first Eco-Fuel run, this resulted in a critical temperature excursion. When agitation failed, due to an overloaded motor, solids settled to the bottom of the vessel. This is where the heating coil is located and, without mixing, extreme temperature gradients can occur. Consequently, the temperature probe, located in the mid-section of the vessel, continually registered below the set point and caused unintentional pasteurization of the digesting mixture. That run was repeated.

A standard inoculum was used to shorten the lag phase of fermentation and to provide each substrate with the same level of microbial activity and distribution. In addition, the inoculum would eliminate the use of digester sewage sludge. The production of gas could then be attributed directly to digestion of MSW and not be confused or masked by the digestion of the more readily biodegradable sewage sludge. In addition to readily biodegradable materials, sewage sludge is normally used in anaerobic digestion to supply microorganisms and nutrients. For this study, microorganisms and nutrients were supplied by an inoculum and inorganic salts.

The frozen standard inoculum was used for all runs except the run with Pompano Beach substrate where 500 g of a freshly prepared (unfrozen) inoculum was introduced. Since this substrate did not undergo pretreatment processing, there was no need to evaluate its biodegradable potential for comparison with the pretreated substrates. In addition, this substrate could not be used as a duplicate control along with the Baltimore substrate because it was processed through fewer steps than typical RDF. Therefore, it was digested to give an estimate of potential total gas yield and profile of the gas production during the anaerobic digestion of MSW. For this purpose, a freshly prepared inoculum was used because it was not known what effect a frozen inoculum would have on the total yield during the process. The frozen inoculum was felt appropriate to inoculate the substrates used for the gas yield comparison because the results were compared on a relative basis. In addition, the runs were made weeks apart and this was the recommended procedure to supply each substrate with a similar inoculum.

Pompano Beach Substrate

Although the Pompano Beach facility processes MSW through a secondary shredder, the sample was taken before that operation. The sample was inoculated with a freshly prepared inoculum, and the run had the shortest lag period (4 days) before appreciable gas production began. This run was extended beyond the initial gas production phase and manifested two gas generation stages. Figure 2 shows the profile of gas production over the entire run. The data provided an indication of total potential gas yield in the fermentation of MSW. The instantaneous gas production rate reached a maximum of 0.013 m^3/kg initial volatile solids per day during the initial production phase. Maximum methane content of the gas was 59.4% measured on the 16th day. The CO_2 was 40.1%. Cumulative gas yield for the entire run was 0.215 and 0.282 m^3/kg, respectively, of dry substrate and volatile solids added to the digester.

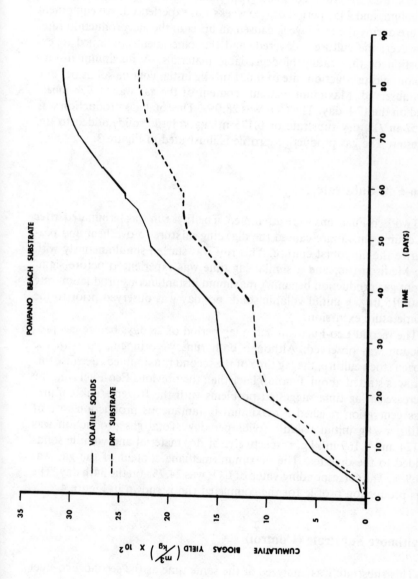

Figure 2. Cumulative gas yield—Pompano Beach substrate. Gas volumes at 18.3°C (65°F), 739 mm Hg, dry.

Madison Substrate

This was the first complete run made with the frozen standard inoculum, and a lag period of two weeks was experienced. An equipment failure during the third week caused an upset in the gas production rate. However, the culture recovered, and the experiment continued to exhaustion of the readily biodegradable materials. A maximum instantaneous gas production rate of 0.021 m^3/kg initial volatile solids per day was observed. Maximum methane content of the gas was 66.9% measured on the 17th day. The CO_2 was 28.0%. The total gas production was 0.152 m^3/kg dry substrate or 0.175 m^3/kg volatile solids added to the digester. The gas production profile is illustrated in Figure 3.

Eco-Fuel Substrate

Two Eco-Fuel runs were attempted. The first run was terminated after an equipment failure caused the digesting mixture to overheat and pasteurize the microbial culture. This run was started simultaneously with the Madison run, and a similar lag time was experienced before significant gas production began. A maximum instantaneous production rate of 0.015 m^3/kg initial volatile solids per day was observed prior to the temperature excursion.

The second Eco-Fuel run had a lag period of 28 days before gas production was observed. Although both runs were inoculated with the frozen stock culture, the lag time of the second run doubled. Because this run was started about a month later than the previous Eco-Fuel run, the increase in lag time suggested problems with the frozen stock culture. Gas generation reached a maximum instantaneous production rate of 0.012 m^3/kg initial volatile solids per day. Total gas production was 0.124 and 0.169 m^3/kg, respectively, of dry material and volatile solids added to the digester. The maximum methane content of the gas was 68.9%. The corresponding value of CO_2 was 24.7% on the 34th day. The gas production profile for the completed run is shown in Figure 4.

Baltimore Substrate (Control)

This substrate was analyzed at the same time as the second Eco-Fuel run. Although the Baltimore substrate was inoculated with the frozen standard inoculum, only a short lag period of about 6 days to reach gas production was experienced. The maximum instantaneous gas produc-

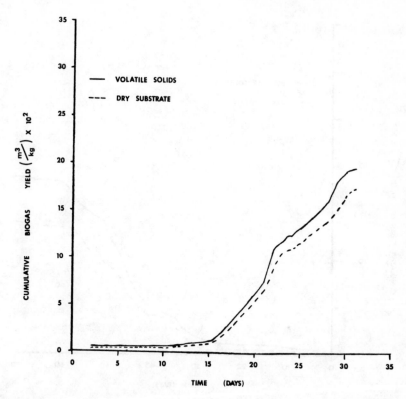

Figure 3.   Cumulative gas yield—Madison substrate. Gas volumes at 18.3°C (65°F) 738 mm Hg, dry.

tion rate observed was 0.006 m³/kg initial volatile solids per day. The maximum methane content of the gas was 61.8%, with a corresponding CO_2 value of 38.0% measured on the 56th day. Gas production was 0.164 and 0.175 m³/kg, respectively, of dry substrate and volatile solids added to the digester. Figure 5 shows the gas production profile.

SUMMARY AND CONCLUSIONS

Pretreatment steps were added in the processing of Eco-Fuel and the Madison substrate. However, there was no increase in gas yield over the control (Baltimore substrate).

Eco-Fuel has a smaller particle size than the Baltimore substrate; however, the particles were reduced through a chemical process. The absence of a gas yield increase, due to smaller particle size, suggested that the

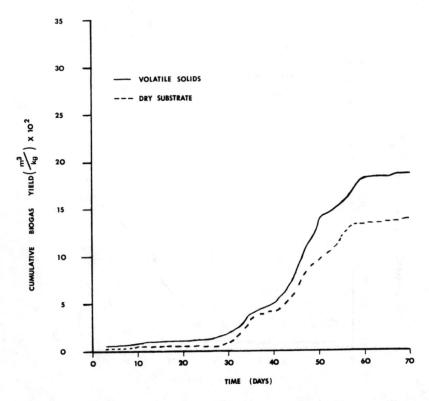

Figure 4. Cumulative gas yield – Eco-Fuel. Gas volumes at 18.3°C (65°F), 738 mm Hg, dry.

chemical process reduced Eco-Fuel biodegradability. In fact, gas production on a dry material basis for Eco-Fuel was lower than that of the Baltimore sample. This, along with the high ash content of Eco-Fuel, suggested that some of the readily biodegradable materials were hydrolyzed in the embrittlement process. It might be possible, however, to enhance Eco-Fuel biodegradability and maintain the desired particle size by modification of the process. Gas yields and solids data for Eco-Fuel and the other substrates are given in Tables I and II respectively.

Eco-Fuel showed the ability to be slurried at higher concentrations than the shredded substrates in a given digester. The 4% solids loading was already at the limit of loading for the shredded materials using these laboratory digesters. During this work, Eco-Fuel was intentionally held at the same loading level for comparison purposes. However, the loading limit for both substrates probably could be increased in a differently designed digester.

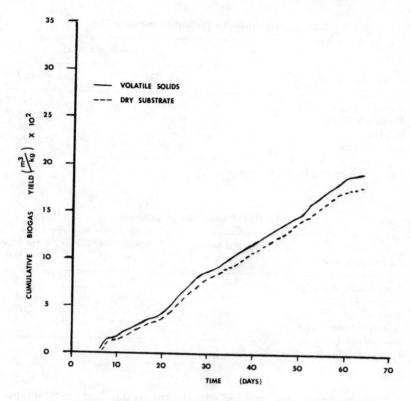

Figure 5.   Cumulative gas yield—Baltimore substrate. Gas volumes at 18.3°C (65°F), 738 mm Hg, dry.

The extended run with substrate from the Pompano Beach facility showed more than one gas production phase in the bioconversion of MSW. In addition, results suggested that gas production in the initial production phase could be increased if a pretreatment method was employed that could convert some of the less biodegradable material to readily biodegradable substances without removing any of the initial volatile solids.

Analyses of gas samples indicate that the methane content of the gas produced in the anaerobic digestion of MSW is about 60–70%. Although the total gas yield is increased when a mixture of sewage sludge and MSW is digested, total gas production from the entire run of the Pompano Beach substrate suggests that at least about 0.28 m^3 of gas can be produced per kg of volatile solids added to the digester when MSW is digested alone. In addition, this yield probably could be improved if key parameters, such as digester loading and particle size, were optimized.

Table I.   Gas Yields for the Initial Production Phase

Substrate Source	Gas Production[a]	
	m³/kg TS	m³/kg VS
Pompano Beach Facility	0.215	0.282
Madison RDF Plant	0.152	0.175
Eco-Fuel	0.124	0.169
Baltimore Plant	0.164	0.175

[a] Based on material added to the digester.

Table II.   Composition of Substrates

Substrate Source	Total Solids (wt %)	Volatile Solids (wt %)[a]	Fixed Solids (wt %)[a]
Pompano Beach Facility	94.1	75.3	24.7
Madison RDF Plant	88.0	86.6	13.4
Eco-Fuel	98.3	74.3	25.7
Baltimore Plant	96.3	91.5	8.5

[a] Percentages based on total solids.

Because of problems observed with the frozen standard inoculum, no firm conclusion can be drawn on enhancement of the gas production rates.

Based on the substrates investigated in this project, shredded MSW from a typical RDF process (primary shredding, magnetic separation, screening, air classification and secondary shredding) appears to be the most attractive substrate for biogasification. However, it is suggested that other pretreatment methods be investigated before making a firm conclusion on the front-end design.

ACKNOWLEDGMENTS

Invaluable support from the following individuals and organizations is gratefully acknowledged: Dr. C. R. Guerra, John Zemkoski and R. J. Terranova of PSE&G; personnel from Linden Synthetic Natural Gas Plant and Harrison Gas Plant; Dr. J. M. Nystrom of Arthur D. Little, Inc.; and personnel from John G. Reutter Associates.

REFERENCES

1. Walter, D. K. and C. Brooks. "Refuse Conversion to Methane (RefCOM): A Proof-of-Concept Anaerobic Digestion Facility," IEEE PES Winter Meeting, New York, NY, (1980), p. 3.
2. Spencer, R. R. "Manual of Procedures for the Operation of Bench-Scale Anaerobic Digesters," prepared for the U.S. DOE under contract EY-76-C-06-1830 (1978) pp. 2, 5.
3. Ghosh, S., and D. L. Klass. "SNG from Refuse and Sewage Sludge by the BIOGAS® Process," IGT Symposium on Clean Fuels from Biomass, Sewage, Urban Refuse and Agricultural Wastes, Orlando, FL, (1976), pp. 123–181.

CHAPTER 10

LANDFILL GAS RECOVERY AT THE
ASCON DISPOSAL SITE – A CASE STUDY

Robert P. Stearns and Thomas D. Wright

SCS Engineers
Long Beach, California

The rapid rise in the cost of energy has prompted increased interest in the recovery and utilization of landfill gas (LFG) at locations throughout the United States. The U.S. Department of Energy (DOE) has estimated that the nation's solid waste landfills generate 200 billion ft³of methane gas annually. Except for a few locations, this potential resource is being lost to the atmosphere. Further, approximately 0.5 million tons of solid waste are added daily to active sanitary landfills in the United States.

The generation of methane gas during anaerobic decomposition of landfilled solid waste is a well-known phenomenon. Landfill gas typically contains 50–60% methane. The balance is composed of carbon dioxide and trace quantities of many other gases. The rate of gas generation will generally be highest during the first few years after solid waste burial, and will tend to decrease with time. The exact details of this time variation are not well known for full-size landfills. Small-scale experiments do not appear to simulate what is found in the field. For lack of a better understanding, it is often assumed that the long-term gas generation rate, after the first few years, can be described by an exponential decay and associated halflife.

Theoretically, the maximum amount of methane which can be produced during the life of a gas-generating landfill is about 0.266 m³/kg of refuse (4.5 ft³/lb). This amount would not, however, be generated in a reasonable time. Moreover, actual recovery will be less than 100%. A maximum recovery of 0.027 to 0.059 m³/kg (1 to 2 ft³/lb) of refuse is considered reasonable at this time.

199

Initial efforts at LFG recovery occurred in Los Angeles County at the Palos Verdes Landfill operated by the Los Angeles County Sanitation Districts in the mid-1960s. From this modest beginning, LFG recovery technology has been applied successfully at seven other landfills and is under active consideration at many other locations.

ASCON SITE DESCRIPTION

The Ascon disposal site is located in the Wilmington area of Los Angeles, California. The site was a former borrow pit and occupies an area of approximately 11.3 ha (38 ac). Household and commercial rubbish, tank bottoms from refining operations and oil field drilling muds have been disposed at the site since 1960 to an average depth of about 18.3 m (60 ft). Soil is scarce at the site, and auto shredder waste is used as daily cover material for the compacted wastes.

A portion of the site was formerly used as a storage area for petroleum coke. Large quantities of water were added to these storage piles, and resulted in perched water and high moisture conditions within the landfill. Filling operations are scheduled to cease in 1980.

FEASIBILITY STUDY

A field test program was conducted during 1976 at the Ascon site by SCS Engineers under contract to the site owner, Watson Energy Systems, Inc. This test program was designed to determine if methane gas could technically and economically be recovered from the site. Three test wells were installed and pumping tests performed over a 3- to 4-month period to determine:

- gas composition as a function of withdrawal rates from the test wells;
- gas flowrates as a function of pressure drop; and
- influence area of withdrawal wells.

During the field test program, preliminary negotiations were conducted with an adjacent Shell Oil refinery for gas sales. Requirements for gas processing and delivery specifications were identified.

Results of the feasibility study indicated that up to 31.4 m³/min (1170 ft³/min) of LFG containing 19.6–21.6 MJ/m³ (500–550 Btu/ft³) could be recovered from the site. This withdrawal rate was estimated to be sustainable for at least six years. User requirements for the LFG were also found to be acceptable—compression to 483 kPa (70 psi) and moisture removal at 4.4°C (40°F).

System design and installation proceeded and were completed in mid-1978. The LFG extraction system as originally installed was comprised of 24 vertical wells drilled to an average depth of 15.2 m (50 ft) with associated header pipe collection system. PVC piping was used throughout the collection system. Wells were perforated for the lower 4.6–6.1 m (15–20 ft) and sealed from the surface with concrete and bentonite clay.

The LFG compressor and cooling equipment utilized rebuilt equipment. A schematic of the gas withdrawal and processing system is shown in Figure 1.

During the placement of extraction wells, a number of unanticipated conditions were encountered. First, landfilled wastes were more compact and had a higher moisture content than indicated by the feasibility testing. Second, as a result of high moisture levels, wells would partially fill with water after drilling. Standing water levels were as high as 9.1 m (30 ft) deep in some wells. This water could be pumped out; however, replacement by seepage occurred with time. Injector pumps were installed in the deeper wells to remove excess moisture.

System startup occurred in August 1978. Great variation existed in gas production rates of wells. Some wells were free-flowing and produced

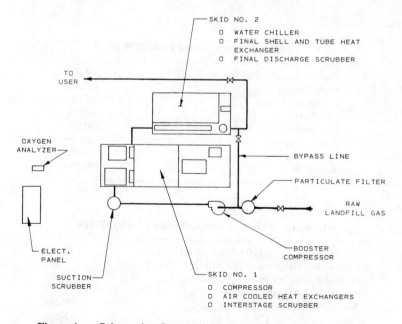

Figure 1.   Schematic of gas withdrawal and processing system.

large quantities of LFG, while others were without positive pressure and yielded little or no gas (even when considerable vacuum was applied). During the ensuing months, additional wells averaging 10.7–12.2 m (35–40 ft) deep were installed to tap more productive areas of the site. Several nonproducing wells were abandoned. A total of 60 wells are now located on the site.

The system has been operating essentially continuously since early 1979, and is currently capable of delivering up to 28.2 normal m³/min (1050 scfm) to the user. Deliveries average about 21.5 normal m³/min (800 scfm or 1.2 × 10⁶ scf/day). No major operating difficulties or maintenance problems have arisen. However, supervision (8 hr/day) was found to be necessary for system adjustments. The system is also monitored by computer and equipped with an alarm system which shuts the system down if problems occur.

ECONOMICS

Tables I and II summarize the capital and operating costs, respectively, associated with the installed system. Annual operating costs, including amortization of capital, average 45% of total installed cost.

Table I. System Capital Costs (1978 $)

Compressor/Gas Chiller	103,000
Wells/Header	376,000
Discharge Pipeline	35,000
Site Work	10,000
Instrumentation/Controls	100,000
Electrical Service	20,000
Engineering	65,000
Total Capital	709,000

Table II. Estimated Annual Operating Costs, 1979 ($)

Electrical Power (150,000 kW/mo at 5¢/kW)	90,000
Compressor Maintenance (5% of capital cost)	5,200
Maintenance Labor (8 hr/day at $15/hr)	43,800
Admin. & Testing ($2500/mo)	30,000
Amortization (7 yr at 12%)	155,400
Total	324,400

Table III presents the estimated annual income from the system. As can be seen, a favorable economic return exists. The sales agreement between Watson Energy Systems, Inc. and Shell Oil assigns the value of the LFG to 70% of the value of No. 6 fuel oil on an equivalent Btu basis. Entitlements are earned by Watson Energy Systems, Inc., under the applicable DOE program.

Table III. Estimated Income from Gas Sales, 1979 ($)

Direct Sales[a]	517,000
Entitlements[b]	137,000
Total	654,000

[a]32,240 normal m^3/day (1.2 × 10^6 scf/day) at 21.0 MJ/normal m^3 (535 Btu/scf) at $2.32/GJ ($2.40 /10^6 Btu).
[b]Estimated based on $0.62/GJ ($0.65/10^6 Btu).

GAS QUALITY

Gas quality at the Ascon landfill has been consistent, with methane concentrations averaging in excess of 50%. Gas obtained from the Ascon landfill is routinely analyzed (bimonthly) by an independent laboratory. A typical result is shown in Table IV.

Extensive analyses of gas obtained from one Los Angeles area landfill has identified more than 65 trace constituents in LFG. Trace components of the gas obtained from Ascon have been identified also. A sample analysis is contained in Table V.

Table IV. Major Constituents – Ascon Landfill Gas

Constituent	Vol %[a]
Methane	55
Carbon Dioxide	42
Hydrogen	0.5
Oxygen	0.2
Nitrogen	1.2

[a]Average of several samples.

Table V.   Trace Constituents – Ascon Landfill Gas[a]

Constituent	Parts per Million
Acetone	32.5
Ethyl Mercaptan	21.1
2-Methyl Furan	6.9
Methyl Ethyl Ketone	5.2
Benzene	5.5
Toluene	20.4
Terpene	12.4
Ethyl Benzene	21.4
Xylene	14.9
Butyl Alcohol	5.2
Alpha Terpinene	11.1
Limonene	26.2
C_3-Substituted Benzenes	9.8
C_4-Substituted Benzenes	7.6
Dichlorobenzene	4.1
2-Ethyl-1-hexanol	6.2
C_4-C_{14} Hydrocarbons	114.2

[a]Sample date: May 15, 1979.

FUTURE FOR LFG RECOVERY

Increases in energy costs have given LFG recovery a needed "shot in the arm." An additional impetus is on its way from the U.S. Environmental Protection Agency (EPA). The Resource Conservation and Recovery Act (RCRA) requirements for controlling migration of LFG as dictated by the EPA sanitary landfill criteria require methane gas concentrations at the disposal site property line to not exceed 5% by volume. Methane gas concentrations in facility structures cannot exceed 1.5% by volume. These requirements will necessitate installation of LFG control systems at most sites. The installed LFG control system may include some of the same facilities (extraction wells, pumps, etc.) required for an LFG recovery system. If the LFG must be removed, many enterprising site owners will actively seek a profitable market for the gas.

Finally, DOE has become increasingly interested in LFG recovery. DOE is supporting a number of projects aimed at improving LFG recovery technology. A number of new projects are likely to be supported under provisions of Public Law 96–126. Legislation supporting LFG recovery has also been introduced at the federal level.

Thus, we can expect more LFG recovery projects in future years. It is hoped that the beneficial effects associated with LFG recovery can dispel some of the negative public reaction to landfilling of our solid wastes, while contributing to our national fuel supply inventory.

CHAPTER 11

EVALUATION OF EMERGING NORTH AMERICAN PYROLYSIS TECHNOLOGY FOR THE CONVERSION OF BIOMASS AND SOLID WASTE TO FUELS

Jonathan K. Tuck

Cal Recovery Systems, Inc.
Richmond, California

Donald R. Deneen

Southern California Edison Company
Rosemead, California

In 1977 Southern California Edison Company (SCE) initiated a study to determine the technical and economic feasibility of producing a fuel gas for electricity generation by combined biological and thermal gasification of select municipal, agricultural and animal wastes [1]. Air pollution emission restrictions, the escalating cost of low-sulfur fuel oil and the limited long-term availability of natural gas were key considerations underlying SCE's interest in the generation of fuel gas from alternative energy sources. Compared to the combustion of solids, gas combustion was relatively clean, required less excess air and offered the opportunity for greater efficiency and control. The gas could also be used as feedstock in the synthesis of methanol, ammonia and petrochemicals.

The study concluded that anaerobic digestion of the putrescible wastes accompanied by thermal conversion of the mechanically separated nonputrescible fraction and dewatered digester sludge appeared to be a feasible means of generating a high-quality fuel gas suitable for use in existing utility boilers. However, it was also concluded that the available database for preliminary design was deficient, and the risks associated with going directly to a commercial scale of operation were unacceptably high. A multiphase program of investigation to establish the database was recommended. If the risks associated with commercialization could be mitigated, the design and construction of a 200-ton/day demonstration facility would follow.

The evaluation of suitable thermal conversion processes represented a key task in this investigation. This chapter summarizes the selection and evaluation of three pyrolysis reactors designed for the conversion of solid wastes and other biomass to alternative fuels. The feedstock used for the testing of reactor performance was refuse-derived fuel (RDF), a predominantly cellulosic material derived from the shredding and beneficiation of municipal solid waste. This feedstock was readily available and represented the most suitable feedstock for a demonstration facility located in an urban area.

THE PYROLYSIS PROCESS

In a pure pyrolysis process, organic material is thermally cracked into a solid char and volatiles in the absence of air. Heat for cracking is usually generated by electrical resistance heating, or by combusting fuel gas either in a firebox surrounding the reaction chamber or in fire tubes inserted into the reaction chamber. In practice, pure pyrolysis is achieved only in batch reactors, where complete removal of air is possible. The operation of more efficient continuous feed units is accompanied by air injection and subsequent partial combustion of volatiles and carbon. As a consequence, the quality of the resultant fuel gas is degraded relative to batch reactor gas by the presence of a large proportion of noncombustible nitrogen and carbon dioxide.

Variables Affecting Pyrolysis

Pyrolysis results in the formation of a solid char, a fuel gas and a condensible fraction consisting of tar, oil and a pyroligneous liquor (a mixture of highly oxygenated aliphatic and aromatic compounds). The quality and yield of these products from a specific reactor design are dependent primarily on the chemical and physical characteristics of the feedstock, the heating rate, the reaction chamber temperature, the solids retention time and the quantity of air introduced into the reaction chamber.

Chemical and Physical Characteristics

Biomass and RDF are characterized by the predominance of cellulose, hemicellulose, other carbohydrates and lignin. Compared to coal, which has a complex polycyclic structure, biomass exhibits high reactivity [2]. The relationship between chemical and physical characteristics and the kinetics and product formation mechanisms of pyrolysis have been studied, but are not completely understood [3,4]. Currently, pyrolysis of the many complex forms of biomass is viewed, for lack of better data, as the sum of the thermal conversion of its three major components: cellulose, hemicellulose and lignin [4].

Available data indicate the reaction between feedstock moisture and the products of pyrolysis is relatively slight at temperatures less than 600°C [5, 6]. At temperatures above 600°C, the volatiles will react with steam to form a synthesis gas [6]. In either case, an excessive amount of moisture in the feedstock requires a significantly greater quantity of heat for vaporization.

Heating Rate

Heating rate is a function of particle size, temperature, residence time and heat transfer technique. Flash pyrolysis, which occurs at a high heating rate (< 6 sec), leads to the generation of more highly oxygenated complex aromatics and unsaturated compounds. Slow heating rates (50°C/min) result in an increase in the char yield and a decrease in tar formation. The reason for this effect is that the secondary reactions which crack the products of pyrolysis occur more easily during a slow heating rate [3,7]. Slow heating rates are usually dominant in large-scale pyrolysis processes.

Reaction Chamber Temperature

The reaction temperature significantly affects product quality and yield. For example, slow heating-rate data indicate that as the reaction temperature is increased from 300 to 500°C (572 to 932°F), the devolatized matter from pyrolysis of poplar wood increases from 10 to more than 70% [3]. Figure 1 shows the effects of reaction temperature on product and energy distribution for poplar wood at a low heating rate [3].

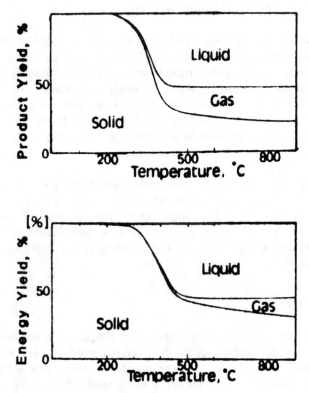

Figure 1.   Product and energy distribution during pyrolysis of poplar
wood: 50°C/min heating rate.

Air Intrusion

The introduction of air into the reaction chamber results in partial oxidation of the products of pyrolysis and the liberation of heat. Although
process heat requirements are reduced and gas yield increased, the
quality of the gas is diminished.

Solids Retention Time

The actual solids residence time is dependent on the speed at which
material is conveyed through the reactor. Optimum residence time is a
function of all the previously mentioned variables; reaction chamber
temperature is the most influential.

Pyrolysis Gas Quality and Boiler Performance

One of the objectives of the reactor evaluation was to identify those reactors which have the potential of producing a fuel gas with a medium range heating value of at least 7.8 MJ/normal m³ (200 Btu/scf). The rationalization for this requirement is discussed by Frendberg [8] and illustrated in Figure 2, which shows the effect of gas heating value on thermal efficiency in a utility boiler. The use of a fuel gas having a heating value lower than 7 or 8 MJ/normal m³ may result in significant derating of the boiler or extensive retrofit modifications. Diminished performance may only be averted if sensible heat in the form of air or fuel gas preheat is also added to the boiler.

SELECTION OF REACTORS FOR EVALUATION

Following the disappointing performance of large-scale, single-unit, thermal conversion plants such as the 200-ton/day Occidental Oil (flash

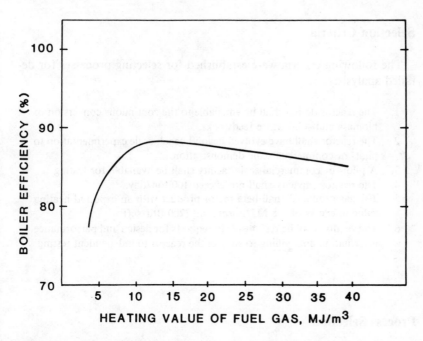

Figure 2. Effect of gas heating value on existing utility boiler efficiency.

pyrolysis), the 200-ton/day Union Carbide (Purox), and the 1000-ton/day Baltimore (Monsanto Landgard) plants, a new approach to thermal plant design seemed appropriate. The approach selected in this study was as follows.

Small- to- medium-capacity reactors would operate in parallel to obtain a 200-ton/day plant design capacity. An economic level of redundancy in the reactor system would be provided to give rated capacity during periods of maintenance and unscheduled repair and allow flexibility of operation unobtainable in a large single-unit system. In particular, a large turndown ratio would be possible without jeopardizing efficiency or product yield and quality. Candidate reactors were to be identified by reviewing the status of biomass and solid waste thermal conversion technology. A number of processes would then be selected for comparative testing and evaluation. Where possible, each reactor type would be tested and evaluated at, or near, commercial scale to minimize risks. In this way, mechanical design and ancillary equipment selection problems and major deviations in conversion efficiency, product characteristics and product yields would be avoided.

Selection Criteria

The following criteria were established for selecting processes for detailed analysis:

1. The reactor design shall be amenable to the continuous conversion of biomass and solid waste feedstocks.
2. The reactor shall have evolved beyond bench-scale experimentation to pilot- or commercial-scale demonstration.
3. A pilot- or commercial-scale facility shall be available for testing.
4. The reactor capacity shall not exceed 100 ton/day.
5. The gas produced shall be a major product with an expected heating value in excess of 7.8 MJ/normal m^3 (200 Btu/scf).
6. The vendor shall be responsive to requests for design and performance information and willing to subject the reactor to independent testing.

Process Selection

Table I describes thermal pyrolysis, gasification and starved-air reactors now under development in North America. Each reactor is identified by developer and location. Reactors developed solely for coal gasification are not included due to the inappropriate database.

Selection of processes for evaluation was undertaken by a system of elimination based on the application of the previously established criteria. Those vendors of processes selected for the evaluation were:

- Pan American Resources Inc. (Lantz Converter)
- Pyro-Sol Inc. (Pyro-Sol Process)
- Total Energy Systems (Total Energy Systems Organic Refining Process)

PROCESS DESCRIPTIONS

The following descriptions of the selected processes were developed from a review of vendor-supplied information.

Total Energy Systems Organic Refining Process

The Total Energy Systems process consists of a low temperature cascading bed reactor and refining train. The system was invented by Atkins to convert biomass and solid waste into char, oil and gas. Total Energy Systems of Florida, Inc., a Los Angeles–based firm, markets the system.

Development of this process began more than eight years ago. Three batch pilot plants, each successively larger than the former, were operated and used to develop the reactor design and test the effectiveness of electrocatalytic refining units. The vendors claim the electrocatalytic refining units crack and reform high-molecular-weight hydrocarbons contained in the gas stream, thus minimizing tar production and enhancing oil and gas production.

A continuous-feed demonstration plant rated at 60 ton/day was constructed in 1979. The plant (Figures 3 and 4) consists of a reactor and five condensation tanks with attached electrocatalytic refining units and bubble tower. The reactor (or cracking unit) is box-shaped and is 1.8 m wide, 2.4 m high and 6.1 m long. The exterior shell is separated from an inner reaction chamber by flues. Natural gas or recycled pyrolysis gas is fired into these flues, and the heat from combustion is transferred through the reaction chamber wall to the feedstock and products of pyrolysis. Four vertically aligned chain-link conveyors, having a total length of 20.7 m, are housed in the reaction chamber. The drive shaft for each conveyor is connected to a variable speed electric motor and gear train. Feedstock is

Table I. Examples of Pyrolysis, Gasification and Starved-Air Reactors for Thermal Conversion of Biomass and Solid Waste Feedstocks

Developer	Location	Reactor Type[a]	Conversion Process[b]	Design Capacity (ton/day)	Target Feedstocks[c]	Primary Products[d]	Gas Quality[e]
Andco, Inc.	Buffalo, NY	VS	SAG	75	MSW, IW, S	C, G	Low
Battelle Columbus Lab	Columbus, OH	FB	R		W	C, G	Low
Biomass Corporation	Yuba City, CA	DFB	SAG	36	Far, W	G	Low
BSP/Envirotech	Belmont, CA	MH	SAG	5-200	RDF, S	C, G	Low
Canadian Industries, Ltd.	Kingston, Ont	FB	SAG	24	MSW, W	C, G	Low
City of Baltimore	Baltimore, MD	RR	SAG	600	MSW, IW	C, G	Low
Dynecology, Inc.	New York, NY	FB	SOG	2	MSW, S, Coal	G	Medium
Energy Incorporated	Idaho Falls, ID	FB	SAG	0.5-1.5	W, FAR	C, G	Low
Energy Resources Co.	Cambridge, MA	FB	SAG	13-20	MSW, FAR, S	C, G, O	Low
Enterprise Co.	Santa Ana, CA	AB	P	50	RDF	C, O, G	Medium
Forest Fuels	Antrim, NH	UMB	SAG	30	W, FAR	G	Low
Garrett Energy Research	Hanford, CA	MH	SAG	4	RDF, M, W	C, G	Low
Intenco	Houston, TX	HMB	P	50	T	C, O	
Kemp Reduction Corporation	Santa Barbara, CA	AB	P	5	RDF, FAR	C, O	Low
Nichols Research and Eng.	Belle Mead, NJ	MH	SAG	5-200	RDF, S	C, O	Medium
Occidental Research	El Cajon, CA	VS	FP	200	RDF, S	C, O, G	Medium
Pan American Resources, Inc.	Upland, CA	RR	P	4	RDF	C, O	
Pyro Conversion, Inc.	Rosamont, CA	AB	P	5-26	T, FAR	C, O	Medium
Pyro Sol, Inc.	Redwood City, CA	HMB	P	50	MSW, IW	C, G	Medium
Q² Corporation	Oakland, CA	VEB	HG	25	T, RDF	G	Medium
Resource Recovery Corporation	Raleigh, NC	VS	P	50	MSW, IW, S	G	Medium
Rotter Gasifier	Portland, OR	DFB	SAG	24	T, FAR, MSW	C, G, O	Medium
Standard Solid Fuels	Kirland, WA	VEB	P,SAG	50	RDF, W, S, IW	G	Low
Tech-Air Corporation	Cordele, GA	VMB	SAG	50	RDF, IW, W	G, O, C	Low

Texas Tech University	Lubbock, TX	FB	R	0.5	M, FAR	C, G	Low
Thagard Research Corporation	Irvine, CA	VEB	HG	10–25	RDF, W, S	C, G	Low/Medium
Tosco Corporation	Los Angeles, CA	RR	P	15	MSW, T	C, O	Low
Total Energy Systems of Florida, Inc.	Los Angeles, CA	HMB	P	60	MSW, W	C, G, O	Medium
Union Carbide Corporation	South Charleston, WV	VS	SOG	200	MSW, IW, S	G	Medium
University of California	Davis, CA	DFB	SAG	18	FAR	C, G	Low
Wright Malta Corporation	Ballston Spa, NY	RR	CG	0.1	MSW, IW, S	C, G	Medium

[a] Reactor type: VMB = vertical moving bed; HMB = horizontal moving bed; DFB = downdraft fixed bed; VEB = vertical entrained bed; UMB = updraft moving bed; RR = rotary reactor; MH = multiple hearth; VS = vertical shaft; FB = fluidized bed; AB = auger bed.

[b] Conversion process: SAG = starved-air gasification; SOG = starved-oxygen gasification; FP = flash pyrolysis; HG = hydrogasification; CG = catalytic gasification; P = pyrolysis; R = reforming.

[c] Target feedstocks: MSW = municipal solid waste; FAR = forestry and agricultural residues; RDF = refuse-derived fuel; IW = industrial waste; S = sludge; M = manure; T = tires; W = woodchips.

[d] Primary products: C = char; G = gas; O = oil.

[e] Gas Quality: low = 0–7.8 MJ/m³; medium = 7.8–15.7 MJ/m³; high = greater than 15.7 MJ/m³.

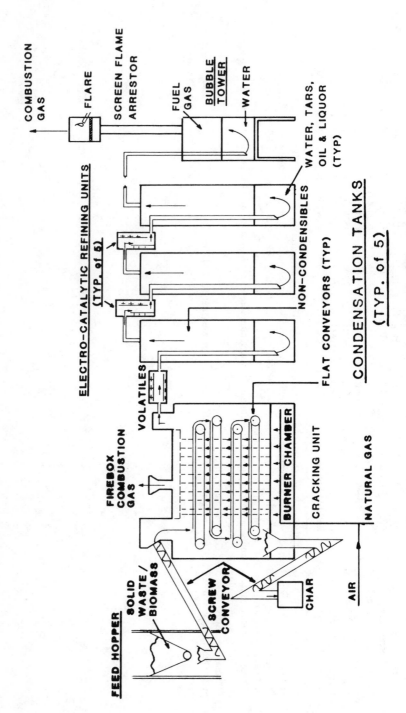

Figure 3. Total Energy Systems demonstration plant schematic.

Figure 4.   Total Energy Systems cracking unit and a section of the refining train.

screw-conveyed from the storage hopper to the top of the reactor, where
it drops onto the first moving conveyor. A 10- to 20-cm layer of material
is formed by the action of a breaker bar. Drying, cracking and devolatization occur as the material is conveyed into zones of increasing temperature during a 20-min retention period. Gravity transfer of material from
one conveyor to another provides some agitation of the feedstock,
thereby exposing unpyrolyzed material to direct heat. Product characteristics and throughput may be optimized by regulating the speed of the
conveyors and the internal temperature.

The volatiles produced by pyrolysis are pressure-driven by the cracking
unit through the condensation tanks where previously condensed oil and
liquor cool and extract the volatiles. Refined gas leaves the fifth condensation unit and is discharged through the bubble tower, which acts as an
air seal. The char produced by the unit is screw-fed to char storage bins.
In a commercial operation, water, oil, tar and particulates which accumulate in the condensation tanks would be removed for oil and tar recovery and wastewater treatment.

Lantz Converter

The Lantz converter is a medium-temperature, rotary-bed reactor invented for biomass and solid waste pyrolysis by Lantz. Pan American Resources, Inc., in Upland, CA, is developing and marketing the reactor. As a commercial unit, the developers offer the reactor in capacities ranging from 50 to 100 ton/day.

The Lantz converter design concept is based on development work initiated in the 1930s. Several commercial units with a 50-ton/day capacity were built and operated between 1963 and 1968. Two units were used by the lumber industry to make charcoal for briquettes. Units were also used at a Ford assembly plant and a naval ammunition depot to dispose of organic solid wastes. Currently, there are no commercial units in operation. However, a 4-ton/day pilot plant is maintained for research and demonstration and a 50-ton/day unit is now being installed in an industrial plant in New York state to demonstrate the generation of fuel gas from solid waste.

The pilot plant is shown in Figures 5 and 6. The 1.3-cm-thick stainless steel cylindrical retort contained within the refractory-lined firebox is approximately 0.6 m in diameter and 2.4 m long, and is lined with helical ribs to assist feedstock agitation and to promote heat transfer. Solid waste, which has been shredded to a nominal size of 3 cm and dried in a rotating tube having direct contact with the firebox combustion gas, is fed by hydraulic ram into the rotating retort, where it is thermally cracked at 430–540°C for a period of 8 to 12 min. From 10 to 30% of the pyrolysis gas is returned to the firebox for combustion in the firebox burners. The flow of gas to the burners is controlled by the temperature in the firebox. Additional heat in the form of natural gas is supplied during startup and when feedstock moisture content is excessive.

Pyro-Sol Process

The Pyro-Sol process consists of a high-temperature, horizontal moving-bed reactor; a gas cleanup train; a boiler; and a turbogenerator. The reactor was invented for pyrolysis of industrial and municipal solid waste by Welty. The process is being developed by Pyro-Sol Inc. and Levin Energy Corporation of Redwood City, CA.

The Pyro-Sol demonstration reactor was developed from pilot tests conducted at Santa Ana, CA, between 1971 and 1974. The pilot reactor was 3.6 m long, 0.3 m wide, and was operated at feed rates varying from 27 to 90 kg/hr. The unit was electrically heated, and material to be

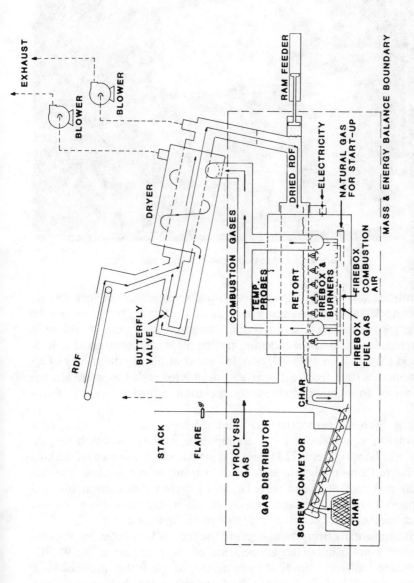

Figure 5. Lantz converter pilot-plant schematic.

Figure 6.   Lantz pilot reactor, dryer and boiler.

pyrolyzed was fed into the system using an overhead air lock device. Feedstock was conveyed and pyrolyzed on a vibrating conveyor and the char discharged via a screw feeder. Based on the results of the pilot reactor tests, a 50-ton/day demonstration plant was constructed at Redwood City, CA, in 1978–1979. A 15-year contract for the supply of an automobile fluff feedstock was also negotiated with Levin Metals, the owner of an automobile shredding operation located adjacent to Pyro-Sol.

The Pyro-Sol demonstration plant (Figures 7 and 8) consists of a pyrolyzer, a gas collection and cleanup train, a boiler and a turbine/generator. The pyrolyzer is 22 m long, 1.8 m wide and 1.8 m high. A shaking conveyor forms the hearth and feedstock transport mechanism. The side walls and roof are lined with 18 cm of ceramic fiber insulation and support 8 radiant heat tubes suspended horizontally over the conveyor. Each radiant heat tube is 20 cm in diameter and 5.8 m long.

Startup entails firing natural gas in the radiant heat tubes for approximately 1 hr to bring the temperature of the pyrolyzer up to 930°C. Shredded feedstock conveyed from the storage bin is then dropped via a three-stage air-lock feeder onto the front end of the shaking conveyor, and the layer of material formed is moved through the pyrolyzer by the

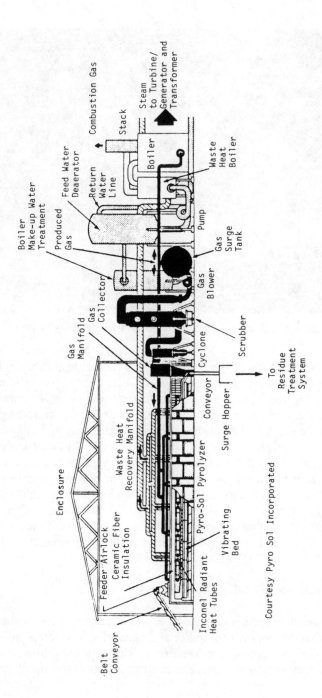

Figure 7. Pyro-Sol demonstration plant schematic.

Figure 8.   Pyro-Sol demonstration plant: 50-ton/day pyrolyzer.

reciprocating action of the conveyor. The radiant heat tubes indirectly transfer the heat from the combustion inside the tubes to the moving bed. During the 5- to 7-min residence time, the conveyor continually agitates the bed to expose unpyrolyzed material to the radiant heat. Volatiles are liberated from the organic material, leaving a char residue, which is discharged by screw conveyor to an airtight bin. Pyrolysis gas exits the pyrolyzer at a temperature of 590°C and a pressure of less than 0.12 kPa. It is passed through a cyclone and wet scrubber to remove particulates, oils and tars, and is then blown into a surge tank. Approximately one-third of this gas is returned to the radiant heat tubes to provide heat for pyrolysis. The remainder is extracted from the surge tank and fired in a boiler to provide steam for electricity generation.

Feedstock Specifications for Pyrolysis

Most vendors require that the biomass and refuse feedstocks be shredded so that the length of the material is reduced to less than 5 cm, and the thickness to less than 0.6 cm. Feedstocks with moisture contents in excess of 20% may require drying. The presence of glass and metal contaminants in the feedstock is acceptable, although removal of these inerts from the feedstock or the char may provide by-product revenue.

Auxiliary Energy and Material Requirements

Electricity and a fuel gas are the only auxiliary inputs required for operation. It is estimated that electricity requirements are less than 25 kWh/ton of feedstock. Natural gas is used for startup and occasionally during operation when the moisture content of the feedstock is too high.

OPERATING CHARACTERISTICS AND TEST DATA

The processes described are in the pilot-plant or demonstration phase of development. Operating experience ranges from a few days to more than a decade in the case of the Lantz converter. Test data are reported and analyzed for the Total Energy Systems Organic Refining Process and Lantz converter. An opportunity to test the Pyro-Sol process did not occur, and the data provided by the vendor were insufficient for adequate analysis.

Total Energy Systems Organic Refining Process

Operating experience for the demonstration plant is limited to an eight-day test in October 1979 and a one-day test in July 1980. The test of October 1979 was undertaken by the vendors using a woodchip feedstock. No quantitative data were collected from this test. Consequently, arrangements were made with the vendor for additional testing, and in July 1980 tests were performed using woodchip and RDF feedstocks.

The test program for RDF involved the continuous operation of the process and the collection of data for the period of performance. Due to technical factors, comprehensive measurements and analyses were not possible. In particular, gas flowrate and volume of condensibles could not be measured and had to be estimated from the vendor's batch tests and national laboratory test data [9]. Measurements collected onsite and by laboratory analysis included:

* RDF feed rate;
* char production rate;
* RDF proximate and ultimate analyses, and heating value;

- char proximate and ultimate analyses, and heating value;
- product gas composition and heating value;
- oil and water content of condensibles;
- elemental analysis and heating value of the oil;
- reaction chamber and product gas temperatures; and
- significant operating characteristics

Test data are summarized in Table II, and the mass and energy balances determined for the test are shown in Figures 9 and 10. The feed rate was set at a level which the vendors considered the maximum possible for trouble-free feeding. It is noted that:

1. The pyrolysis reactor operated in a 200–370°C temperature range.
2. The test feed rate was less than 10% of that reported by the vendors as the rated capacity.
3. The nitrogen content of the fuel gas was excessive.
4. The high volatile content of the char indicated that devolatization was not complete.
5. The firebox gas requirement was unacceptably high for the test feed rate.

Lantz Converter

The existing pilot reactor has been operational since 1961. Test data have been collected since 1963 and are summarized for RDF in Table III. The tests in August 1979 were arranged and observed by SCE. Mass and energy balances for the second test conducted on August 22, 1979 are shown in Figures 11 and 12. It is noted that:

1. There is a considerable proportion of nitrogen in the pyrolysis gas.
2. At steady-state operating conditions, less gas is required for heating the retort and firebox, resulting in an increase in overall reactor efficiency. In the August 1979 test, the calorific yield of the dry gas available for distribution increased from 4.6 at the beginning to 7.4 MJ/kg of RDF after 3 hr of operation. The heating value of the gas also increased from 5.6 MJ/normal m³ (143 Btu/scf) to 10.3 MJ/normal m³ (264 Btu/scf) over the same period.
3. The heating value of the pyrolysis gas generated by the pilot reactor appears to increase with increase in refuse feed rate.

The energy balance indicates that almost 12.6 GJ (12 MBtu) of pyrolysis gas is generated per ton of RDF. After satisfying the process heat requirement, about

Table II.   Summary of Data from Total Energy Systems
Process Test Conducted on July 2, 1980, RDF Feedstock

Item	Value
Feedstock Characteristics	
Heating value (MJ/kg dry)	16.86
Moisture content (wt %)	25
RDF feed rate, (kg/hr)	167
Test period (hr)	4.6
Fuel Gas Characteristics	
Composition (mol %)	
N_2	71.3
CO_2	21.7
C_1	0.8
C_2	0.4
C_3	5.8
C_{6+}	0.03
Heating value (MJ/m^3)	6.34
Miscible Condensate Characteristics	
Organic content	1–5
Heating value (MJ/kg)	
Minimum	17.63
Maximum	39.55
Char Characteristics	
Ash content (wt %)	47.7
Volatile content (wt %)	29.0
Heating value (MJ/kg)	14.66
Production rate	
(kg/hr)	87.5
(kg/metric ton RDF)	523
Temperatures (°C)	
Top of reactor	304
Bottom of reactor	399
Gas	
Minimum	204
Maximum	232
Firebox Natural Gas Requirement	
Flowrate (m^3/hr)	60.0
Applied heat rate (GJ/hr)	2.24
Applied heat rate (GJ/metric ton RDF)	13.35

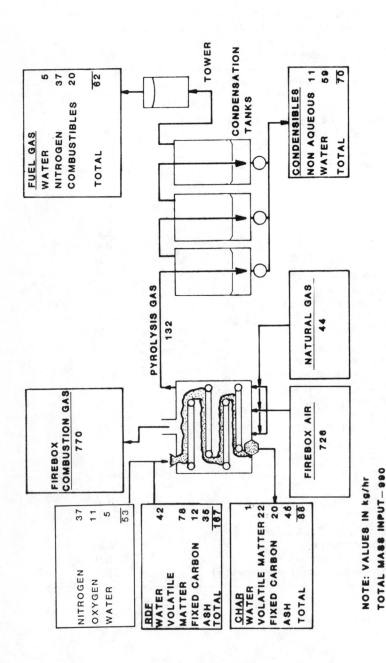

Figure 9. Mass balance for Total Energy Systems process test (July 2, 1980).

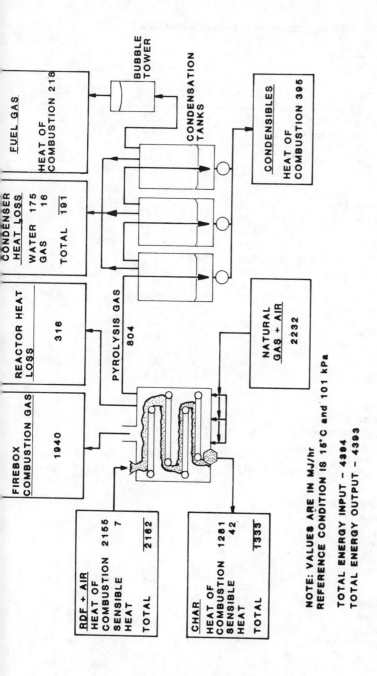

Figure 10. Energy balance for Total Energy Systems process test (July 2, 1980).

Table III. Summary of Lantz Converter Data for
Solid Waste Feedstock

Item	Test Date				
	8/22/79	8/22/79	9/24/79	9/24/79	11/2/76
Dry refuse heating value (MJ/kg)	15.75	15.75	NA[a]	NA	19.42
Refuse moisture content (wt %)	6	6	NA	NA	NA
Feed rate (kg /hr)	107	107	184	184	159
Startup to sample time (hr)	0.5	3	0.5	0.5	0.5
Sampling location[b]	D	D	D	R	R
Pyrolysis gas characteristics					
Composition (mol %)					
CO_2	7.35	13.20	7.1	15.0	24.8
O_2	11.20	0.70	6.2	0.7	3.4
N_2	63.89	53.47	57.8	37.0	31.3
CO	8.93	16.10	15.4	27.6	26.6
CH_4	2.42	4.64	4.5	6.7	6.1
C_2H_6	0	0	NA	NA	NA
C_2H_4	0.27	1.05	NA	NA	0.84
C_2H_2	2.04	4.51	2.3	3.4	1.36
C_3	0.60	0.79	0.3	1.3	1.32
C_4	0.28	0.35	0.4	0.6	0.7
C_5	0.17	0.17	0.2	0.4	0.3
C_6	0.19	0.30	0.2	0.2	0.5
C_7 and higher	0.06	0.12	0.3	0.4	
H_2	2.60	4.60	4.7	6.7	2.8
Heating value					
(MJ/normal m³)	5.6	10.4	8.6	13.1	11.2
(MJ/metric ton refuse)	4648	7785	NA	NA	NA
Flowrate[c]					
(normal m³/hr)	94	85	NA	NA	NA
(normal m³/metric ton refuse)	882	797	NA	NA	NA
(kg/hr)	116	104	NA	NA	NA
Organic condensate characteristics					
Heating value					
(MJ/kg)	32.54	27.80	NA	NA	33.38
(MJ/metric ton refuse)	569	697	NA	NA	NA
Flowrate					
(kg/metric ton refuse)	17.5	25.2	NA	NA	0.1
(kg/hr)	1.9	2.7	NA	NA	0.02
Particulate characteristics					
Heating value					
(MJ/kg)	18.99	18.06	NA	NA	NA
(MJ/metric ton refuse)	1034	1394	NA	NA	NA
Flowrate					
(kg/metric ton refuse)	54.6	78.8	NA	NA	NA
(kg/hr)	5.8	8.4	NA	NA	NA

Table III. continued

Item	Test Date				
	8/22/79	8/22/79	9/24/79	9/24/79	11/2/76
Char characteristics					
Ash content (wt %)	72	72	NA	NA	NA
Heating value					
(MJ/kg)	6.38	6.38	NA	NA	NA
(MJ/metric ton refuse)	1975	1975	NA	NA	NA
Flowrate					
(kg/metric ton)	312	312	NA	NA	NA
(kg/hr)	33.4	33.4	NA	NA	NA

[a]NA = not available.
[b]D = fuel gas distributor, R = inside retort.
[c]Does not include firebox fuel gas.

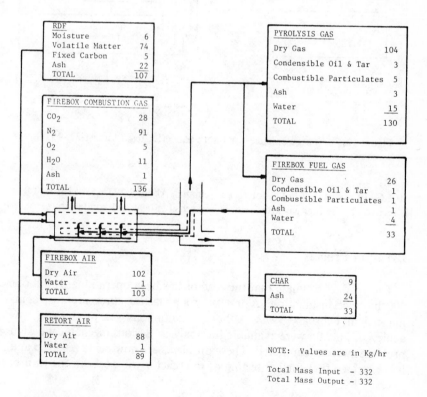

Figure 11. Mass balance for Lantz converter test (August 22, 1979).

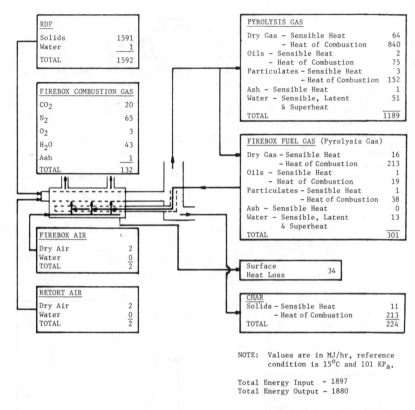

Figure 12.   Energy balance for Lantz converter test (August 22, 1979).

10.1 GJ (9.6 MBtu) of hot pyrolysis gas remains. About 7.1 GJ (6.7 MBtu) of this is recoverable as fuel gas after removal of sensible heat and condensibles.

Pyro-Sol Process

The vendors estimate that the reactor has been operational for about 300 hr since installation. A test program, lasting three months, commenced in February 1979. Although continuous gas production was achieved, no data were available for analysis. Serious mechanical design problems were encountered. These problems are now being resolved and it is expected that further testing of the reactor will occur in the future.

Optimized Process Performance

To facilitate an understanding of pyrolysis objectives, optimized mass and energy balances have been estimated for pyrolysis of a standardized RDF (SRDF) feedstock having a heating value of 17.4 MJ/kg of dry solids (7500 Btu/lb) and a moisture content of 20%. These estimates (Figure 13) are for a generic process operating at steady-state conditions.

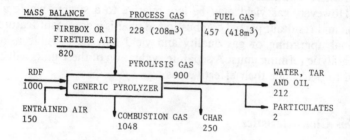

NOTE: VALUES IN Kg/t RDF
RDF MOISTURE = 20%

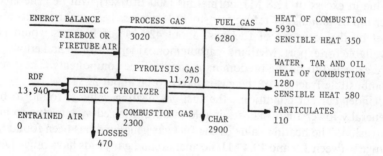

NOTE: VALUES IN MJ/t RDF
REFERENCE CONDITION IS 15°C, 101 KP$_a$

Figure 13.   Mass and energy balances estimated for optimized pyrolysis of SRDF in a generic process.

Vendor data, national laboratory research data [9] and thermodynamic considerations were used to develop the estimates. The balances assume that:

- Operation of the generic reactor is at a temperature in excess of 650°C.
- Efficient heat transfer between the firebox or fire tubes and the feedstock occurs.
- Air intrusion is minimized by using more effective char removal systems.

Under these optimized conditions, gas and char quality are exceptional. However, gas yield may be low relative to a system where air intrusion and resultant partial combustion of char is a dominant feature. Additional upgrading of gas quality and yield may be possible using electrocatalytic refining units. A separate evaluation of these units will be required to determine their effectiveness.

Fuel Gas Characteristics

For optimized pyrolysis conditions in a generic reactor, it is estimated that more than 620 normal m³ (21,900 scf) of fuel gas, having a heating value in excess of 11.8 MJ/normal m³ (300 Btu/scf), will be generated per ton of biomass or refuse feedstock. If process heat requirements are met by the recycling of fuel gas, about 420 M³ of fuel gas would be available for export. Methane, carbon monoxide, hydrogen and ethylene are expected to be the predominant combustible components. The noncombustible fraction of the gas is likely to be about 35%.

Under current conditions, the fuel gas heating value is found to be generally lower than that attainable in an optimized system due to air intrusion. The heating value of the fuel gas generated has been found to range between 3.9 and 10.4 MJ/normal m³ and gas yields have ranged up to 950 m³/metric ton of feedstock.

By-product Characteristics

Associated with pyrolysis gas are particulates and organic condensibles representing as much as 7 and 2% by weight, respectively. In terms of energy content, particulates may represent about 14% and the organic condensibles about 7% of the gaseous energy flow. Between 150 and 300 kg of char is generated per ton of RDF. The char has low volatile and high ash content, and a heating value ranging between 5.8 and 11.6

MJ/kg. Sensible heat is also a by-product of pyrolysis. It may be recovered from the firebox combustion gas by using the gas to either dry the incoming RDF or generate steam in a waste heat boiler.

By-product Refinement and Utilization

The char may be steam gasified to produce synthesis gas or, after metals and other inert components are removed, converted to char briquettes suitable for firing in industrial boilers. Other possibilities include use as an activated carbon for wastewater treatment and generation of carbon feedstock for the chemical industry.

In a commercial plant, pyrolysis gas tars may be recovered by wet scrubbing, clarifying the scrubber water and then centrifuging the underflow. The overflow from this process will consist of a solution of highly oxygenated hydrocarbons which may be extracted by distillation or solvent extraction methods. However, recovery of these hydrocarbons is complicated not only by their miscibility in water, but also by their corrosiveness and possible mutagenic properties. Possible uses of recovered tar and oil include returning it to the reactor for cracking, or marketing it as a fuel or petrochemical feedstock.

Thermal Efficiencies

Based on the previously described mass and energy balances, the pyrolysis gas thermal efficiency (PGTE) and fuel gas thermal efficiency (FGTE) have been calculated for each test and the optimized conditions in an idealized generic process. The efficiencies are not directly comparable since the test feedstocks have different characteristics. In particular, RDF moisture content is 6% for the Lantz test and 25% for the Atkins test. The efficiencies shown in Table IV are defined as follows:

$$PGTE = \frac{H_p}{H_{fs} + H_{pg}} \times 100$$

$$FGTE = \frac{H_f}{H_{fs}} \times 100$$

where H_p = pyrolysis gas calorific and sensible heat (MJ/hr)
H_f = fuel gas calorific heat (MJ/hr)
H_{fs} = feedstock calorific heat (MJ/hr)
H_{pg} = calorific and sensible heat of process (or recycled) gas (MJ/hr)

Table IV.   Gas Thermal Efficiencies for Test
and Optimized Conditions

Item	TES Test[a]	Lantz Test[b]	Generic Reactor[c]
Pyrolysis Gas Thermal Efficiency (PGTE) (%)	18	79	66
Fuel Gas Thermal Efficiency (FGTE) (%)	10[d]	53	43

[a] Total Energy Systems test, July 2, 1980; from Figure 10.
[b] August 11, 1979; from Figure 12.
[c] Optimized for minimum air intrusion and medium-Btu gas production from pyrolysis of SRDF from Figure 13.
[d] Not appropriate for comparative purposes since natural gas and not fuel gas was used for reactor heating.

The PGTE is the efficiency of the pyrolyzer before process gas removal and before gas cleanup or cool-down. The FGTE may be shifted by the use of electrocatalytic refining units (which one vendor claims can increase the fuel gas yield in low temperature pyrolysis by as much as 50%) or by the introduction or exclusion of air. Gas cleanup may be avoided where close coupling of a boiler and pyrolyzer is possible. In addition to an increase in thermal efficiency, a reduction in capital and operating costs is achieved by close coupling.

It should be noted that the thermal efficiencies shown in Table IV do not take into account the energy value of the char and by-product tar and oil, which in some cases, may be the desired primary product.

ANTICIPATED PROCESS DEVELOPMENT SCENARIOS

Discussions with the vendors and a review of performance of each reactor has led to the following scenarios for development.

Total Energy Systems Organic Refining Process

Since testing of this reactor has been conducted at less than the rated capacity, the effect of feed rate scale-up is of immediate concern. It is possible that at the rated capacity, gas quality will be improved as air intrusion and the resultant pyrolysis gas dilution become less significant. However, the conversion efficiency may deteriorate significantly as a result of a thicker bed and/or decreased retention time brought about by the higher feed rate. The volatile content of the char is also likely to increase even higher than the 29% level determined to date. If, by changing internal design, the temperature of pyrolysis can be raised, devolatilization of the feedstock will improve, and more pyrolysis gas and a less volatile char may be produced at the rated reactor capacity.

Total Energy Systems believe that the tests undertaken so far do not provide sufficient information for a conclusive evaluation. Modifications are being implemented which will improve the performance and efficiency of the process.

Lantz Converter

Currently, Pan American Resources is installing a 50-ton/day reactor with dryer at an industrial site in the state of New York. It is expected that the reactor will be pyrolyzing refuse to provide fuel gas for a steam boiler. The retort temperature will be in the 650–700°C range, and the reactor is to be fitted with a variable drive motor to provide control of the retort's rotational speed. This will allow adjustments to bed depth, retention time and liberation of volatiles. Steps are also being taken to minimize air infiltration through the feeder, the char discharge mechanism, and the interfacing between the retort, feeder and gas distributor.

Pyro-Sol Process

Vendor testing has shown that the conveyor inside the pyrolysis reactor fails to perform satisfactorily. The vertical component of the oscillating movement of the conveyor causes internal pressure fluctuation (0.4 kPa) resulting in overheating of the underside of the conveyor and some loss of char to the underlying plenum. A new conveyor having a horizontal reciprocating movement (zero pitch) has been selected by the vendor as a replacement. Since this conveyor produces no vertical agitation of the feedstock, the efficiency of heat transfer to the feedstock may not be as great as with the former conveyor. Further development and testing is planned using a new 75-ton/day pyrolyzer which will be 50% wider than the 50-ton/day unit.

COST AND REVENUE ESTIMATES

Cost estimates for installing a turnkey system capable of processing 100 ton/day of feed range from $2 to 4 million. Amortization costs for such a plant are approximately $8–16/ton of RDF pyrolyzed (30 years at 12.5%). The annual operating and maintenance costs (excluding costs for gas cleanup, electricity generation and residue disposal) range between $12 and $21 per ton. Revenue for the gas is estimated at $20/ton of RDF if it is converted to electrical power (at $0.06 /kWh) or $30/ton if used to displace natural gas (at $4/GJ). Revenues from the sale of recovered metals, oil and carbon cannot be estimated since specifications have not been developed.

CONCLUSIONS

While the commercial feasibility of pyrolysis has yet to be demonstrated, the concept remains attractive. Installation of a biomass or waste pyrolysis plant represents a means of retrofitting existing oil- or gas-fired boilers at a cost less than that of a new solid-fuel fired boiler. In addition, the production of a medium-Btu gas as opposed to the low-Btu gas of air-blown gasification processes will ensure the maintenance of boiler rating. Because of their relatively low reaction chamber temperature, pyrolyzers also have the advantage of not requiring front-end separation of the metals and glass contained in refuse feedstock. These materials may be separated from the char should this appear profitable.

Of the three process evaluated, the Lantz converter appears to be the closest to commercial demonstration. The most significant aspects of the reactor design are that the feedstock is continually agitated, the retort is maintained at an acceptable temperature and the ram feed mechanism is relatively simple and amenable to improvements which would reduce air intrusion. The Total Energy Systems and Pyro-Sol processes also have potential; however, both require considerably more testing and evaluation. In the case of the Total Energy Systems process, a critical review of reactor design is desirable.

ACKNOWLEDGMENTS

Mr. Ira J. Wright of Brown and Caldwell provided insight and direction during the course of this study, and was largely responsible for the project's inception. Ms. Carol Murray helped arrange reactor testing, procure laboratory services and meticulously analyze test data. We also acknowledge the cooperation provided by the reactor vendors. In particular, we thank Mr. Lyle Atkins of the Wallace-Atkins Oil Corporation and Ms. Mabbie Igleheart, Mr. Edward Bales and Mr. John Trescot of

Total Energy Systems, Inc.; Mr. William Fio Rito and Mr. Dae Lantz III of Pan American Resources Company; and Mr. James Welty and Mr. Harvey Oberg of Pyro-Sol Systems Inc. We hope that the results of this study will provide opportunities for further development of their processes. Finally, we would like to thank the various laboratories, particularly Truesdail Laboratory Inc. in California and E. W. Saybolt Laboratory and Co., Inc. in Texas, for services rendered in what may have been difficult circumstances.

REFERENCESS

1. Brown and Caldwell Consulting Engineers. "Microbial Production of Methane From Refuse," prepared for Southern California Edison Company (1978).
2. Wan, E. I., and M. Cheng. "A Comparison of Thermochemical Gasification Technologies for Biomass," in *Symposium Papers, Energy from Biomass and Wastes,* (Washington, DC: Institute of Gas Technology 1978), pp. 781–814.
3. Rensfelt, E. et al. "Basic Gasification Studies for Development of Biomass Medium-Btu Gasification Processes," in *Symposium Papers, Energy from Biomass and Wastes* (Washington, DC: Institute of Gas Technology, 1978), pp. 465–494.
4. Solar Energy Research Institute. "A Survey of Biomass Gasification, Volume 1 – Synopsis and Executive Summary" (1979).
5. Sanitary Engineering Research Laboratory. "Comprehensive Studies of Solid Wastes Mangement," University of California, Berkeley (1972).
6. Antal, M. J. "Synthesis Gas Production from Organic Wastes by Pyrolysis/Steam Reforming," in *Symposium Papers Energy from Biomass and Wastes,* (Washington, DC: Institute of Gas Technology, 1978), pp. 495–523.
7. Boyd, M., C. Anderson, A. DeVera and M. Hawley. "Pyrolysis and Gasification of Hybrid Poplar SPP," paper presented at the 1979 Annual AIChE Meeting, San Francisco, CA, November 1979.
8. Frendberg, A. M. "Performance Characteristics of Existing Utility Boilers When Fired with Low Btu Gas," paper presented at the Electric Power Research Institute Symposium, April 1974.
9. Sanner, W. S., C. Ortuglio, J. G. Walters and D. E. Wolfson. "Conversion of Municipal and Industrial Refuse into Useful Materials by Pyrolysis," Bureau of Mines R.O.I. 7428 U.S. Dept. of the Interior (1970).

CHAPTER 12

GASIFICATION OF FEEDLOT MANURE IN A FLUIDIZED BED: EFFECTS OF SUPERFICIAL GAS VELOCITY AND FEED SIZE FRACTION

K. Pattabhi Raman, Walter P. Walawender and L. T. Fan
Department of Chemical Engineering
Kansas State University
Manhattan, Kansas

Agricultural wastes, such as feedlot manure, are one class of materials that are being investigated for possible utilization as a supplemental energy resource and/or a future source of chemical feedstocks. Feedlot manure is a low-sulfur material that is renewable and available in significant amounts in certain areas. Manure can be converted into useful products by anaerobic digestion, atmospheric pressure gasification or liquefaction. Of these, atmospheric pressure gasification appears to be the most economically attractive [1], Contacting devices, such as the fixed bed, the moving bed, the entrained bed and the fluidized bed can be used for gasifying manure. Reed provides an excellent review of these options, the fundamental principles and the current state of the art for the developing gasification technologies [2]. From the standpoint of gas production, fluidized beds are highly desirable because of their high heat transfer characteristics and their capabilities for maintaining isothermal conditions.

A survey of the literature on the gasification of manure indicates that the available experimental data are somewhat limited. Most investigators have only examined the influence of temperature. Burton carried out two experimental runs with dried cow manure in a fluidized bed reactor [3]. The reactor used was 0.38 m (15 in.) in diameter and employed an inert matrix of sand as the bed material. Hot fluidizing gas for the reactor was generated by combusting methane or propane. The reactor operating temperatures used for the two runs reported were 1041° and 1022°K. Smith et al. published partial oxidation data obtained in a moving bed reactor [4]. The experiments were conducted in a 0.05-m (2-in.)-diameter reactor and used recycled product gas and air as the gas medium. Data were obtained for a temperature range of 894–950°K. Walawender and Fan [5] presented preliminary pilot-plant data on the fluid bed gasification of manure in a reactor similar to the one used by Burton [3]. They conducted tests over a temperature range of 1000–1100°K under conditions where the superficial velocity and feed rate were varied. One conclusion presented was that the feed rate variations are not important.

Bench-scale operating data were obtained by Halligan et al. [6] in a 0.05-m (2-in.)-diameter reactor which was operated in a partial combustion mode using steam and air. The reactor was externally heated with electrical heaters and the data were obtained between 977 and 1069°K. Mikesell et al. [7] reported limited data on the flash pyrolysis of steer manure in a multiple hearth reactor. The operating temperatures for these experiments were between 873 and 1023°K. Beck et al. [8] presented partial oxidation data on manure obtained in a pilot plant reactor. Steam and air were used as the fluidizing medium in the 450-kg/day pilot plant. The reactor used was 0.15 m (6 in.) in diameter and had an axial temperature variation of about 500°K. Data on the reactor offgas were presented for an average temperature of about 870°K in the reactor. Howard et al. conducted a comparative study on the gasification of a variety of biomass materials (including manure) in a 0.5-m-i.d. fluid-bed pilot plant [9]. They examined the influence of fluidization velocity and reactor loading on the gasifier performance. They also developed a preliminary model to describe the liquid yield as a function of temperature. Raman et al. examined the influence of temperature on the gas composition, yield and heating value in a pilot-scale reactor fluidized with a mixture of flue gas and steam [10]. Results were presented for conditions of fixed feed size, feed rate and superficial velocity.

Raman [11] also developed a model to describe fluid-bed gasification and applied it to manure to estimate the yield of gas, liquid and solid as well as the gas composition. Although the model is a preliminary one, it does a reasonable job of predicting the experimental results.

To properly design a system for the gasification of manure or other biomass, it is necessary to develop a systematic database which includes the effects of operating temperature, feed size, superficial gas velocity and perhaps other variables on the gasification characteristics. These would be most useful if obtained on a pilot-plant scale. The objectives of the present work were to conduct gasification experiments with manure in a fluidized bed reactor and to assess the influence of the feed size fraction and superficial gas velocity on: (1) product gas composition; (2) higher heating value of the product gas; and (3) product gas volumetric yield. The operating temperature was also varied in the experiments to examine these influences as a function of temperature.

METHODS

Facilities

The pilot plant facility used for the gasification of manure is shown schematically in Figure 1. The pilot plant consisted of the following seven components: (1) reactor, (2) screwfeeder, (3) cyclone separator, (4) venturi scrubber, (5) afterburner, (6) control and instrumentation panel, and (7) gas sampling train.

The reactor was constructed from heat resistant stainless steel 310 alloy. The reactor proper had an i.d. of 0.23 m (9 in.) with an expanded freeboard of 0.41 m (16 in.) i.d. A burner with a duty of 47.5 MJ/hr (45,000 Btu/hr) located at the bottom of the reactor (plenum) generated the

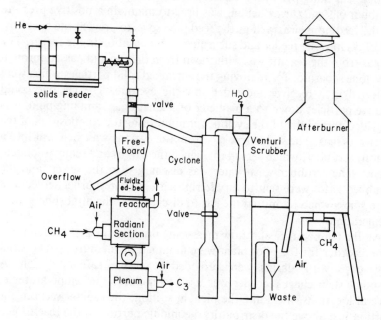

Figure 1. Flow scheme of pilot plant.

gas for fluidization by the combustion of propane under starving air conditions. Water was injected into the plenum section as necessary to maintain the temperature below 1250°K and to supply additional gas for fluidization. A sampling port was provided at the plenum section to permit monitoring of the composition of the fluidizing gas. Supplemental heat (as needed) for operation was transferred across the walls of a radiant jacket surrounding the reactor. A burner with a duty of 105.5 MJ/hr (100,000 Btu/hr) supplied heat to the jacket using natural gas as the fuel. The distributor plate for the reactor was made from a 3-mm thick 316 stainless steel plate and had 844 holes of 15 mm diameter. The reactor was well insulated with a minimum of 0.1 m of Kao Wool and had adequate temperature and pressure measuring elements located at various strategic points. An inert matrix composed primarily of silica sand was used as the bed material. Approximately 45 kg of sand with a mean particle size of 0.55 mm was used to give a static bed height of 0.6 m (24 in.). An overflow pipe for withdrawing solid samples from the bed was provided on the reactor as shown in Figure 1.

The solids to be gasified were fed into the bed through a feed pipe of 0.075 m (3 in.) diameter, which discharged the feed just above the expanded bed surface. The feed material was delivered to the feed pipe from a sealed hopper with a variable speed screw feeder. A purge stream of about 0.36 m³/hr of helium was used to maintain a positive pressure on the feed hopper as well as the feed pipe so as to prevent the backflow of offgas into the feeder and subsequent condensation in the feeder. The offgas from the reactor was withdrawn from the top and passed through a cyclone separator for removing the entrained solid particles which were collected in a receiver located below the cyclone. The cyclone could remove particles down to a diameter of 5 μm. A gas sampling point was provided at the inlet of the cyclone for monitoring the composition of the reactor offgas. The solids-free gas from the cyclone was then sent into a venturi scrubber, which served to quench the offgas and remove condensables. The scrubber wastewater was discharged to the sewer, and the scrubbed gases were sent to an afterburner. The afterburner served as a flare stack which permitted the gas to discharge to the atmosphere after incineration.

All the temperature and flow measuring instruments and the temperature recorder for the pilot plant were mounted on a control panel. Control loops with alarms were provided to ensure safe operation. A 12-point strip chart recorder was used to monitor the temperatures at several locations, including the plenum section, the radiant section, the portion just above the distributor, the middle portion of the reactor and the freeboard section.

A sampling train was constructed to collect samples of the plenum gas as well as the reactor offgas. The sample stream was passed through a series of glass condensers and condensate receivers permitting the separation of condensables from the stream. The cooled sample gas was passed through a wet test-meter, and then through a sample bottle, and subsequently was incinerated.

Feed Material Preparation

The manure used was collected from paved feedlots at Kansas State University's Beef Research Center. The manure had a moisture content of about 80% and was subsequently flash-dried to reduce moisture to about 8%. The dry manure was sieved to obtain three size fractions, namely; − 2 + 8 mesh (0.45 cm), − 8 + 14 mesh (0.19 cm) and − 14 + 40 mesh (0.09 cm). The ultimate analyses of the three sizes of manure are presented in Table I.

Procedure

The reactor was initially heated to the desired operating temperatures using both the plenum and radiant burners. The temperatures in various parts of the reactor were monitored to establish a stable starting condition. The propane used in the plenum burner was burned under starving air conditions to ensure an oxygen-deficient atmosphere in the reactor. Two or three gas samples were taken from the plenum section for analysis before a run was initiated. Over the course of the sampling period, condensate was collected for a measured volume of the burner

Table I.   Elemental Analyses of the Feed (Dry Basis)

	− 2 + 8 mesh (wt %)	− 8 + 14 mesh (wt %)	− 14 + 40 mesh (wt %)
C	45.3	42.9	44.0
H	6.1	6.0	5.6
N	3.2	3.4	3.5
O (by difference)	27.8	33.6	32.1
Ash	17.6	14.1	14.8
Total	100.0	100.0	100.0
Moisture	4.8	9.4	6.6

gas (saturated at the metering conditions) to determine the water content of the fluidizing gas.

Manure was introduced into the reactor at a continuous, prespecified rate, and the temperature profile of the reactor was closely monitored. Samples of the reactor offgas were taken with the simultaneous collection of condensate. Run durations were 30 min to 1 hr. Feeding was then terminated and the char collected in the cyclone was weighed. Samples of the cyclone char were reserved for analysis. After the completion of each run, the char retained in the reactor was burned with excess air, and the ash produced was elutriated from the bed and collected in the cyclone. A sample of the ash generated was also reserved for analysis.

The flowrates of the propane, air and injection water were noted during each run. The solid feed rate was determined by the difference in weights of solids in the hopper before and after the experiment. For each of the runs, the gas samples were drawn after flushing the sample bottles for about 5 min. The volumetric flow of gas through the wet test-meter and the pressure and the temperature of the wet test-meter were noted. The condensates were measured volumetrically.

Chemical Analysis

Gas analysis was accomplished using a Packard Model 417 Becker Gas Chromatograph equipped with thermal conductivity detectors. The gas components of interest included H_2, CO, CO_2, CH_4, C_2H_4, C_2H_6, C_3H_6, N_2 and O_2. Column packings used were a 5Å molecular sieve for the separation of H_2, O_2, N_2, CO and CH_4, while the remaining components were separated using a column of Porapak Q with a short lead section of Porapak R to shift the retention of water. The chromatograph was operated isothermally at $80°C$ with helium as a carrier gas. The instrument was calibrated with puchased calibration mixtures. Solid materials were analyzed with respect to their elemental composition (C, H, N, O) using a Perkin-Elmer Model 240 Elemental Analyzer. The ash analysis was performed according to the standard ASTM procedure in a muffle furnace and the moisture content was determined by drying the samples in an oven for 3 hr at $373°K$. The composition of the solids and gases given in this study represent the average of at least two determinations.

Operating Conditions

Gasification experiments were conducted by varying the operating temperature, feed size fraction and gas superficial velocity. The arithmetic average between the bed temperature and freeboard temperature was taken as the operating temperature of the reactor. In all the experiments, the freeboard temperature was less than the bed temperature. The maximum temperature difference observed in this study was 120°K and the average temperature difference was 80°K. A summary of the operating conditions is presented in Table II.

Data Analysis

Mass balance calculations were first performed on the plenum burner using the analysis of the dry plenum gas, the condensate collected and the flowrates of propane, air and injection water. The flowrate of the dry burner gas entering the reactor was computed by performing a nitrogen balance around the burner. An overall nitrogen balance on the reactor was then used to evaluate the dry offgas flowrate with the aid of the offgas analysis. For computing the amount of gas produced from the manure, it was assumed that the burner gas did not significantly take part in the reactions. The yield and the composition of the dry product gas were computed from the difference between the dry offgas and the dry burner gas. Product gas represents the dry gas that results from the gasification of the feed.

From the condensate collected for a unit volume of the burner gas, the total water content of the burner gas was computed using the volumetric flowrate of the dry burner gas. Similarly, the condensate associated with the dry offgas was computed from the volumetric flowrate of the dry offgas and the condensate data obtained for a unit volume of the dry offgas.

To complete the overall material balance around the reactor, it was necessary to know the total amount of char produced. Since a portion of the char was retained in the bed, it was necessary to establish a procedure for evaluation of the total char generated. Attempts were made to estimate the char in the bed by performing an inert balance on the ash pro-

Table II. Summary of Operating Conditions

Feed Size Fraction	−2 + 8 Mesh	−8 + 14 Mesh	−14 + 40 Mesh
Feed Rate (kg/hr)	11.0–17.2	11.9–30.2	5.1–31.8
Reactor Temperature (°K)	900–980	800–1040	800–1040
Superficial Gas Velocity (m/sec)	0.31–0.37	0.33–0.45	0.33–0.45
Injection Water Rate (kg/hr)	2.0–2.5	2.0–3.5	2.0–3.5

duced during combustion and the char. This method was not very satisfactory, since substantial amounts of the ash were carried past the cyclone to the scrubber and drain. From experimental observations, it was found that for any run, the elemental analysis of the cyclone char and the char retained in the reactor agreed with each other closely. Hence, the char retained in the bed was assumed to have the same composition as that of the cyclone char. The total char produced was estimated using the ultimate analysis of char and feed coupled with an ash balance on the reactor. This procedure was subsequently checked with a pelleted feed material, whose char could be separated from the inert solids in the bed. The check indicated that the inert balance was a satisfactory approach. A detailed example of the material balance procedure has been presented elsewhere [10].

RESULTS AND DISCUSSION

Approximately 100 experiments were conducted, 45 each for the -8 $+ 14$ mesh and $-14 + 40$ mesh fractions and the remainder for the -2 $+ 8$ mesh fraction. For each run, material balance calculations were performed to evaluate the quantity and the composition of the product gas. Material balance closure ranged from 80 to 115%, with most runs closing to better than 90%. The higher heating value of the product gas was calculated from its composition and the heating values of the individual components. The effects of superficial gas velocity and the feed size fraction were assessed from the results obtained.

Product Gas Composition

To examine the influence of gas superficial velocity on the concentrations of the individual components of the product gas, data obtained at different superficial velocities were plotted against the operating temperature for a given feed size fraction. These plots, presented in Figures 2 and 3 (for the $-8 + 14$ mesh and the $-14 + 40$ mesh sizes) showed a minimal scatter ($\pm 1\%$) indicating that for the range investigated, the gas superficial velocity did not have a discernible influence on the product gas composition. In these two plots as well as in the subsequent ones, the actual data points are not shown for the sake of simplicity.

The effect of the size fraction used on the composition of the product gas can be assessed by comparing Figures 2 and 3. For a given operating

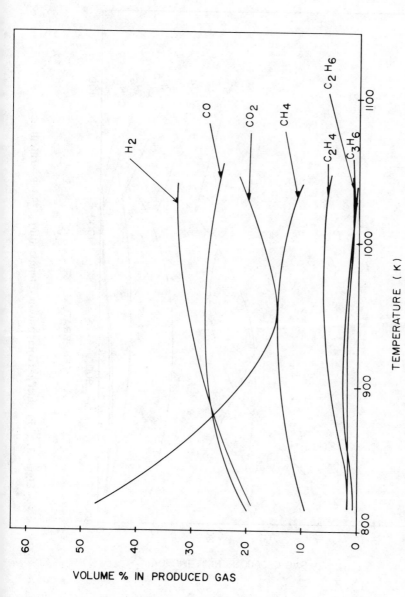

Figure 2. Product gas composition *vs* temperature, − 8 + 14 mesh size.

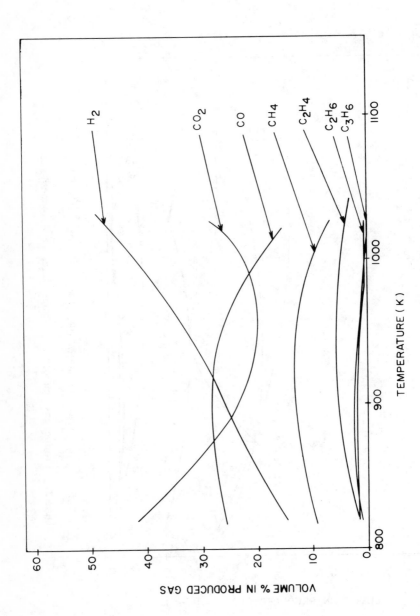

Figure 3. Product gas composition vs temperature, − 14 + 40 mesh size.

temperature, comparison shows that the concentrations of C_2H_6 and C_3H_6 are very close to each other for the two feed size fractions. The concentrations of CH_4, C_2H_4 and CO_2 show similar trends in both cases. Their numerical values are in good agreement with each other up to an operating temperature of about 950°K. Beyond this temperature, the differences are more pronounced. The concentrations of H_2 and CO complement each other in the two plots. A higher value of H_2 concentration is offset by a lower value of CO concentration and vice versa. It can also be seen that the concentration of CO_2 goes through a minimum in the two figures with the numerical values for the two size fractions being distinctly different.

Figure 2 shows that for the $-8 + 14$ mesh size fraction, the concentration of H_2 in the product gas varied between 19 and 33% and that of CO varied between 20 and 25%, reaching a maximum of 27.5% at 950°K. In Figure 3, it can be seen that for the $-14 + 40$ mesh size fraction, the concentration of H_2 varied between 15 and 50% and that of CO between 25 and 15%, reaching a maximum of 28.3% at 900°K. These two figures suggest that there is a distinct difference in the concentrations of CO, H_2 and CO_2 from the two size fractions. Limited data for the $-2 + 8$ mesh size fraction did not show an appreciable difference from the results for the $-8 + 14$ mesh fraction.

Heating Value

In Figure 4, the higher heating value of the gas produced at different gas superficial velocities is plotted against the operating temperature. It can be seen that the heating values go through definite maxima and then diminish. As in the case of the product gas composition, for a given feed size fraction, the gas superficial velocity did not have a significant influence on the heating values. The deviation observed was ±0.8 MJ/std m³ (±21.5 Btu/scf).

By comparing the two curves in Figure 4, the effect of the size fraction on the heating value of the product gas can be assessed. For the size fraction of $-8 + 14$ mesh, the heating value increases from 8.30 to 19.75 MJ/std m³ (223 to 530 Btu/scf) and then diminishes to 14.30 MJ/std m³ (384 Btu/scf) over the temperature range studied. In the case of the -14 to 40 mesh fraction, the heating value increases from 10.43 to 18.20 MJ/std m³ (280 to 489 Btu/scf), and then decreases to 12.3 MJ/std m³ (330 Btu/scf). These data indicate that the feed size fraction may have a marginal influence on the heating value of the product gas. The data were also compared with a limited number of data obtained for a $-2 +$

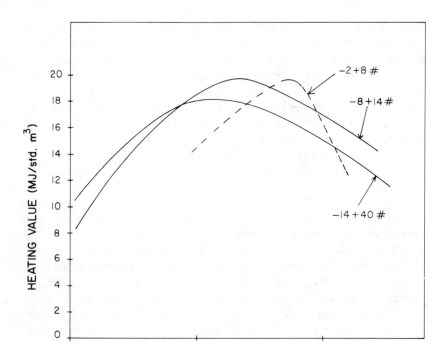

Figure 4.   Gas higher heating value vs temperature.

8 mesh fraction of manure as shown in Figure 4. A comparison of the heating value curves for the three size fractions indicates that the peak of the heating value curve shows a shift to the right as the size fraction becomes larger. Also, as size increases, the peaks become narrower.

Product Gas Yield

Figure 5 presents plots of the volumetric yield of dry product gas (on a dry ash-free basis) vs temperature for the different size fractions. The volume is expressed as standard cubic meters which pertains to dry gas at 288°K and 101.3 kPa. The data points for a given size fraction showed a fair amount of scatter. The scatter was such that bands of ±0.25 std m³/kg about the lines shown in Figure 5 were needed to contain the data for a given size fraction. There were no discernible trends in the data to suggest that superficial velocity variations were responsible for the

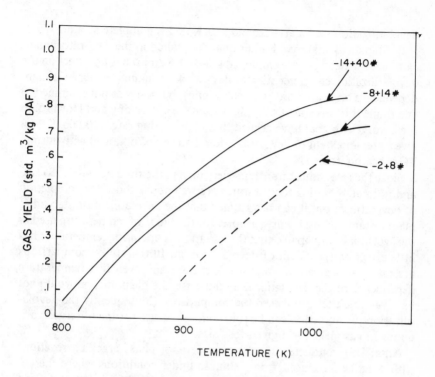

Figure 5. Gas yield vs temperature.

scatter observed. The average yield of dry product gas ranged from 0.13 std m³/kg (2.1 scf/1b) at 820°K to 0.86 std m³/kg (13.8 scf/1b) at 1020°K for the −14 + 40 mesh fraction. For the −8 + 14 fraction the average yield ranges from 0.04 std m³/kg (0.6 scf/1b) at 820°K to 0.72 std m³/kg (11.5 scf/1b) at 1020°K. A limited amount of data for the −2 + 8 size fraction are also presented in Figure 5. The comparison shows a definite tendency for higher gas yields with smaller feed size fraction.

A simple conceptual model for the gasification of manure can be envisioned to consist of the following steps: (1) devolatilization of the solid to form char and volatile matter; (2) thermal cracking of volatiles to produce light components and char (carbon deposition) and gas-phase water-gas and steam-hydrocarbon reactions. The yield of total volatiles in the first step will dictate the level of gas yield that can be obtained from the solid feed. The extent of thermal cracking and other gas phase reactions of the volatiles is determined by their time-temperature history. These reactions will establish the final ratio of gas to liquid and the gas composition.

Thermogravimetric studies on manure have indicated that the

devolatilization step starts around 420°K and is complete around 770°K [12]. Statistical analysis of additional data taken in this laboratory indicate that the heating rate employed (40–160°K/min) has no effect on the devolatilization characteristics. Antal's work on manure indicates a slight dependence on heating rate (5–140°K/min), but this was not examined to determine if it was statistically significant [13]. Anthony and Howard, in their work with coal have argued that high heating rates (10,000°K/sec) give a greater extent of devolatilization than can be obtained with normal TGA heating rates [14].

Since the rate of heat transfer is very high in the fluid bed (1000°K/sec) and normal operating temperatures are well above those for completion of devolatilization, it can be assumed that the devolatilization step takes place instantaneously. Furthermore, for the range of temperature used in this work, it is anticipated that the variation in operating temperature has little effect on the extent of devolatilization. In this study, comparisons are made at a given operating temperature and even if a temperature dependence of the devolatilization did exist, the phenomenon would not be a variable that influenced the comparison. Consequently the devolatilization phenomenon can be ruled out as a cause for the observed variations in gas yield for a given feed size.

Antal [13] conducted studies on the vapor-phase cracking reactions with volatiles produced from cellulose under conditions where the devolatilization phenomenon was held constant. He found that at a given temperature, the amount of each component in the product gas was affected by the residence time. His results indicated that for residence times up to about 5 sec, the yield of the components such as CH_4, C_2H_6, C_2H_4 and C_3H_6 increased dramatically, and beyond 5 sec, the effect was much less pronounced. He also found that the amount of each component increased with temperature and, in the case of both C_3H_6 and C_2H_6, the amounts of each passed through maxima as temperature increased.

In the present work, the residence time of the gas in the reactor was calculated to be approximately 6 sec. The variation in this value over the experimental range was about ±1 sec. Since this value is more than 5 sec and the variation observed is not extensive, it can be expected that the gas residence time does not have a significant effect on the data. This is corroborated by the experimental observation that the superficial velocity of the gas, which is related to the residence time, did not have a significant influence on the gas composition and heating value for a given feed size fraction and a given gasification temperature. However, the yields obtained with a given feed size fraction and temperature showed a significant scatter, which is far beyond the bounds of variations that can be expected on the basis of the variations in the time-temperature history that the product gases experienced.

One plausible explanation for the observed behavior might lie in possible variations in the feed makeup as a consequence of segregation effects between batches. This is supported by the observation made during the test program that the gas yield data for a given batch of manure were consistent but varied from batch to batch for a given feed size and operating temperature. An examination of the elemental analyses from the different feed batches did not indicate significant variations in elemental composition. Since manure consisted of a mixture of stalks, hulls and other plant materials, it was next decided to examine possible variations in the cellulose content of these components. Whistler and Smart indicate that a considerable variation in cellulose content exists for different parts of a plant as well as between different types of vegetation [15]. For example, leaves contain 10–20% cellulose, stalks 40–50%, hulls 35% and cobs 40%. Consequently segregation phenomena between batches could give rise to feeds with different cellulose content.

The influence of cellulose content on the devolatilization characteristics of biomass materials was then examined. Howard et al. reported on the maximum oil yield obtainable from different biomass materials [9]. The maximum oil yield can be related to the extent of devolatilization that will take place for a given material. In their work with paper, sawdust and mixtures of the two, it was observed that the maximum oil yield increased in the order sawdust, mixture, paper. The cellulose content increases in the same order. Their study does not relate this observation to the gas yield, unfortunately.

To examine this dependence further, limited data on the cellulose content, TGA analysis and gas yields for cellulose, paper hardwood, softwood, manure and coal were compared. Table III presents the summary of the TGA results obtained by Antal [13] for cellulose, paper and wood, and by Howell [12] for manure and coal. The cellulose content and rela-

Table III.   **Devolatilization Characteristics of Different Materials**

Materials	Total Devolatilization (wt %)	Relative Gas Yield	Cellulose Content (wt %)
Cellulose	90	11	100
Paper	85		
Cherry (hardwood)	80		
Pine (softwood)	70		58
Cane (sorghum)		7	35–50
Manure	55–60	5	
Coal	30	3	0

tive gas yield for some of these materials are also presented for comparison. The relative gas yield is for 970°K with the result for cellulose from Antal [13] and the remaining values from this laboratory. As these limited data indicate, it appears that the TGA results, gas yields and cellulose content show the same trends implying that increasing cellulose content may correlate with increasing devolatilization and subsequent gas yield. Walawender et al. presented a comparative study on the gasification of various carbonaceous materials over a range of temperatures [16]. The results clearly show that the gas yield increases as the cellulose content of the material increases. Beck et al. report similar results for their partial oxidation studies on wood and manure [17]. They also attribute the increase in gas yield to an increase in the cellulose content of the feed material. This apparent correlation needs to be explored further and quantified if possible.

It is quite possible in this work that, between batches, the cellulose content of the manure feed could have been different due to segregation. This difference could very well be responsible for the scatter observed in the product gas yield for a given feed size fraction. In view of this, caution should be exercised in interpreting the influence of particle size on the gasification characteristics of biomass. For the different feed size fractions, variations in the material makeup were evident. The $-2 + 8$ mesh size fraction consisted of hulls and undigested grain. The $-8 + 14$ mesh size fraction was spherical in shape and had a small amount of undigested grain, whereas the size fraction of $-14 + 40$ mesh was comprised primarily of fine strands.

The difference in heating value and yield observed for different size fractions could well be due to variations in material makeup alone but it cannot be ruled out that particle size may also have some influence. Maa and Bailie [18] in their study on cellulosic materials theorized that for particle sizes less that 0.2 cm in diameter, pyrolysis is reaction controlled and the particle size has no influence. In the present study, the size fraction $-14 + 40$ mesh (0.09 cm) falls below this value, the size fraction $-8 + 14$ mesh (0.19 cm) is marginally below, while the size fraction $-2 + 8$ mesh (0.45 cm) is above the 0.2-cm size stipulated by Maa and Bailie. Consequently for this study, size effects should not be important for the smallest size fraction, but may be intruding for the other two sizes, especially the largest size.

CONCLUSIONS

Gasification studies were conducted with different size fractions of manure in a fluid-bed reactor. The effects of gas superficial velocity and feed size fraction on the gasification were studied at different operating temperatures. Superficial gas velocity did not appear to have a significant influence on the composition and heating value of the product gas for the range of variation. The feed size fraction did have an influence on the composition, heating value and yield of the product gas. The observations indicate that the yield increases and the maximum in the heating value curve shifts to the left as the size fraction becomes smaller. In the conduct of the experiments, considerable scatter was observed in the gas yield obtained with different batches of feed for a given operating condition. A possible explanation for this behavior is offered which suggests that segregation phenomena between batches of feed and subsequent variations in the cellulose content of the batch may be primary factors influencing the observed scatter. The apparent correlation between the cellulose content and the gas yield needs further examination.

ACKNOWLEDGMENT

This is Contribution No. 80-380-A, Department of Chemical Engineering, Kansas Agricultural Experiment Station, Kansas State University, Manhattan, Kansas.

REFERENCES

1. Engler, C. R., W. P. Walawender and L. T. Fan. "Synthesis Gas from Feedlot Manure," *Environ. Sci. Technol.* 9:1152 (1975).
2. Reed, T. B., Ed. *A Survey of Biomass Gasification, Vols. 1–3*, SERI-TR-33-239 (Golden, CO: Solar Energy Research Institute, 1979).
3. Burton, R. S. "Fluid Bed Gasification of Solid Waste Materials," MS Thesis, University of West Virginia, Morgantown, WV (1972).
4. Smith, G. L., C. J. Albus and H. W. Parker, "Products and Operating Characteristics of the TTU Report," paper presented at the 76th National American Institute of Chemical Engineers Meeting, Tulsa, OK, March 1974.
5. Walawender, W. P., and L. T. Fan. "Gasification of Dried Feedlot Manure in a Fluidized Bed — Preliminary Pilot Plant Tests," paper presented at the 84th National American Institute of Chemical Engineers Meeting, Atlanta, GA, February 1978.
6. Halligan, J. E., K. L. Herzog and H. W. Parker. "Synthesis Gas from Bovine Wastes," *Ind. Eng. Chem. Proc. Descrip. Devel.* 14:64 (1975).
7. Mikesell, R. D., D. E. Garrett and D. C. Hoang. "A Thermal Process for Energy Recovery from Agricultural Residues," paper presented at the ACS Symposium on Advanced Thermal Processes for Conversion of Solid Wastes and Residues, Anaheim, CA, March 1978.

8. Beck, S. R., W. J. Huffmann, B. L. Landeene and J. E. Halligan. "Pilot Plant Results of Partial Oxidation of Cattle Feedlot Manure," *Ind. Eng. Chem. Proc. Descrip. Devel.* 18:328 (1979).
9. Howard, J. B., R. H. Stephens, H. Kosstrin and S. M. Ahmed. "Pilot Scale Conversion of Mixed Wastes to Fuel, Vol. 1," Project Report to U.S. EPA (1979).
10. Raman, K. P., W. P. Walawender and L. T. Fan. "Gasification of Feedlot Manure in a Fluidized Bed Reactor. The Effect of Temperature," *Ind. Eng. Chem. Proc. Descrip. Devel.* 19:623 (1980).
11. Raman, K. P. "Fluidized Bed Gasification of Carbonacious Solids," PhD Dissertation, Kansas State University, Manhattan, KS (1980).
12. Howell, J. A. "Kansas Coal Gasification," MS Thesis, Kansas State University, Manhattan, KS (1979).
13. Antal, M. J., H. L. Friedman and F. E. Rogers. "Kinetic Rates of Cellulose Pyrolysis in Nitrogen and Steam," paper presented at the Fall Meeting, Eastern Section of the Combustion Institute, Hartford, CT, November 1977.
14. Anthony, D. B., and J. B. Howard. "Coal Devolatilization and Hydrogasification," *Am. Inst. Chem. Eng. J.* 22:625 (1976).
15. Whistler, R. L., and C. L. Smart. *Polysaccharide Chemistry* (New York: Academic Press, Inc., 1963).
16. Walawender, W. P., K. P. Raman and L. T. Fan. "Gasification of Carbonaceous Materials in a Fluidized Bed Reactor," in *Proceedings, Bio-Energy '80 World Congress and Exposition* (Washington, DC: The Bio-Energy Council, 1980), p. 575.
17. Beck, S. R., and M. J. Wang. "Wood Gasification in a Fluidized Bed," *Ind. Eng. Chem. Proc. Descrip. Devel.* 19:312 (1980).
18. Maa, P. S., and R. C. Bailie. "Influence of Particle Sizes and Environmental Conditions on High Temperature Pyroylsis of Cellulosic Materials—I (Theoretical)," *Comb. Sci. Techno.* 6:1-13 (1973).

CHAPTER 13

USE OF CORN COBS FOR
SEED DRYING THROUGH GASIFICATION

Stanley L. Bozdech

DEKALB AgResearch, Inc.
DeKalb, Illinois

Coupled with discoveries of the earliest uses of corn in the western hemisphere are indications that very early, the cob was in general use as fuel for cooking and heating. Later, the early immigrant farmers of the United States, especially those in the plains where wood was scarce, found that cobs could cook their meals and heat their homes. The petroleum age came; harvest systems changed; the cob was left to rot in the field.

Today the seed corn industry in its normal operations generates enough dry cobs as a by-product to dry the seed. After many failures with direct cob combustion, the industry has turned to gasification, an old technology, which can efficiently convert cobs to clean, controllable heat for drying seed and grain.

The hybrid seed corn industry is unique. Its objectives, including the seed conditioning procedures, are in no way similar to those in the grain handling industry. Nowhere is this more apparent than in harvesting and drying operations. Seed corn is harvested early and with high moisture content. It is also dried on the cob in large batch bins. Temperature of the drying air must be maintained below 42°C to preserve the vitality of the germ.

As early as 1972, DEKALB recognized cobs as a natural fuel for its seed drying operations. It takes 3.76 GJ to dry a ton of seed, but the cobs from that ton contain about 4.89 GJ. Cobs are good fuel, and have been conditioned with the seed. The market value of the cobs at that time at the plant site was about $10/ton. This has since gone down to $5/ton. Natural gas costs are 10 times that amount; propane 20 times. The cost differential could provide the capital for a cob conversion system, although dryers are used only five weeks per year. Each dryer consumes 200,000 m³ of natural gas during that time. Readily available, cheap fuel is definitely an advantage. However, the short season and low annual fuel consumption places constraints on capital costs. Marginal fuel supply dictated an equipment development plan which would make use of cob combustion gases as the drying medium. Without heat exchangers, which are wasteful, the drying stream had to be relatively free of particulate matter and toxic gases. The batch drying system itself is a good filter for particulates.

In 1973, because of the high price of fuel oil in Europe, DEKALB's French associate, RAGT, invested in an incineration system for cobs and piped the combustion gases to the dryers. Today it is still used, but with tubular heat exchangers which must be cleaned weekly. Its high particulate output, without the heat exchanger, is shown in Table I, along with the particulate outputs of other units. DEKALB felt that since the French had done so well with everything but particulate, surely some technique could readily solve that problem. Two direct combustion units were installed in seed plants in 1976: one a sophisticated incinerator with afterburners, which performed poorly; the other a torroidal unit for burning fine material in suspension, which, in addition to high grinding costs, discharged particulate as fast as the other units. Additionally, both units slagged badly. Both were complex, expensive and difficult to operate. In addition to the financial incentive, fuel interruption, and therefore drying interruption, would expose a seed crop to frost and wipe it out completely.

Table I.   Particulate Emissions of Cob Drying Systems

Unit	Particulates (kg/metric ton fuel)
French Incinerator	10.05
Suspension Burner	2.43
Hearth Incinerator	1.26
Gasifier	0.11

DEKALB GASIFICATION SYSTEM

The elemental and physical similarity between wood and cobs became more apparent as our work progressed (Table II). Through these studies and direct contact with European groups working on wood gasification, a system was specifically designed to dry grain with cobs (Figure 1). It meets the criteria we had established at the outset. Although it requires more management than natural gas or oil-fired equipment, the technology is attractive enough to be widely acceptable. It is characterized by:

- continuous operation
- automated output control
- five-to-one turn-down ratio
- simple and direct operating techniques
- safe operation
- clean, efficient heat output

It also meets our economic requirements:

- low capital costs
- reasonable operating costs
- serviceability
- complete heat release
- complete utilization of the heat
- no fuel preparation costs
- favorable fuel prices

DEKALB's gasifier system is an atmospheric, up-draft, negative-pressured system, powered by a single fan which discharges regulated, heated

Table II.   Typical Analysis of Cobs and Oak Wood

	Cobs (wt %)	Oak (wt %)
Carbon	44.96	50.49
Hydrogen	6.10	6.59
Nitrogen	2.42	
Oxygen	44.77	42.77
Chlorine	0.29	
Ash	1.46	0.15
Moisture	0.55	
Higher Heating Value (MJ/kg)	16.78	20.49

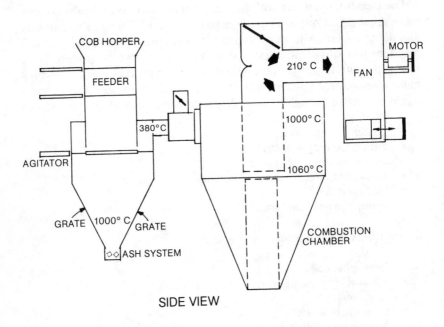

SIDE VIEW

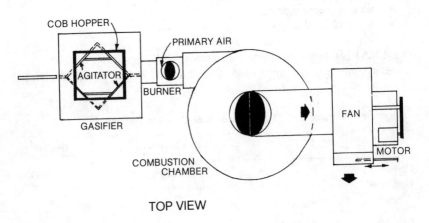

TOP VIEW

Figure 1. Cob-fueled grain drying system.

gases to a seed dryer. Starting at the right (Figure 1), a valve at the discharge of the fan modulates the complete system starting with the production of the gas in the gasifier. Dry cobs as they come from the sheller, whole or in pieces, are fed on demand into the system through alternating slide gates, which keep the system sealed. At the reaction zone, with the reduction in cob size, channeling has a tendency to occur. However, the agitator keeps the bed uniformly packed. The grates are perforated stainless steel, and in operation, are generally protected by a layer of ash. At 1000°C, there is usually some indication of slagging, which appears as soft clumps of ash and carbon. While slagging is not completely eliminated, the horizontal, rotary powered, ash-removal system breaks up and removes most slagged materials.

Producer gas leaves the system at 380°C, carrying tars that often cause serious problems in updraft units. In some earlier tests, tars going through the burner as droplets were not totally consumed and left a sticky coating on fans and drying equipment. Also, on extremely cold days, primary air at the burner has a tendency to condense the tars in the burner head. This can be overcome by feedback from the combustion chamber to heat the ambient air. The primary air adjustment is manual. Once set, it responds to the pressure variations in the system and maintains a constant air-to-fuel ratio. The design of the combustion chamber offers the gas dwell-time in the high-temperature environment to promote complete combustion. In addition, the combustion enclosure provides the first chance to blend ambient air with combustion gases. Adjustment to 210°C is accomplished by a modulating valve controlled by a sensor at the fan inlet. This fan discharges to a mixing chamber at the intake of the fan that supplies drying air at temperatures less than 42°C to the corn bins. A thermostat controls the damper in the discharge of the fan to provide a constant, correct amount of heat.

After 1000 hours of running time on a 1600-MJ/hr pilot unit, and many hours on a full-size 6300-MJ/hr unit, sufficient information has been generated to support the cob gasification concept and continuation of the program.

The particulate emissions are now controllable without a heat exchanger. The cobs produced from the dried seed satisfy fuel needs for drying the seed. Gas quality is good and it burns well. Analyses of gas samples taken just ahead of the burner from two typical runs show some variations, but heat content stays within the range of 4.79–5.21 MJ/m³ (Table III). This provides satisfactory ignition and combustion.

Although the system satisfies the definition of attractive technology in many respects, it is not a push-button system. One person can operate two units if they are close together. The fuel is bulky and cannot be piped

Table III. Gas Composition

	Sample 1 (mol %)	Sample 2 (mol %)
Nitrogen	56+	54+
Oxygen	1.61	0.66
Carbon Dioxide	8.00	7.40
Hydrogen	7.50	8.00
Carbon Monoxide	24.72	27.10
Methane	1.99	2.80
Cob Moisture (wt %)	11	11
HHV (MJ/standard m^3)	4.79	5.21

from a tank; special handling equipment is necessary. Also, the system does not utilize all the thermal potential of the cob; sensible heat from the process, which in some systems can escape to the atmosphere, is captured by placing the unit in the inlet air stream to the drying fan. Mechanically, the gasifier is simple. Most of the annual maintenance problems will probably come from deterioration of refractory during the 47 weeks of downtime.

The fuel as it comes from the seed dryer requires no further drying nor does it need classification by size or grinding. Cob prices are expected to stay low since demand for cob products has been severely depressed by substitutes with lower collection and processing costs.

ECONOMICS

Table IV shows comparative costs of a retrofit gasifier system and other DEKALB systems that are already installed. Cob handling equipment is included in capital cost.

The additional cost of using fossil fuels is high and will probably increase at a fast pace. On new installations where the capital cost for propane storage and firing equipment, or for natural gas firing equipment, must be considered, the figures are much more favorable for cob gasification.

Gasifiers for cobs can protect the seed industry from sudden interruptions of drying fuel supplies. These interruptions, if they occur at certain times, could wipe out a complete year's seed production. DEKALB intends to start equipping its seed corn plants with cob gasifiers. The system developed to date is not perfect, but in actual use, improvements can be made. The gasifier system is particularly adapted to retrofitting on grain dryers. If cobs are an available fuel, the gasifier concept can also be used to provide fuel gas for internal combustion engines.

Table IV. Comparative Seed Drying Costs, 1980 Dollars

	Natural Gas	Propane	Cob Gasification
Capital Cost			141,000
Depreciation Charge			11,280
Operating Days	34	35	35
Operating Labor	960	960	3,200
Operation Electrical	116	116	1,920
Cobs ($0.32/1,000 MJ)			2,500
Propane ($5.17/1,000 MJ)		38,150	
Methane ($3.32/1,000 MJ)	24,500		
Maintenance	400	400	1,100
Tons Dried	2,000	2,000	2,000
$/Ton	12.98	19.80	10.00

SECTION 3

HYDROLYSIS
AND EXTRACTION

CHAPTER 14

REVIEW OF RECENT RESEARCH ON THE DEVELOPMENT OF A CONTINUOUS REACTOR FOR THE ACID HYDROLYSIS OF CELLULOSE

Jean-Paul Franzidis and Andrew Porteous

Open University
Milton Keynes, Buckinghamshire
United Kingdom

The reaction to be considered is the hydrolysis of the cellulose in materials such as newspaper, refuse or agricultural residues to produce glucose, which may then be fermented to alcohol (ethanol). Glucose is the only product of cellulose hydrolysis, but the reaction takes place very slowly by the action of water alone, so acid catalysts and high temperatures are used to accelerate the process.

Unfortunately, these conditions also favor the rapid decomposition of the glucose formed into various other (less desirable) products. The basic problem in developing an effective and economical cellulose hydrolysis process is, and has always been, to overcome the difficulty of hydrolyzing the cellulose while minimizing the decomposition of the glucose product.

This review, after briefly touching on the reasons for the difficulty encountered in hydrolyzing cellulosic materials and the advantages to be gained from cellulose hydrolysis, summarizes the history of commercial dilute acid hydrolysis processes, and sets out the details of recent research into developing a continuous dilute acid hydrolysis reactor for the treatment of cellulosic wastes. Processes using concentrated acids at low temperatures have also been developed for the hydrolysis of cellulose, and research is continuing in this field too; but this work is outside the scope of this chapter.

THE DIFFICULTY OF HYDROLYZING CELLULOSIC MATERIALS

It is hardly surprising that cellulosic materials make such intractable substrates for chemical and biological reactions — it is this very property which makes cellulose a particularly valuable natural polymer, so useful in the building, textile and communications fields. The reasons for the highly unreactive nature of cellulosic materials are twofold, and relate to the crystalline structure of cellulosic molecules, and the presence of another natural polymer, lignin.

The molecular structure of cellulose has been the subject of intense study over the last 30–40 years [1], and it has been found that a high degree of crystallinity exists in the cellulose macromolecule, making it extremely resistant to chemical or biological attack. Cellulose is a long-chain molecule, composed of repeating units of anhydrous glucose residues. It is found in nature as the principal cell-wall material of higher plants. Cotton is the purest natural form (>90%), while the wood of coniferous and deciduous trees contains between 40 and 50% cellulose (Table I).

Within the cell wall, cellulose occurs mainly in the form of fibrils which are visible under the optical microscope. The electron microscope shows that these fibrils are composed of still finer microfibrils, which are thought to consist of numerous nearly parallel cellulose molecules,

Table I.   Composition of Some Cellulosic Materials[a]

	Approximate Composition (% dry weight)		
	Cellulose	Hemicellulose	Lignin
Coniferous Wood	40–50	20–30	25–35
Deciduous Wood	40–50	30–40	15–20
Cotton	94	2	0
Bagasse	40	30	20
Nut Shells	25–30	25–30	30–40
Corn Cobs	45	35	15
Corn Stalks	35	25	35
Wheat Straw	30	50	15
Paper	85–99	0	0–15
Newsprint	50	20	30
Sorted Refuse	60	20	20

[a] Adapted from various sources.

bound to each other laterally by hydrogen bonds, and so exactly ordered in places that they form crystalline regions [2]. Heterogeneous reactions occur slowly in these regions, as the close packing of the cellulose chains does not permit easy penetration by reagent molecules.

Other regions exist within the cell wall in which the cellulose chains are less ordered and may even be tangled—these are the paracrystalline or amorphous regions. This fact is indicated by X-ray diffraction and other evidence [3]. In these regions, heterogeneous reactions take place rapidly, as the reaction sites are readily accessible to reagent molecules. However, although the proportion of amorphous-to-crystalline cellulose varies between one cellulosic material and the next, it is generally small, particularly in the case of naturally occurring ("native") cellulose.

The second important factor limiting the availability or "accessibility" of cellulose for reaction is the presence of lignin. Lignin is another major constitutent of lignocelluosic materials, and makes up about 15–35% of typical wood cell walls (see Table I). It is a random, complex three-dimensional polymer of phenylpropane residues which is deposited in an amorphous form surrounding or "encrusting" the cellulose microfibrils [4]. Lignin and cellulose molecules together form an interpenetrating system of high polymers [5] so dense that not even water molecules can enter. Consequently, native cellulose is essentially inert in the biological and chemical processes, and some form of physical or chemical pretreatment such as fine grinding or swelling in strong alkaline solution is absolutely necessary to get it to react [6].

It is probably as well at this point, to complete the picture, to mention the third major constituent of cellulosic materials, hemicellulose. This is the generic name for another group of carbohydrates associated with the cell wall, which, unlike cellulose, *is* readily accessible to the action of dilute acids and alkalis. This group is composed mainly of heteropolymers, often branched, of various pentose and hexose sugars, with xylose, mannose, glucose, galactose and arabinose being the most predominant. Hemicellulose makes up between 20 and 40% of the composition of wood, and is also found in significant amounts in agricultural residues such as cotton hulls, sunflower husks, corncobs and other nonfood raw materials. In contrast to cellulose, which is hydrolyzed with difficulty, and lignin, which for all practical purposes remains insoluble during acid hydrolysis, hemicellulose is very rapidly hydrolyzed to its constituent sugars by the action of dilute acids, even at moderate temperatures.

During acid hydrolysis of cellulosic materials, then, the acid molecule, being small, penetrates the lignin, and rapidly hydrolyzes the hemicellulose and the noncrystalline regions of the cellulose. Attack then begins on the crystalline cellulose, at a rate that is almost a hundred times slower than the initial rate of hydrolysis [7]. It is this, the rate of reaction of crystalline cellulose, which may be speeded up by the action of strong acids at high temperatures and pressures.

MOTIVATION FOR CELLULOSE HYDROLYSIS

Having noted the difficulty with which cellulosic materials react, it is worthwhile considering the advantages to be gained from a process which will accomplish the desired hydrolysis effectively, efficiently and economically. The reasons are again twofold.

First, there is a large and ever-increasing amount of cellulosic material being generated as waste, which is becoming more and more of a problem in terms of waste disposal. Urban refuse, agricultural and food waste, manure, municipal sewage, and wood waste from the timber industry all contain significant amounts of cellulose. Both the pollution problems these wastes create, and the necessary cost incurred in disposing of them safely, have led scientists to think in terms of turning them to some advantage, by converting them through biological or chemical reaction mechanisms (e.g., acid hydrolysis) to useful products.

Second, the end products of cellulose hydrolysis could turn out to be very useful indeed, particularly in the light of present concern over the limited availability and unstable price of oil and products derived therefrom. Glucose produced by acid hydrolysis may be converted to ethanol by fermentation, and then used either blended with gasoline or on its own as a liquid fuel. Alternatively, it may be used as a chemical feedstock to produce ethylene by dehydration, and thence a whole range of other chemicals presently derived from oil. This is not as far-fetched as it might seem. Before the advent of petrochemicals, ethanol produced by fermentation was used extensively in the manufacture of ethylene and butadiene, and Sweden, to give one example, had an important chemical industry during and a few years after the Second World War based on ethanol derived from wood [8]. Recently, many papers have appeared in the literature outlining a whole range of possible reaction schemes for the production of ethanol-based chemicals [9–12].

REVIEW OF EARLY COMMERCIAL
DILUTE ACID HYDROLYSIS PROCESSES

Attempts to develop commercially feasible processes for the acid hydrolysis of cellulose go back a long way, at least to the year 1900. It will be profitable to review very briefly some of these early processes based on the use of dilute acids, as present-day research is in many ways a logical development of this early work.

Then, as now, the problem consisted in obtaining as high a yield of fermentable sugars as possible, by trying to minimize the decomposition of the glucose formed during the hydrolysis. To have an idea of the kind of success achieved by these early workers, it should be remembered that wood has been the traditional raw material of the hydrolysis industry, and on average, wood contains about 45% cellulose based on the original dry wood weight. The conversion of cellulose to ethanol is expressed in chemical terms, with the number below referring to the molecular weight, by:

$$(C_6H_{10}O_5)_n + nH_2O \xrightarrow{\text{hydrolysis}} nC_6H_{12}O_6 \xrightarrow{\text{fermentation}} 2nC_2H_5OH + 2nCO_2$$

cellulose	glucose	ethanol
162	180	92

Thus, 100 kg of dry wood should on average contain 45 kg of cellulose which may be converted to 50 kg of glucose, and hence to 25.6 kg of ethanol or 34.0 liters of 95% ethanol (for a ton of dry wood, the values would be 510 lb of alcohol or 82 gal of 95% ethanol [13]).

In the southern United States, between 1910 and 1922, there were two plants producing some 19,000–26,500 liters of 95% ethanol per day (5000–7000 gal) from southern yellow pine sawmill waste [14,15]. The process used was known as the "American" or one-stage process, also called the Simonsen method [16]. This was a batch hydrolysis with 0.5% sulfuric acid under steam pressure of about 912 kPa (9 atm). The conversion was accomplished in 15 min, but even in this short period of time, much of the sugar formed was decomposed. Other process losses reduced the yield further to an overall net 9.2 liters of 95% ethanol per 100 kg of wood (22 gal/ton). This yield was too low to be able to maintain competitive production when the local sawmills that supplied the plants were closed down, and raw materials had to be transported long distances from other areas at considerable cost. Simultaneously, the price of blackstrap molasses, the alternative source of industrial fermentation alcohol at that time, was lowered substantially, and these pioneering hydrolysis plants were forced out of operation.

But it was not too many years later that a German, Heinrich Scholler, developed a new technique that resulted in a very marked improvement in the yields of alcohol obtainable from the acid hydrolysis of wood. By allowing successive batches of hot dilute sulfuric acid (0.8%) to percolate through a stationary column of compressed wood waste at temperatures which increased from 120 to 180°C (a method known as "pulse percolation"), and removing the sugars from the reaction zone immediately as they were formed, decomposition of glucose was reduced, and yields of 21.7–24.2 liters of alcohol were obtained per 100 kg of oven-dry wood (52–58 gal/ton) [17,18]. The pressure in the percolator began at 1.14 MPa (165 psi) and reached 1.24 MPa (180 psi) in the final batch. The total extraction time was 13–20 hr. With some modifications, this process was used extensively in Germany before and during the Second World War. In 1945 the combined capacity of Scholler plants in Germany was 10 million liters of alcohol per year [17].

In America, the Madison Wood Sugar Process, developed by the U.S. Forest Products Laboratory towards the end of World War II, was a further refinement of the Scholler process and represented an improvement in terms of alcohol yield. By allowing the dilute acid to flow continuously through a charge of wood, with continuous removal of hydrolysate, decomposition was reduced even further because the sugars were in contact with the acid for a shorter time. Using 0.5–0.6% sulfuric acid at 150–180°C, 26.9 liters of 95% alcohol were obtained per 100 kg of dry, bark-free Douglas fir wood waste (64.5 gal/ton) in under 3 hr of hydrolysis time [19].

A full-scale saccharification plant based on the Madison Process was designed and built at Springfield, OR, with an estimated capacity of 15 million liters of ethanol per year. But unfortunately, it never had a chance to operate commercially. The end of the war, and with it the shortage of alcohol, combined with the advent of cheap naphtha and synthetic ethanol from ethylene derived from petroleum and natural gas, led to the premature shutdown of the Oregon plant, and the abandonment of work on the acid hydrolysis of cellulose for over 20 years in the Western hemisphere.

As far as is known, only in the controlled economy of the USSR has acid hydrolysis of cellulosic materials established itself firmly at the industrial level. There are thought to be over 40 Soviet wood hydrolysis plants in existence, with a combined annual output of some 400,000 tons of wood sugar [20]. The process used is one of periodic percolation [21], in which liquid flows continuously through a stationary bed of solid, as in the Madison process. Modifications have been made in the design of the percolator vessel to improve the hydrodynamic flow conditions, assuring both a horizontal and vertical flow of liquid through the percolator. A great deal of fundamental research has been conducted in the Soviet Union during the last three decades directed at developing improved methods for the hydrolysis of wood and agricultural wastes. But because the published results are in Russian, and are often not readily available, this research has been largely ignored outside that country.

APPLICATION OF CELLULOSE
HYDROLYSIS TO REFUSE TREATMENT

Turning to the recent revival of interest in the acid hydrolysis of cellulosic materials in America and other Western countries, this was sparked off largely by the suggestion of Porteous [22] that an acid hydrolysis technique might be an answer to the refuse disposal problem of large cities. He proposed, as a method of reducing the amount of refuse commited to sanitary landfill, that the 50–60% cellulosic content, made up of paper and putrescibles, be separated out and converted to sugar by acid hydrolysis, and then fermented to ethanol.

To make the process economical, he suggested that the hydrolysis be carried out in a continuous plug-flow reactor, at very much higher temperatures than ever before (220–230°C) and at very small contact times (of the order of a minute). This would have the effect of reducing the capital outlay considerably, as the reactor would be roughly 200 times smaller in size than the equivalent batch or semibatch reactor handling the same throughput [23]. A continuous-flow reactor together with short residence times would also have the advantage of eliminating the blockage problems so frequently encountered in the wood hydrolysis industry.

The idea of carrying out the hydrolysis of cellulose at such high temperatures and short residence times was a logical extension of the fundamental research done by Saeman [24] at the U.S. Forest Products Laboratory more than 20 years previously. As part of the work that went into the development of the Madison process, Saeman studied the kinetics of cellulose hydrolysis at 170, 180 and 190°C with catalyst concentrations of 0.4, 0.8 and 1.6% sulfuric acid. He found that, despite the fact that the hydrolysis takes place heterogeneously during the entire course of the reaction, as long as the cellulosic material he was investigating was ground into fine particles of 0.85 mm or less (−20 mesh), the experimental data for glucose yield could be well represented by the simple equation describing first-order consecutive reactions:

$$A \xrightarrow{k_1} B \xrightarrow{k_2} C \tag{1}$$

where A, B and C are cellulose, glucose, and glucose decomposition products, respectively, and k_1 and k_2 are the first-order reaction rate constants.

The rate equations for this system are:

$$dC_A/dt = -k_1C_A \tag{2}$$

$$dC_B/dt = k_1C_A - k_2C_B \tag{3}$$

$$dC_C/dt = k_2C_B \tag{4}$$

In this reaction scheme, the concentration of cellulose (C_A) decreases continuously, the concentration of glucose decomposition products (C_C) increases continuously, while the concentration of glucose (C_B) passes through a maximum (Figure 1) with time (t). Strictly speaking, these formulas apply only to homogeneous materials, but Saeman found that they also held good for cellulose hydrolysis as long as the cellulose concentration was expressed as potential sugar. He determined the rate constants k_1 and k_2 under various reaction conditions, and found that they

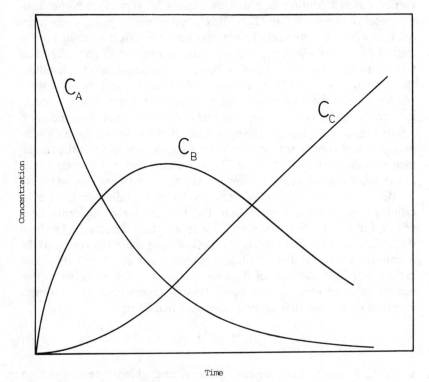

Figure 1.   Concentration-time curves for substances A, B and C in consecutive first-order reactions.

were related to the isothermal reaction temperature and acid catalyst concentration in the usual way:

$$k_1 = P_1 C_a^m \exp(-E_1/RT) \tag{5}$$

$$k_2 = P_2 C_a^n \exp(-E_2/RT) \tag{6}$$

where P_1, P_2, m, n = constants
C_a = sulfuric acid concentration
E_1, E_2 = activation energies
R = universal gas constant
T = absolute temperature

Naturally, all of these relationships applied only to the hydrolysis of the crystalline cellulose. At a reaction temperature of 170°C or more, amorphous cellulose is converted to glucose almost instantaneously by the action of dilute acids (a fact which had to be taken into account in the calculations). With any first-order consecutive reaction, the maximum value of component B (glucose) is a function only of the ratio of the rate constants k_1/k_2, and Saeman found that this ratio increased with increase in reaction temperature and/or acid concentration. Higher yields of glucose were obtained from the hydrolysis of Douglas fir cellulose when the reaction temperature was increased from 170 to 190°C, and the sulfuric acid concentration from 0.4 to 1.6% (Figure 2).

What Porteous [22] did was to assume that the data collected by Saeman for the hydrolysis of cellulose in wood chips would be applicable to the cellulosic material in refuse. This seemed a reasonable assumption, as paper makes up the greatest proportion of the cellulosic content of refuse, and paper is made from wood pulp. Newsprint in particular, which is derived from wood subjected to a purely mechanical pulping process, has essentially the same composition as wood. So, with the aid of the Arrhenius rate equation, Saeman's kinetic parameters were extrapolated to much higher reaction temperatures. Then, using these predicted values of k_1 and k_2, it was calculated that the hydrolysis of the cellulosic material in refuse should be achieved with 55% conversion of cellulose to glucose in 1.2 min using 0.4% sulfuric acid. An economic evaluation of a plant designed around this idea suggested that ethanol production from the wastepaper content of refuse should have a strong profit potential [22].

However, the assumption that Saeman's kinetic parameters could be extended accurately to much higher temperatures and for a different substrate still needed to be tested in practice. Four years later, the predictions were verified. Fagan, working at the Thayer School of Engineering, Dartmouth College, carried out batch tests similar to those of Saeman,

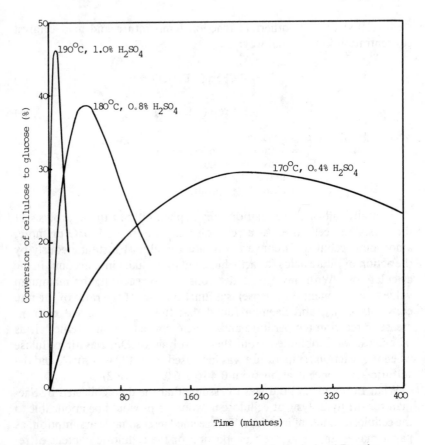

Figure 2.   Predicted glucose yields for isothermal hydrolysis of Douglas fir cellulose (after Saeman [24]).

but at much higher temperatures, and showed that paper cellulose does hydrolyze to sugar, and that the kinetic model of two consecutive first-order reactions may still be applied [25]. He hydrolyzed small samples of ground paper or refuse under nonisothermal conditions at temperatures of up to 240°C, and derived a similar set of curves for the hydrolysis of paper as Saeman had obtained for the hydrolysis of wood (Figure 3). He calculated that in an isothermal reaction, a maximum conversion of cellulose to glucose of 52% could be obtained at 230°C using 1% sulfuric acid. The temperature of 230°C was about the practical limit for hydrolysis, as the reaction time was only 20 sec [26]. It now remained to develop a continuous reactor.

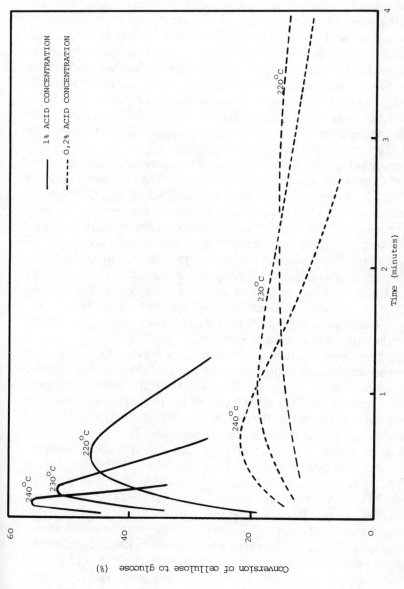

Figure 3. Predicted glucose yields from isothermal hydrolysis of paper cellulose (after Fagan et al. [26]).

DEVELOPMENT OF A CONTINUOUS REACTOR

The problems associated with the development of a continuous reactor for the acid hydrolysis of cellulose are numerous — as is evidenced by the fact that work in this field is still in progress almost 50 years after the first patents were issued [27,28]. The problems relate to the extreme reaction conditions of temperature and pressure required to hydrolyze the cellulose, the necessity of removing the glucose product swiftly from the reaction zone once it is formed and the need to use materials which will be resistant to the highly corrosive conditions of strong mineral acid at high temperatures. Physical problems of solids handling and carrying out a controlled heterogeneous reaction at high solid/liquid ratios only add to the complications.

The advantages to be gained from continuous operation, however, are equally numerous, and so the work continues. The capabilities required from a continuous process include, among other things [9,29]: continuously conveying a high-solids cellulosic slurry to the reactor; increasing the temperature and pressure in the reactor to the optimal hydrolysis conditions; adding the required amount of acid catalyst just before reaction (to minimize corrosion); effecting the reaction in the optimum reaction time to obtain maximum glucose yield; quenching the products of the reaction as swiftly as possible to minimize glucose decomposition; recovering the glucose efficiently and continuously from the hydrolysate. A simple schematic diagram of the type of system which would satisfy these requirements is shown in Figure 4.

The develpment of a flow reactor along these lines has for several years been the subject of investigation at the Thayer School of Engineering, Dartmouth College. Initial problems with the slurry pump limited the solids concentration that could be used to 1% [30] and blockages of the needle valve at the exit of the reactor also caused delays in the experimental program. But recently, a bench-scale plug-flow reactor was designed, constructed and successfully operated to study the kinetics of the continuous acid hydrolysis of cellulosic substrates [31]. Slurries of up to 13.5% Solka-Floc of particle size finer than 74 μ (-200 mesh) or newsprint of particle size finer than 0.2 mm (-65 mesh) were pumped at about 300 ml/min through a heating unit of 6 kW capacity which consisted of 26 m of concentrically coiled tubing wrapped with electrical heating tape. The slurry temperature was raised from 30 to 240°C, while the fluid velocity was maintained at a sufficiently high level to prevent any settling of slurry particles. In a specially designed mixing tee, shaped like a venturi, a controlled flow of concentrated sulfuric acid was injected into the preheated slurry to make up an acid concentration

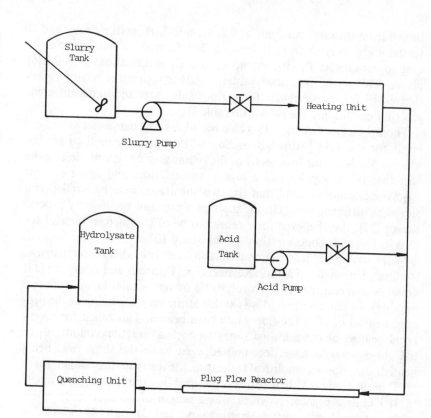

Figure 4. Flow diagram for continuous hydrolysis reactor.

between 0 and 2%. The acidified slurry then entered the reactor, made of Carpenter 20 cb3 stainless steel, and mounted vertically to prevent the settling of slurry particles. The reactor was in fact a series of tubes of various lengths which could be interchanged to vary the residence time. On leaving the reactor, the hydrolysate passed through a 1.8-m-long capillary tube of 0.7 mm i.d. This was found necessary to maintain a pressure differential of 3.45–4.14 MPa (500–600 psi) across the reactor. The hydrolysate was cooled as it passed through the capillary by means of a cooling water jacket. The cooling was accomplished in a fraction of a second, from 240 to 50°C.

The results of the experimental work show that 50–55% conversion of cellulose to glucose could be obtained repeatedly from the Solka-Floc slurries when operating in the temperature range 235–240°C at a residence time of 0.22 min with 1% sulfuric acid. These and other results obtained from the same substrate at 0.5, 1.5 and 2.0% sulfuric acid, and at temperatures varying from 180 to over 240°C, were used to estimate the kinetic parameters P_1, P_2, m, n, E_1 and E_2 in Equations (5) and (6) above. This set of parameter values, together with other sets of values derived by other researchers for various substrates and under different reaction conditions, are shown in Table II.

Thompson and Grethlein [31] also found that the parameter values derived for the acid hydrolysis of Solka-Floc could be used to predict glucose yields for the hydrolysis of the cellulose in newsprint, despite the fact that the newsprint had a higher hemicellulose and lignin content than the Solka-Floc, and that the two substrates were ball-milled and sieved at different sizes. On the basis of a previous preliminary process design [32], they believed their results to be of commercial interest for slurries of solids concentrations greater than 10%.

Work similar to that of Thompson and Grethlein is being conducted at the Open University in Milton Keynes, U.K. Porteous and Anderson [33] designed and constructed a hydrolysis rig of very similar layout with one notable difference — the acid was added at the very beginning of the process, instead of after the slurry had been heated. This called for a very rapid heating of the acidified slurry to optimal reaction conditions, or the glucose would have decomposed even while the slurry was being heated. This was accomplished by heating the slurry in three stages: (1) to 90°C in the feed tank, before it was pumped into the system; (2) from 90 to 180°C in a preheater, which was a length of concentrically coiled tubing wrapped with electrical heating tape; and (3) from 180 to 240°C in a heating section, which was a length of straight tubing similarly fitted with heating tape. As the hydrolysis of the crystalline cellulose proceeds very slowly below 180°C, it was only the heating between 180 and 240°C which had to be done very rapidly. The results published so far [33] have been of experiments with knife-milled filter paper and computer paper. Owing to problems with blockages and inadequate temperature control, glucose yields were low and not very consistent. More recently, however, we have overcome some of these difficulties, and conversions of cellulose to glucose of up to 39% have been obtained in the temperature range 230–240°C, using acid concentrations as low as 0.1%. But slurry concentrations have also had to be low. To date, an upper limit of 1% solids has generally applied. Work is in progress to improve this.

A conceptual design and economic evaluation of a rather different reactor system was put forward by Meller [34]. This was a multistage process intended for the hydrolysis of paper pulp, and consisted of four reactor tubes in series with a screw press after each unit. Hot dilute sulfuric acid would be added at each reactor inlet, to heat the charge to

Table II.   Kinetic Parameters for Hydrolysis of Cellulose and Glucose as Determined by Various Researchers

	Experimental Conditions	P_1 (min^{-1})	m	E_1 (cal/mol)	P_2 (min^{-1})	n	E_2 (cal/mol)	Reference
Douglas Fir	170–190°C; isothermal batch	1.73×10^{19}	1.34	42,900	2.38×10^{14}	1.02	32,870	24
Kraft Paper	180–230°C; nonisothermal batch	28×10^{19}	1.78	45,100	4.9×10^{14}	0.55	32,800	26
Solka-Floc	180–240°C; isothermal plug-flow	1.22×10^{19}	1.16	42,500	3.79×10^{14}	0.69	32,700	31
Paper Pulp	Not reported	6.15×10^{18}	1.275	48,660	1.42×10^{14}	0.997	37,540	55
Pinus radiata Sawdust	190–220°C; nonisothermal batch	1.95×10^{19}	1.45	41,440	2.15×10^{14}	0.9	32,470	67
Lignocellulose Residue	205–228°C; isothermal batch	2.79×10^{19}		44,500	1.43×10^{17}		39,900	35

the desired temperature, and the acidified slurry would be transported through the reactor tube in the required residence time by a screw conveyor. The screw press after each reactor would squeeze most of the hydrolysate from the pulp, thereby removing from the reaction system most of the sugars formed in that reactor tube. The unreacted solids, now containing about 50% moisture, would be mixed with a fresh batch of hot acid at the inlet of the next reactor, and so the process would continue. In the first reactor, only the hemicellulose portion of the feed would be hydrolyzed, so this reactor was designed to be operated at a lower temperature and shorter residence time than the others, which were meant for the hydrolysis of the crystalline cellulose. The remaining reactors were to be operated at 200°C, with a residence time of about 10 min in each. At the end of the process, the unreacted lignin would be used as fuel in the steam boiler. This process was stated to be commercially available from the Black Clawson Company of Middletown, OH [34], and this company was consulted for information on costs, operating labor and power requirements (which were used in the economic evaluation). But on inquiry, this system was found to be obsolete [21], and it is not known whether the process was ever actually built and tested.

Another system, incorporating a screw pump for feeding slurry continuously through a preheater into a hydrolysis reactor, was designed and constructed for operation at moderate temperatures and fairly long residence times by Pohjola [21] and his associates [35,36] at the Technical Research Center of Finland. They developed their 5.5-liter flow reactor for the hydrolysis of various cellulosic materials which are difficult to hydrolyze by percolation methods because of their poor permeability. The preheater was designed in such a way that the temperature of the reaction mixture could be raised by external or direct steam heating, with heat transfer being assisted by means of an internal rotating shaft fitted with scrapers. Product from the reactor was discharged into a flashing chamber by a periodically operating ball valve [21]. A typical substrate, chosen to test the performance of the apparatus, and to check the reproducibility of kinetic data from batch reactor runs, was slightly decomposed moss peat, which contained a high proportion of cellulose in the amorphous form. The optimum reaction temperatures had been determined by Russian investigators [37] to be 135–175°C, at a reaction time between 10 and 30 min. The peat was ground in a Wiley mill to a particle size of 1 mm or less, and made into a slurry with dilute sulfuric acid. The final liquid/solid ratio was 8 or 9, and the acid concentration was 0.5%.

Results are reported for three runs in which the reaction temperature was maintained in the upper range of the recommended values

(169–173°C) and the residence time varied between 10 and 20 min. Glucose yields were steady at 39–40% of theoretical, and compared very favorably with results obtained from hydrolysis in a batch reactor. The flow reactor was said to operate very steadily throughout the experiments. Unfortunately, though it seems that other experiments were carried out with this reactor, these results are not reported, and in fact this group of researchers has published no more work in this field. It is also not known whether their optimisation studies directed at developing a larger scale reactor were ever carried out.

Another group of researchers which has taken its work to the small pilot-plant scale is in the New York University Department of Applied Science. Working in association with Werner and Pfleiderer, manufacturers of extruder equipment, Rugg and Brenner [38] developed a twin-screw hydrolysis reactor, which has been used to treat up to 90 kg/hr (200 lb/hr) of sawdust 95% solids) or newspaper (10% solids) continuously, with high conversion of cellulose to glucose [39]. The reactor was operated at high temperatures (about 240°C) and pressures (3.45 MPa) and at very low residence times (about 10–20 sec). Normally, a machine of this type would be used in the plastics industry to melt, mix and squeeze polymers through a die to form products of various sizes, shapes and colors. In its application as a hydrolysis reactor, sawdust or a slurry of newspapers was fed into the machine instead, and carried forward through the barrel by the twin co-rotating screws. The shearing action of the screws performed a pretreatment function by breaking up the cellulose, making it more accessible to reaction. At the same time, excess water was pressed out and removed, leaving a very dense cellulosic slurry which formed a plug in the screws and thereby prevented the backflow of material as the pressure built up toward the front of the barrel. All the time, heat was applied to the slurry in the barrel indirectly by super-heated steam [40].

When the optimal conditions of pressure and temperature were reached, at a point shortly before the reactor exit, dilute sulfuric acid was injected into the slurry. The residence time of the slurry in the reactor from that point onward was only a matter of seconds before the entire reaction products were discharged from the end of the barrel through a high-pressure valve. The results of the tests carried out indicate that 50–60% conversion of cellulose to glucose could be obtained from the hydrolysis of sawdust with 1–2% sulfuric acid; similar tests gave 40–50% yields from newsprint. One of the significant advantages of the particular extruder used was that it had a modular design with interchangeable screws and barrel sections. These could be altered to suit the specific conditions required at different points along the reactor length (e.g., high shear action, increased pressure generation). In addition, as the acid was introduced into the slurry so late in the process, just before the reactor exit, only the forward sections of the barrel needed to be made from corrosion-resistant materials. The development of this extruder reactor was the culmination of a five-year research program at New York University and represents an important step toward the establishment of an industrial continous hydrolysis process. Of particular significance are the high throughputs, the good yields of glucose, and the capacity of the reactor to treat slurries with high solid/liquid ratios. The theoretical framework seems to be lacking, but experiments are said to be under way with the express aim of developing accurate material and energy balances, and optimizing the reaction conditions for maximum glucose yield [38].

Finally, a word ought to be said about the research work presently being conducted in the USSR on developing continuous hydrolysis reactors. Being the only country in the world at the moment to have a full-fledged hydrolysis industry, treating wood and agricultural wastes by percolation hydrolysis, it is only natural that there has been much activity there in recent years aimed at developing continuous processes for cellulose hydrolysis. Hardly a month goes by without some reference in the *Chemical Abstracts* to this ongoing research. A review of this subject was made some years ago by Pohjola [21].

It appears that most of the Russian work is directed at developing continuous countercurrent percolator reactors. Percolation hydrolysis, in which a fluid flows continuously through a stationary bed of solid, suffers from the disadvantage that the volume of the solid bed shrinks to half soon after the hydrolysis begins. This leads to underutilization of the reactor volume, and makes operation and control difficult as the process parameters are continuously varying. Then too there are the usual problems with batch processes of time lost in charging, heating, and discharging each batch. These problems would not exist in continuous percolation.

According to Efimov [41], systematic studies aimed at the development of continuous hydrolysis processes began in Russia as long ago as 1947. The earliest apparatus types considered were continuous percolators which were fed with material from the top, but since 1965, when systematic theoretical work began to find the fundamental physicochemical relationships of continuous percolation, efforts in process development have been directed toward countercurrent percolation [21]. Tests with a bottom-fed 440-liter pilot percolator gave 45–52% yields of sugar based on dry wood; the capacity of the reactor was 80 kg of dry wood/hr. The kinds of parameters studied were the hydrolysis rate constants, coefficients of mass transfer, diffusion of sugar from wood shavings of various dimensions, effective contact time between solid and liquid in continuous hydrolyzers, geometry of the reaction vessel, and effects of particle size, solution viscosity, temperature gradients, liquid volume to solid volume ratios, time of hydrolysis and stability of the operating conditions [42-45]. As was stated above, however, all these published results are in Russian and lately in the *Chemical Abstracts;* many of these papers have been listed with only the title translated. So it is thus likely that a great wealth of published work in this field is being completely ignored.

PRETREATMENT METHODS TO ENHANCE CELLULOSE HYDROLYSIS

The difficulty of hydrolyzing cellulosic materials, owing to the crystallinity of the cellulose and the presence of lignin, has been described above. It is on account of this "inaccessibility" of cellulose to chemical or biological attack that such drastic reaction conditions are required to convert cellulose to glucose, conditions which result in the destruction of the glucose itself and severely limit the maximum sugar yields obtainable from a simple batch hydrolysis.

Considerable effort has been expended over the years to develop various physical and chemical pretreatment methods which would substantially reduce the crystallinity of the cellulose macromolecule, and/or break the lignin-carbohydrate bond, thereby making the cellulose more readily available for reaction. If the hydrolysis of the cellulose could be carried out effectively at much lower temperatures than presently required for useful reaction rates, decomposition of glucose would cease to be a problem. Alternatively, during hydrolysis at elevated temperatures, if the accessibility of the cellulose to reaction were increased, the ratio of the rate constants k_1/k_2 would be higher, and greater maximum yields of glucose would be obtainable.

The entire subject of cellulose pretreatment was reviewed several years ago by Millett et al. [6]. The chemical methods considered were swelling with alkaline agents, such as sodium hydroxide or ammonia; delignification or the use of delignified residues as raw materials; and steaming. Physical pretreatments reviewed included fine grinding; irradiation by gammas rays or high-velocity electrons; thermal treatment at high or low temperatures; and compression. All of these methods have proved effective to varying degrees (depending on the conditions of hydrolysis and the particular substrate used) in improving cellulose hydrolysis.

Brenner et al. [9], during batch experiments with newspaper, showed that hydropulping followed by low-level electron irradiation (5–10 Mrads) was more effective as a pretreatment method than Wiley-mill grinding, industrial grinding, hydropulping with a ferric ion/hydrogen peroxide soak or hydropulping per se. Conversions of cellulose to glucose of 50% or more could be obtained after hydropulping/irradiation pretreatment, from hydrolysis at 230°C with 1% sulfuric acid at very short reaction times of 10–20 sec.

Pohjola et al. [35] carried out batch hydrolyses on a lignocellulosic residue from a furfural process. After the steam digestion of birch chips, during which the hemicellulose fraction was converted to furfural, the unreacted solids were blown out of the reactor under their own pressure. This blow-out had the effect of partly destroying the fibrous structure of the particles, and the aim of the investigation was to determine whether in the process the material became more amenable to hydrolysis with dilute acids. Experiments were carried out in a 2-liter autoclave at temperatures up to 240°C with a sulfuric acid concentration of 0.5%. Comparison of the results with those of Converse et al. [30] for waste paper showed that the yields for the lignocellulose residue were considerably higher.

The influence of fine grinding on the acid hydrolysis of cotton linters, newsprint, and two species of wood was studied recently by Millett et al [46]. Samples were first ground to pass a 0.5-mm screen, then milled for up to 2 hr in a vibratory ball mill. Subsequent hydrolyses were carried out at 180°C with 0.1 N sulfuric acid at times ranging from 4 to 160 min. The results showed that the rates of hydrolysis of all four substrates increased substantially as the time of milling was increased.

Research into pretreatment methods for cellulose hydrolysis has also led to the development of a new process which it is claimed will give quantitative yields of sugars from cellulosic materials. The Tsao-Purdue process relies on the use of chemical solvents to destroy the crystallinity of the cellulose by extracting it from a cellulosic residue, and then reprecipitating it in an amorphous form which is much easier to hydrolyze at less stringent reaction conditions [47,48]. The solvents suggested are Cadoxen (a mixture of cadmium oxide and ethylenediamine), ferric tartrate or concentrated sulfuric acid. This reaction scheme is in fact the basis of a quantitative saccharification technique [49], but the idea of using this established knowledge for the development of a commercial process is new. However, although paper studies have been made of the process [50,51], and results have been reported for several different agricultural products such as corn stalks, alfalfa, orchard grass, tall fescue and sugarcane bagasse, it is not known whether this scheme has ever been tested outside the laboratory.

CONTINUOUS ACID HYDROLYSIS
AS A PRETREATMENT PROCESS

While some kind of pretreatment step is required to make cellulosic materials more susceptible to hydrolysis by dilute acids, the hydrolysis reaction itself has the effect of opening up the structure of the macromolecule, removing the soluble carbohydrates, and leaving the residue more porous and more readily accessible to microbial or enzymatic attack. It is in this sense that Klee and Rogers [52] considered acid hydrolysis as a pretreatment to the anaerobic digestion of municipal waste. Other researchers in the United Kingdom [53] developed a tubular reactor for hydrolyzing activated sludges to glucose, as a first step in the production of single-cell protein. The slurry was first heated to 310°C in an oil bath and then injected into the reactor together with sulfuric acid. The hydrolysate was neutralized and centrifuged to separate the dilute sugar solution from the solids.

Perhaps the most interesting recent development along these lines is the work of Knappert et al. [54] at the Thayer School of Engineering, Dartmouth College. The same continuous plug-flow reactor that was designed and constructed by Thompson and Grethlein [31] for their study of the high-temperature hydrolysis of cellulose (described above) was used to carry out a partial acid hydrolysis of four cellulosic materials as a pretreatment to enzymatic hydrolysis. Solka-Floc, newsprint, corn stover and oak chips were made into slurries containing 5% solids by weight and hydrolyzed for 0.22 min at temperatures ranging from 160 to 220°C, and acid concentrations of 0.4–1.2%. These pretreated substrates were then subjected to enzymatic hydrolysis of 50°C for up to 48 hr. The results showed that the native substrates in particular – the oak and the corn stover – gave much improved yields of glucose after pretreatment compared with enzymatic hydrolysis without pretreatment. Quantitative conversion of cellulose to glucose was obtained for oak which had been pretreated with acid at only 189°C. On the other hand, newsprint, which gives a much higher yield of glucose on enzymatic hydrolysis than does oak (when neither of them are pretreated), was not so much improved after pretreatment, except if the temperature of acid hydrolysis was about 220°C, then quantitative yields were obtained after just 24 hr of enzymatic hydrolysis. Solka-Floc did not improve at all after pretreatment.

These results are interesting, as they represent a significant improvement in the rate and extent of enzymatic hydrolysis for certain substrates when subjected to a prior partial acid hydrolysis. The possibility arises of other methods being developed to convert the cellulose after a pretreatment step of this kind.

ECONOMIC CONSIDERATIONS

High-temperature dilute acid hydrolysis of cellulosic materials was originally proposed with a clear economic motive in mind—deriving profit from the disposal of urban refuse. Since then, the oil crisis and the appreciation of the need to conserve vital resources has lent another dimension to the whole issue of deriving energy from biomass, and researchers have studied the hydrolysis of a whole range of cellulosic materials. Ultimately, economic considerations prevail, and money will only be forthcoming to build a large-scale continuous hydrolysis plant when it can be shown that such a plant will be operated successfully, and at a profit.

Over the years, many paper analyses of various designs of hydrolysis plant have appeared [22,30,32,38,50,55,56]. It is very difficult to compare one with another, as there is great disparity between them on such factors as the size of the plant, the nature of the process, the starting material and its assumed price, the nature of the product (this can vary from a dilute glucose solution to anhydrous ethanol), whether there is any credit assumed for by-products, and just how the authors have scaled up from their laboratory data. Production costs have been estimated at anything from $0.60 to $6.00/gal of ethanol, although recently there does seem to be a consensus that ethanol may be produced at a price comparable to that of alcohol derived from grain [57].

The most important consideration in the economics of cellulose hydrolysis is the cost of raw material. It was this factor which caused the early closure of the pioneering hydrolysis plants in the southern United States, and it is still vital for economical operation that there be large supplies of cheap cellulose within close proximity to a hydrolysis plant, hence the application to municipal refuse. In large cities, refuse is a very substantial source of cellulose, which is already collected and might even be available at negative cost. A plant designed to treat wood waste or agricultural residues would have to be located very near a large source of cellulose to be profitable.

The cost of pretreatment is another significant economic consideration. Grinding, which is the most common form of pretreatment in the laboratory, is a very energy-intensive and costly unit operation. The Tsao-Purdue solvent pretreatment process stands or falls economically on the recovery of the solvent. The advantage of selecting a raw material already in a form which is easy to hydrolyze, or which responds readily to pretreatment, has been discussed [58].

The third major economic problem is the cost of alcohol recovery. Present technology requires that the fermentation of glucose be carried out in dilute solution, and the ethanol product separated from ten times or more its own volume of water. At the moment, distillation is the only method used to accomplish this separation at considerable energy cost. Ways of improving the fermentation process to give higher concentrations of alcohol have been reported [59,60]. It is also possible to use solid residues of the conversion process, such as unreacted cellulose and lignin to provide energy for distillation, and this has been taken into account in some economic analyses. There is still much scope for further improvement in this area.

A related issue is the need for a hydrolysis reactor to be able to treat a feed with a high solid/liquid ratio to obtain a high concentration of sugar in the hydrolysate. If the hydrolysate had first to be concentrated from 1 or 2% glucose to 10% prior to fermentation, the additional distillation costs could well be prohibitive. Grethlein [32] has calculated that an acid hydrolysis plant treating a 30% newsprint slurry would be energy self-sufficient if the lignin and unreacted cellulose were burned to provide process heating and electrical energy. A similar plant capable of handling only a 10% slurry would have to purchase additional energy.

Another factor is that the price of ethanol from alternative sources is continually rising. Practically all synthetic alcohol in the United States is produced from ethylene derived from petroleum or natural gas, and the price of ethanol has increased sharply in the last decade to keep up with increases in the price of oil. Ethanol produced by hydrolysis of cellulose is becoming more competitive, and this trend is irreversible. "Ultimately the true cost of a limited exhaustible material will become greater than that of a renewable resource" [12].

Moreover, in the past, the relative stability in the price of oil was a major factor in its being the preferred prime raw material for the synthesis of a wide variety of chemicals. Now that the price and availability of oil have become so uncertain, cellulosic wastes which are locally available at predictable prices pose a serious challenge as a rival feedstock for the chemical manufacturing industries [59].

A further very important factor favoring cellulose hydrolysis is the possibility of obtaining by-product credits. Lignin may be burned as solid fuel, but it also has potential as a chemical feedstock in its own right in the production of phenol and phenolic polymers [12], the manufacture of coke [61], or a wide range of other uses [62]. The recovery of hemicellulose too would provide significant cost benefits. The products of hemicellulose hydrolysis are xylose and furfural, for which an established market already exists. Many different methods have been outlined for the extraction of hemicellulose [48,63–65] which could be used in conjunction with cellulose hydrolysis to form an integrated process by which all the major components in a cellulosic substrate would be converted to useful products.

To put all of this into quantitative terms, alcohol from grain was sold in 1980 at about $0.41/liter ($1.57/gal), while the cost of alcohol from natural gas was $0.46/liter ($1.80/gal) [66]. In comparison, the developers of the New York University extruder reactor claim that alcohol could be produced by their method of hydrolyzing cellulose from newspapers and sawdust for a little as $0.21/liter ($0.80/gal) [66]. Elsewhere, they have estimated the capital cost of a hydrolysis plant treating 1816 metric ton/day (2000 ton/day) of dry sawdust at $67.4 million in 1980 [38]. In this analysis, the cost of the extruder reactors alone was $17.5 million. With a 50% cellulose content in the sawdust, a conversion of cellulose to glucose of 60%, and of glucose to ethanol of 48%, the annual production of anhydrous ethanol was calculated at 112.3×10^6 liters (29 Mgal). Then assuming that the lignocellulose residue would be used to generate steam for the process, and that a by-product credit would be obtained for carbon dioxide produced during the fermentation, an estimated production cost of $0.28/liter ($1.08/gal) of anhydrous ethanol was arrived at. The cost of sawdust feedstock, taken to be $33/metric ton ($30/ton), represented 64% of this production cost.

This illustrates very clearly the importance of raw material cost on the economics of cellulose hydrolysis. In another economic evaluation, Converse and Grethlein [56] compared the hydrolysis of refuse with that of wood in a single-pass reactor, operated at 230°C and using 1% sufuric acid. The plant was designed to treat 804 metric ton/day (885 ton/day) of solids. Refuse, which was assumed to contain 61% cellulose, was priced at $11/metric ton ($10/ton), while wood was assumed to contain 41% cellulose and cost $33/metric ton ($30/ton). The significant difference in raw material prices is reflected in the cost of ethanol production: $0.19–0.22/liter ($0.75–0.84/gal) from refuse, compared with $0.40–0.44/liter ($1.55–1.70/gal) from wood.

Finally, in the United Kingdom, a recent committee studying the production of fuel alcohol from biomass [57] concluded that sorted refuse was the only material which appeared to be energetically and economically feasible as a substrate for acid hydrolysis of cellulose to glucose.

DISCUSSION

Continuous acid hydrolysis of cellulose in a plug-flow reactor has yet to reach the commercial stage. Of the research which has been carried out in the last 15 years, much has been at the level of small batch tests.

Only recently, in the last three or four years, have laboratory-scale continuous reactors been designed, constructed and operated, with maximum glucose yields in the region of 50-60%.

These studies have been almost exlusively concerned with pure substrates. It still remains to determine the kinetics of refuse hydrolysis in a continuous reactor. The only work which has been done to date with refuse is that of Fagan et al. [26], who hydrolyzed very small samples (25 g) under nonisothermal conditions, and Converse et al. [30], who had problems in maintaining a constant temperature and steady flowrate in their flow reactor. It is most important to obtain accurate kinetic data for the hydrolysis of refuse, as economic considerations point to this as the most suitable raw material for a commercial hydrolysis process. Determining the kinetic parameters will be no easy task, however, as refuse is an extremely unhomogeneous product. In fact, on account of this, it might well be that the kinetic model of Saeman does not apply in this instance, or it might have to be modified to take into account the proportions of the constituent cellulosic materials and the effect of impurities. Experimental results to date in this field are extremely sketchy, and these possibilities are not ruled out. Even with pure substrates, such as Solka-Floc, there is a wide scatter between experimental data points and values predicted from the kinetic model [31]. This could of course, be put down to experimental error, but it is nevertheless essential that these areas be investigated thoroughly and the kinetics be understood properly, to create a reliable basis on which to design and cost a full-scale hydrolysis plant for this application.

Work also needs to be done on the hydrolysis of much larger particles. Until now, research in this area has been limited by the small scale of experimental rigs. Initial experiments by Guha et al. [67], who compared the hydrolysis of 2.5-mm prehydrolyzed woodchips with sawdust, showed that diffusional effects were significant with the larger particles. The work of Soviet wood chemists, who have devoted much attention to diffusional studies over the years (because of their use of percolator reactors) ought to be studied before their results are duplicated in other laboratories.

Scope exists too for improving the yield of glucose from high-temperature hydrolysis. At the moment, the practical limit for glucose conversion is 50-60% because of the very short reaction times required to produce maximum yields at high temperatures. The Scholler and Madison Processes, in comparison, were able to convert about 80% of the available cellulose to glucose. A variation of the multistage hydrolysis system [34] could be tried, although on account of the substantial glucose decomposition which occurs at high temperatures, a cheaper and perhaps more profitable alternative might be a single-stage reactor with a recycle stream.

Finally, the whole concept of "total biomass utilization" deserves closer scrutiny. Hemicellulose may be extracted in a prehydrolysis step and lignin used as fuel or chemical feedstock. Consideration should also be given to recovering the glucose decomposition products present in the hydrolysate in a concentration of about 20% of the original cellulose. Hydroxymethylfurfural, the product of glucose decomposition, itself breaks down into levulinic and formic acids. These organic acids, together with the sulfuric acid catalyst, could probably be taken up on a weak-base ion exchange resin and recovered separately as salts by selective regeneration. This might assist even more in the economics of the process.

SUMMARY AND CONCLUSIONS

Hydrolysis of cellulose by dilute acids at high temperature has been described as a method of converting present wastes such as refuse into useful products. The difficulties involved in carrying out this reaction, relating to the crystallinity of the cellulose and the presence of lignin, have been outlined together with some methods of pretreatment which make the cellulose more available for reaction. Recent progress towards developing a plug-flow reactor, capable of converting cellulosic slurries with high solids content continuously to glucose has been reviewed.

On the basis of research undertaken during the last decade, there is much room for optimism that acid hydrolysis will have a role to play in the treatment of cellulosic wastes. The original predictions that hydrolysis at high temperatures and very short reaction times would result in the conversion of 50–60% of the available cellulose to glucose have been verified. Reactors have been developed in which this process can be carried out under carefully controlled conditions. Some of the reactors are capable of handling newsprint slurries of 10% solids content and more, and the extruder reactor can treat sawdust which contains only 5% moisture.

Economic analysis suggests that, in the immediate future, urban refuse is the most suitable raw material for a commercial-scale hydrolysis plant. With this in mind, much more research ought now to be concentrated on the hydrolysis of this substrate, and in particular on the kinetics of the process, which should be determined accurately, to allow proper mass and energy balances to be drawn up. On this basis, a more reliable design and cost estimate of a full-scale hydrolysis plant could be made.

REFERENCES

1. Shafizadeh, F., and G. D. McGinnis. "Morphology and Biogenesis of Cellulose and Plant Cell-Walls," *Adv. Carbohyd. Chem.* 26:297–349 (1971).
2. "Cellulose," in *McGraw-Hill Encyclopedia of Science and Technology, Vol. 2,* 4th ed., (New York: McGraw-Hill Book Co., 1977), pp. 662–663.
3. Segal, L. "Derivatives of Cellulose. Effect of Morphology on Reactivity," *High Polym.* 5(Pt.5):719–739 (1971).
4. Fratzke, A. R., U. Gunduz and G. B. Oguntimein. "Composition, Purification, and Enzymatic Hydrolysis of Corn Cob Xylan. A Literature Search," NTIS Pub. PB-246 895 (1974).
5. Cowling, E. B. "Physical and Chemical Constraints in the Hydrolysis of Cellulose and Lignocellulosic Materials," *Biotechnol. Bioeng. Symp.* 5:163–181 (1975).
6. Millett, M. A., A. J. Baker and L. D. Satter. "Pretreatments to Enhance Chemical, Enzymatic and Microbiological Attack of Cellulosic Materials," *Biotechnol. Bioeng. Symp.* 5:193–219 (1975).
7. Sharples, A. "Degradation of Cellulose and Its Derivatives. Acid Hydrolysis and Alcoholysis," *High Polym.* 5(Pt.5):991–1006 (1971).
8. Jullander, I. "Early Experiences from the Swedish Wood-Based Chemical Industry," *Appl. Polym. Symp.* 28:55–60 (1975).
9. Brenner, W., B. Rugg and C. Rogers. "Utilization of Waste Cellulose for Production of Chemical Feedstocks via Acid Hydrolysis," in *Clean Fuels from Biomass and Wastes,* (Chicago, IL: Institute of Gas Technology, 1977), pp 201–212.
10. Humphrey, A. E. "Economical Factors in the Assessment of Various Cellulosic Substances as Chemical and Energy Resources," *Biotechnol. Bioeng. Symp.* 5:49–65 (1975).
11. Edwards, V. H. "Potential Useful Products fom Cellulosic Materials," *Biotechnol. Bioeng. Symp.* 5:321–338 (1975).
12. Goldstein, I. S. "Potential for Converting Wood into Plastics," *Science* 189:847–852 (1975).
13. Miller, D. L. "Ethanol Fermentation and Potential," *Biotechnol. Bioeng. Symp.* 5:345–352 (1975).
14. Sherrard, E. C., and F. W. Kressman. "Review of Processes in the United States Prior to World War II," *Ind. Eng. Chem.* 37(1):5–8 (1945).
15. Plow, R. H., J. F. Saeman, H. D. Turner and E. C. Sherrard. "The Rotary Digester in Wood Saccharification," *Ind. Eng. Chem.* 37(1):36–43(1945).
16. Nikitin, N. I. *The Chemistry of Cellulose and Wood,* J. Schmorak, trans. (Jerusalem: Israel Program for Scientific Translations, 1966), p. 561.
17. Locke, E. G., J. F. Saeman and G. K. Dickerman. "The Production of Wood Sugar in Germany and its Conversion to Yeast and Alcohol," F.I.A.T. Final Report No. 499 (London: H. M. Stationery Office, 1945).
18. Harris, E. E., E. Beglinger, G. J. Hajny and E. C. Sherrard. "Hydrolysis of Wood. Treatment with Sulphuric Acid in a Stationary Digester," *Ind. Eng. Chem.* 37(1):12–23 (1945).
19. Harris, E. E. and E. Beglinger. "Madison Wood Sugar Process," *Ind. Eng. Chem.* 38(9):890–895 (1946).
20. Humphrey, A. E. "The Hydrolysis of Cellulosic Materials to Useful Products," *Adv. Chem. Ser.* 181:25–53 (1979).

21. Pohjola, V. J. "Dilute-Acid Hydrolysis of Cellulosic Materials. I. Review on Continuous Processes," *Mat. Proc. Technol.* Vol. 15 (1977).
22. Porteous, A. "Towards a Profitable Means of Municipal Refuse Disposal," ASME Pub. 67-WA/PID-2 (1967).
23. Porteous, A. "The Recovery of Fermentation Products from Cellulose Wastes via Acid Hydrolysis," *Octagon Pap.* 3:17–57 (1976).
24. Saeman, J. F. "Hydrolysis of Cellulose and Decomposition of Sugars in Dilute Acid at High Temperature," *Ind. Eng. Chem.* 37(1):43–52 (1945).
25. Fagan, R. D. "The Acid Hydrolysis of Refuse," ME Thesis, Thayer School of Engineering, Hanover, NH (1969).
26. Fagan, R. D., H. E. Grethlein, A. O. Converse and A. Porteous. "Kinetics of the Acid Hydrolysis of Cellulose Found in Paper Refuse," *Environ. Sci. Technol.* 5(6):545–547 (1971).
27. Bergius, F., and F. Koch. "Saccharifying Ground Wood or Similar Vegetable Matter," German Patent 709,758 (July 17, 1941); from *Chem. Abs.* 37:3605 (1943).
28. Dreyfus, H. "Hydrolysing Cellulose," British Patent 424,970 (February 28, 1935).
29. Grethlein, H. E. "Chemical Breakdown of Cellulosic Materials," *J. appl. Chem. Biotechnol.* 28:296–308 (1978).
30. Converse, A. O., H. E. Grethlein, S. Karandikar and S. K. Kuhrtz. "Acid Hydrolysis of Cellulose in Refuse to Sugar and its Fermentation to Alcohol," NTIS Pub. PB-221 239 (1973).
31. Thompson, D. R., and H. E. Grethlein. "Design and Evaluation of a Plug Flow Reactor for Acid Hydrolysis of Cellulose," *Ind. Eng. Chem. Prod. Res. Dev.* 18(3):166–169 (1979).
32. Grethlein, H. E. "Comparison of the Economics of Acid and Enzymatic Hydrolysis of Newsprint," *Biotechnol. Bioeng.* 20:503–525 (1978).
33. Porteous, A., and J. Anderson. "Progress in the Recycling of Cellulosic Wastes by Acid Hydrolysis," paper presented at the International Solid Wastes Association Congress, London, June 18, 1980.
34. Meller, F. H. "Conversion of Organic Solid Wastes into Yeast. An Economic Evaluation," U. S. Public Health Service Publication No. 1909 (1969).
35. Pohjola, V. J., M. Pulkkinen and P. Perttilä. "Dilute-Acid Hydrolysis of Cellulosic Materials. III. Kinetics of a Lignocellulose Residue," *Mat. Proc. Technol.* Vol. 17 (1977).
36. Pohjola, V. J., A. Salminen and P. Perttilä. "Dilute-Acid Hydrolysis of Cellulosic Materials. IV. Experiments on Moss Peat," *Mat. Proc. Technol.* Vol. 18 (1977).
37. Chalov, I. V. et al. U.S.S.R. Patent 283,939 (June 23, 1969).
38. Rugg, B., and W. Brenner. "Utilization of the New York University Continuous Acid Hydrolysis Process for Production of Ethanol from Waste Cellulose," in *Proceedings of the Bio-Energy '80 World Congress and Exposition* (Washington, DC: The Bio-Energy Council, 1980), pp. 160–162.
39. "New Process Converts Cellulose Waste into High Btu Alcohol Fuel," *Reuse Recycle* 10(3):1–4 (1980).
40. "Continuous Cellulose-to-Glucose Process," *Chem. Eng. News* 57:19–20 (1979).
41. Efimov, V. A. "The Present State of Research on the Development of Continuous Processes for the Hydrolysis of Plant Materials with Dilute Sulphuric Acid," *Chem. Abs.* 68:88 297q (1968).

42. Belyaevskii, I. A. "Theory of Continuous Hydrolysis Methods," *Chem. Abs.* 78:59 898 (1973).
43. Starostina, V. A., and I. A. Belyaevskii. "Hydrolysis of Wood Shavings," *Chem. Abs.* 78:59 905n (1973).
44. Molchanova, M. N., I. A. Belyaevskii and V. A. Efimov. "Residence Time of Raw Material in Continuous-Operation Apparatus," *Chem. Abs.* 78:5 635 f (1973).
45. Molchanova, M. N., V. A. Efimov and I. A. Belyaevskii. "Effect of the Density of the Raw Material on Hydrolysis in Continuous-Operation Counter-current Apparatus," *Chem. Abs.* 80:49 455t (1974).
46. Millett, M. A., M. J. Effland and D. F. Caulfield. "Influence of Fine Grinding on the Hydrolysis of Cellulosic Materials—Acid vs. Enzymatic," *Adv. Chem. Ser.* 181: 71–89 (1979).
47. Ladisch, M. R., C. M. Ladisch and G. T. Tsao. "Cellulose to Sugars: New Path Gives Quantitative Yield," *Science* 201:743–745 (1978).
48. Hsu, T. A., M. R. Ladisch and G. T. Tsao, "Alcohol from Cellulose," *Chemtech* 10(5):315–319 (1980).
49. Saeman, J. F., J. L. Bubl and E. E. Harris. "Quantitative Saccharification of Wood and Cellulose," *Ind. Eng. Chem. Anal. Ed.* 17:35–37 (1945).
50. McKee, A. G. and Co. "Preliminary Engineering and Cost Analysis of Purdue/Tsao Cellulose Hydrolysis (Solvent) Process," U.S. Dept. of Energy Publication HCP/T4641-01 (1978).
51. Jones, J. L., S. M. Kohan and K. T. Semrau. "Mission Analysis for the Federal Fuels from Biomass Program. Volume VI. Mission Addendum," U.S. Dept. of Energy Publication SAN-0115-T4 (Menlo Park, CA: SRI International, 1979).
52. Klee, A. J., and C. J. Rogers. "Biochemical Routes to Energy Recovery from Municipal Wastes," *AIChE Symp. Ser.* 74(181):32–37 (1978).
53. Stafford, D. A. and L. Kane-Maguire. "Extraction of Biologically Useful Materials from Sewage Sludges," in *Processes for Chemicals from Some Renewable Raw Materials* (London: The London and South-Eastern Branch of the Institution of Chemical Engineers, 1979), pp. 29–34.
54. Knappert, D., H. Grethlein and A. Converse. "Partial Acid Hydrolysis of Cellulosic Materials as a Pretreatment for Enzymatic Hydrolysis," *Biotechnol. Bioeng.* 22:1449–1463 (1980).
55. Alpay, H. E., M. K. Naushahi and H. Sawistowski, "Conversion of Cellulosic Part of Municipal Waste into Industrial Ethanol," *Pak. J. Sci. Res.* 31(1–2):67–73 (1979).
56. Converse, A. O., and H. E. Grethlein. "Acid Hydrolysis of Cellulosic Biomass," in *Proceedings of Third Annual Biomass Energy Systems Conference* SERI/TP-33-285, (Golden, CO: Solar Energy Research Institute, 1979), pp. 91–95.
57. Emery, A. N., and C. A. Kent. "Fuel Alcohol Production from Biomass," in *Energy from the Biomass,* Report No. 5 (London: The Watt Committee on Energy, 1979), pp. 46–59.
58. Lipinsky, E. S. "Perspectives on Preparation of Cellulose for Hydrolysis," *Adv. Chem. Ser.* 181:1–23 (1979).
59. Finn, R. K. "The Prospects of Fermentation Alcohol from Hydrolyzed Cellulose," *Biotechnol. Bioeng. Symp.* 5:353–355 (1975).
60. Flickinger, M. C. "Current Biological Research in Conversion of Cellulosic Carbodydrates into Liquid Fuels: How Far Have We Come?," *Biotechnol. Bioeng.* 22(Suppl. 1):27–48 (1980).

61. Longo, W. P., and J. S. A. Neto. "Hydrolysis of Cellulosic Materials in Brazil," in *Proceedings of the Bio-Energy '80 World Congress and Exposition* (Washington, DC: The Bio-Energy Council, 1980), pp. 409–411.
62. Falkehag, S. I. "Lignin in Materials," *Appl. Polym. Symp.* 28:247–257 (1975).
63. Funk, H. F. "Recovery of Pentoses and Hexoses from Wood and other Material Containing Hemicellulose, and Further Processing of C_5- and C_6- Components," *Appl. Polym. Symp.* 28:145–152 (1975).
64. Kalninsh, A. Y., and N. A. Vedernikov. "Utilization of Hardwood as a Chemical Raw Material in Latvian SSR," *Appl. Polym. Symp.* 28:125–130 (1975).
65. Lee, Y. Y., C. M. Lin, T. Johnson and R. P. Chambers. "Selective Hydrolysis of Hardwood Hemicellulose by Acids," *Biotechnol. Bioeng. Symp.* 8:75–88 (1978).
66. "Cheap Alcohol from Cellulose," *Business Week* (January 14, 1980).
67. Guha, B. K., G. E. Fowler and A. L. Titchener. "Engineering Evaluation of Chemical Conversion of Wood to Liquid Fuel Alcohol," in *Proceedings of the Alcohol Fuels Conference* (Sydney, Australia: Inst. Chem. Eng. N. S. W. Group, 1978), pp. 8/6–8/11.

CHAPTER 15

STARCH HYDROLYSIS FOR
ETHANOL PRODUCTION

Gerald B. Borglum

Biotechnology Group
Miles Laboratories, Inc.
Elkhart, Indiana

Ethanol and ethanol-gasoline mixtures have been considered for use as fuel since the early days of the automobile. The abundant and less expensive petroleum supply precluded extensive use of ethanol as fuel, and only in the last few years has the general public become aware of and concerned about the dwindling and increasingly expensive petroleum supplies. Interest in extending gasoline supplies with ethanol-gasoline mixtures has increased greatly.

STARCH SOURCES

Ethanol has been produced by anaerobic yeast fermentation of simple sugars since early recorded history. These fermentations used the natural yeast found on fruits and the sugars of these fruits to produce wines. Beer fermentations made use of the amylases of germinating grain to hydrolyze the grain starches to fermentable sugars. Current practices utilize bacterial and fungal amylases to efficiently hydrolyze grain or tuber starch to glucose for fermentation to ethanol.

Starch is a polymer of glucose, having mainly the structure shown in Figure 1. The glucose units are joined by glycosidic bonds between carbon one, the reducing group, and carbon four for nearly all the glucose-to-glucose bonds. The glucosidic bonds are all alpha configuration.

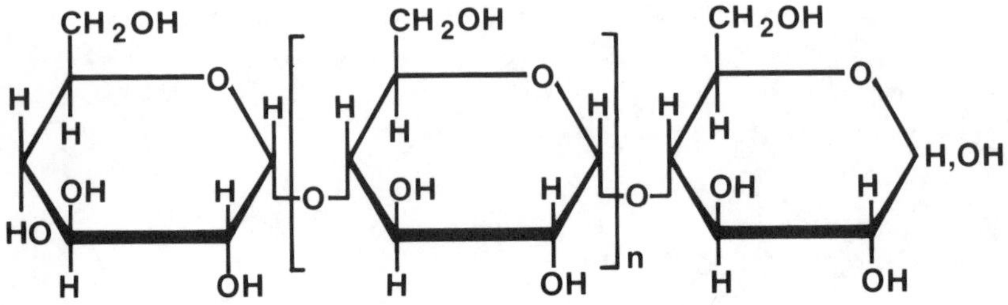

Figure 1.   Diagrammatic representation of amylose. Subscript n has a value of about 220–3700 [2]. Glucose units are linked by α-1,4 glycosidic bonds.

Cereal grain starch is normally a mixture of two types of polymers: amylose, a linear glucose polymer which is represented in Figure 1, and amylopectin, a branched polymer. The branch points in amylopectin are alpha 1,6-bonds. Cereal grains can vary in starch content due to the cultivar analyzed and to the growth environment. The typical compositions of common industrial starch sources are given in Table I. In the United States, corn is the major starch source used commercially. Refined or crude starch, or ground whole corn, can be used as substrate. However, plants constructed for fuel alcohol production typically use ground whole corn.

Starch granules contain both amylose and amylopectin polymers which are associated by hydrogen bonds between carbohydrate hydroxyl groups or through water hydrate bridges between polymer chains. The associated polymers have a highly ordered structure resulting in the observed birefringence when the starch granule is viewed under polarized light. The highly ordered structure is resistant to hydrolysis by bacterial and fungal amylases. Subjecting an aqueous starch suspension to heat disrupts the hydrogen bonds and permits hydration of the polymers, causing the starch granule to swell and lose birefringence under polarized

Table I.   Typical Composition (% dry wt) of Starch Sources

	Corn [1]	Sorghum [1]	Wheat [2]	Potato [3]
Starch	71.5	74.1	74.4	75
Protein	9.9	11.2	14.5	10
Fat	4.8	3.7	1.9	0.3
Ash	1.4	1.5	2.0	4.5
Other[a]	12.4	9.5	7.2	10.2

[a]Fiber, pentosans, sugars.

light. This irreversible swelling is starch gelation which facilitates its attack by enzymes. Typical gelation temperature ranges are: corn, 62–70°C; sorghum, 68–78°C; wheat, 59.5–64°C; and potato, 58–66°C [4]. If a gelatinized starch solution is allowed to cool, the amylose polymers aggregate to form a gel and partially crystallize. The crystallization is called retrogradation; the crystalline portions are resistant to enzymic attack resulting in a lower yield of ethanol.

ENZYMIC HYDROLYSIS

The gelatinization of starch requires high temperatures. Thus, the starch substrate is processed in high solids slurries to minimize energy costs. Refined starch concentrations are typically 25–40 wt % dry solids, and ground whole corn slurries of 20–30 wt % dry solids are used. Gelatinization greatly increases the viscosity. A thermostable bacterial alpha-amylase is added to the slurry before heating to the gelatinization temperature to partially hydrolyze the starch polymers and to reduce viscosity.

Microorganisms produce two types of amylase: endo-amylases (alpha-amylases), which attack alpha-1,4 bonds of starch polymers at random points along the polymer chain, and exo-amylases (glucoamylase, β-amylase), which hydrolyze units of glucose or maltose from the nonreducing end of the starch polymer. Alpha-amylases are produced by bacteria, fungi and germinating cereal grains, but only the bacterial enzymes exhibit the high temperature stability needed for commercial starch liquefaction. Bacterial alpha-amylase is a metal ion–containing protein and requires a small amount of calcium ion during use for maximum stability. The enzyme from *Bacillus licheniformis* exhibits highest activity at 90°C, pH 5.5–7.0 [5]. The enzyme is typically used at 90–100°C, pH 6.0–6.5. Glycosidic alpha-1,4 bonds are hydrolyzed exclusively. The alpha-1,6 branch points are not attacked, but do not halt enzyme attack on the rest of the polymer. A representation of alpha-amylase attack on amylopectin is presented in Figure 2.

Starch produces a blue color with iodine; the enzymic hydrolysis yields successive color changes to purple, red and finally yellow. At the stage the hydrolysis is allowed to proceed in the industrial process, the products are dextrins (oligosaccharides) and small amounts of glucose and maltose. Hydrolysis to the stage that no color change takes place with iodine results in oligosaccharides containing less than 18 glucose units [6]. When the starch hydrolysis has proceeded to the extent that an 18 DE (dextrose equivalent reducing sugar value expressed as percent of total solids) is reached, the liquefied starch will not retrograde on standing. The hydrolysis is often continued only to about 14 DE [7,8]; however,

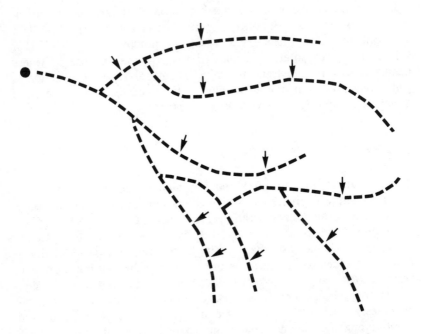

Figure 2. Alpha-amylase action on amylopectin represents the reducing terminus of the polymer; ———, glucose units joined by α-1,4 glycosidic bonds; ⟍⟋, glucose units joined by α-1,6 glycosidic bonds; arrows represent α-amylase attack on α-1,4 glycosidic bonds to form dextrins.

the liquefied starch must be further hydrolyzed with glucoamylase without delay to avoid retrogradation.

The progress of starch hydrolysis in terms of DE as a function of time [8] is presented in Figure 3. The substrate was a 38.8 wt % refined starch with 0.1 wt % DSB (dry starch basis) Taka-Therm® added. It was gelatinized with a laboratory-scale jet cooker at 130–135°C. At plant scales, the higher steam pressures available allow closer temperature control, and 0.02 wt % DSB alpha-amylase is recommended. The enzyme added at this stage serves only to reduce the starch viscosity, and the enzyme is denatured by the high temperature. Enzyme was added to the cooked starch at the levels indicated to hydrolyze the starch to dextrins at 95–100°C.

Liquefied starch is hydrolyzed to glucose with fungal glucoamylase. Commercial glucoamylases are produced by selected strains of *Aspergillus niger, Rhizopus* sp. and *Endomyces* sp. [9]. The *A. niger*

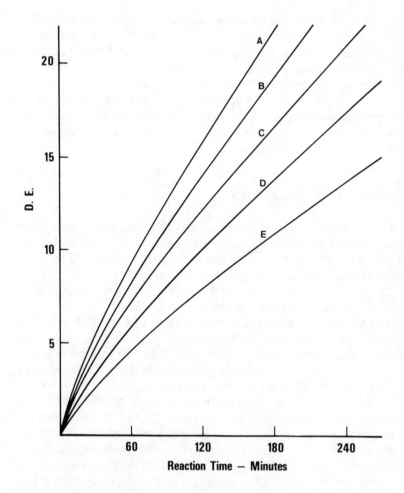

Figure 3. Starch liquefaction DE versus time. Refined starch, 38.8 wt %, jet-cooked at 130–135°C with 0.1 wt % DSB, Taka Therm for gelatinization and then incubated at 95–100°C with additional Taka-Therm at levels of (wt % DSB), (A) 0.13%, (B) 0.11%, (C) 0.09%, (D) 0.07%, (E) 0.05% [8].

enzyme is used extensively in the United States; its optimum use conditions of low pH and high temperature are advantageous for minimizing the growth of microorganisms during saccharification. The use of *A. niger* glucoamylase is described in this paper.

The commercial enzyme product (Diazyme® L-100) contains two glucoamylase isozymes and a minor amount of alpha-amylase [10–12]. The alpha-amylase has optimum activity at more acid pH values than bacterial alpha-amylase, but is less thermoduric than the bacterial enzyme. It has the beneficial effect of aiding the hydrolysis of amylopectin to glucose [12]. The glucoamylase contains no metal ions and requires no cofactors for activity. The optimum conditions for activity are pH 3.8–4.5 and 60°C. The enzyme is an exo-glucosidase that attacks glucose polymers at the nonreducing terminus. The enzyme hydrolyzes both alpha-1,4 and alpha-1,6 bonds; a nearly quantitative yield of glucose is produced from starch.

STARCH LIQUEFACTION

Three methods are used industrially to gelatinize and liquefy the refined or crude starch slurry: atmospheric batch, pressure batch and continuous liquefaction. The terms refer to the method of heating the starch slurry to gelatinize the starch. With refined starch, slurry concentrations of 25–35 wt % solids are used for the atmospheric batch process; 25–40 wt % solids can be processed by the other two methods. Ground corn or other ground grain mashes are more difficult to process, and a solids concentration of 20–30 wt % is recommended. Holt et al. [13] recommend holding a ground corn slurry at 60°C for 1.5 hr to allow the starch time to absorb water. The refined or crude starch slurry is adjusted to pH 6.0–6.5 with a lime slurry. The lime provides calcium ion in addition to pH adjustment; sufficient calcium ion for enzyme stability may be provided by pH adjustment for refined starch slurries. Ground corn and other grains contain phytic acid, which reacts with calcium at pH 6 and higher pH to form insoluble calcium phytate, making the calcium ion unavailable to the alpha-amylase. For crude starch slurries, calcium ion to 500 ppm may be needed [13]. Lime slurry can be added to give 500 ppm calcium ion in the slurry and sulfuric acid is used to titrate the slurry to pH 6.0–6.5.

Atmospheric Batch

Alpha-amylase is added to the substrate slurry to give a ratio of 0.15% enzyme product (Taka-Therm) to dry starch content (0.15 wt % DSB). The slurry is continuously agitated and heated in a jacketed tank. The rate of heating is controlled to reach 90–95°C in minimum time but slow enough to allow the enzyme time to act on the gelatinizing starch and avoid too rapid thickening which would prevent efficient stirring. The mash is held at 90–95°C until 14 DE is reached or an iodine test gives a light red to yellow color indicating complete liquefaction. About 1.5 hr at 90–95°C is required.

Pressure Batch

Slurries having higher percent solids can be processed at higher than atmospheric pressures. A portion of the alpha-amylase (0.05 wt % DSB) is added to the agitated starch slurry initially to liquefy at the gelatinization temperature. The slurry is heated to 72°C, held if necessary until a viscosity break (decrease) occurs, and then heated to 140–163°C and held for 15 min. The mash is flash-cooled to 90–95°C. The enzyme added initially does not survive this heat treatment, and an additional 0.1% wt % DSB alpha-amylase is added to complete the liquefaction as in the atmospheric batch liquefaction process.

Continuous

Continuous starch cooking by injecting steam into a flowing stream of starch slurry (jet cooking) is a simple, efficient and economical method for achieving the benefits of pressurized starch gelatinization without the expense of pressure vessels [14,15]. Batch liquefaction is described below; however, the liquefaction vessel could be designed to permit passage of the substrate at the proper rate so liquefaction is complete at the bottom of the vessel (see Blanchard [16] for an example of a continuous process).

The jet cooking device can be a type designed to inject steam for heating liquids or be designed specifically for cooking starch slurries. There are several manufacturers of steam jet heaters; the jet heater used in the apparatus described below is a Hydro Heater® manufactured by Hydro Thermal Corporation, Milwaukee, WI. The jet starch cooking apparatus is diagrammed in Figure 4 [17]. The mixing tank holds the starch slurry with 0.02 wt % DSB alpha-amylase added. The steam injection rate and starch slurry pump rate are adjusted to obtain a temperature of 140–163°C. A suitably sized coil connects the jet heater and throttle valve to give a 5-min hold period. The gelatinized and partially thinned starch is flash-cooled to 90–95°C in the jacketed receiver vessel (liquefaction vessel). Additional alpha-amylase, 0.13 wt % DSB, is added to complete the liquefaction as described for the above liquefaction procedures. The three procedures are summarized in Table II.

There are advantages and disadvantages for the three processes. The

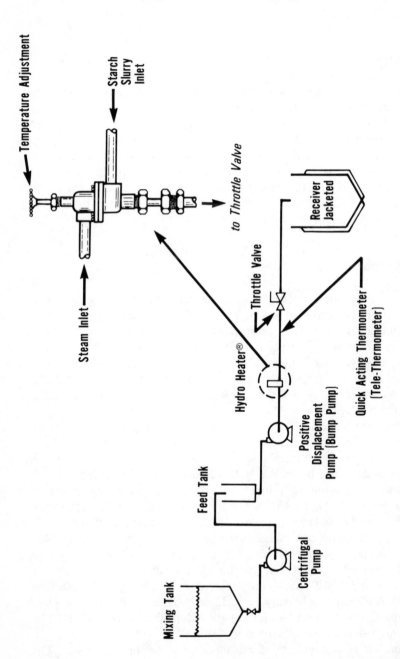

Figure 4. Jet starch cooking apparatus. The length of pipe connecting the Hydro Heater and throttle valve is chosen to yield a 5-min hold period; a quick acting thermometer is attached to monitor temperature. The gelatinized starch is flashed to 90–95°C, atmospheric pressure, in the jacketed receiving tank [17].

Table II.   Taka-Therm Starch Liquefaction

Atmospheric Batch	Pressure Batch	Continuous
0.15 wt % DSB	0.05 wt % DSB	0.02 wt % DSB
Heat to 90–95°C	Heat to 140–163°C (Hold at 72°C, if needed); flash-cool to 90–95°C; add 0.1 wt % DSB Taka-Therm	Jet cook to 140–163°C; flash-cool to 90°C; add 0.13 wt % DSB Taka-Therm
Hold at 90–95°C, 30–90 min	Hold at 90–95°C, 30–90 min	Hold at 90–95°C, 30–90 min

atmospheric batch process does not require pressure equipment and high-pressure steam, which are cost advantages. On the other hand, complete gelatinization and solubilization of high cornstarch concentrations are difficult to accomplish in a short processing time at the lower atmospheric batch process temperature. The jet cooking process permits the use of high solids slurries, more efficient energy use and can be modified for continuous liquefied starch production.

SACCHARIFICATION AND FERMENTATION

At the completion of the liquefaction step, the mash is diluted with water and stillage to 19 wt % solids, cooled to 60°C, and titrated to pH 4–4.5 with acid. Glucoamylase is added to a concentration of 0.22 ml/ 100 g DSB (enzyme concentration recommended for Diazyme L-100 [5]. After 1–2 hr at 60°C, the mash is cooled to 30°C and yeast is added for the fermentation. In this hydrolysis time, sufficient glucose is produced to give 40 DE, sufficient fermentable sugar for the start of the yeast fermentation. The glucoamylase action is considerably slower at 30°C, but the rate of hydrolysis is great enough to keep the yeast supplied with dextrose. An alternative method is to completely saccharify the mash (to at least 95 DE in 36–72 hr) and then dilute, cool and ferment the glucose syrup. The concurrent saccharification and fermentation process described above is complete in about 48 hr, the typical fermentation time using a glucose syrup. Thus, there is no apparent advantage to completely saccharify the starch before fermentation.

The yeast requires a source of nitrogen and vitamins [18]. The ground whole corn supplies nitrogen as proteins and peptides and may supply the required vitamins. Stillage from the ethanol distillation of the fermentation beer is added as part of the dilution water to provide vitamins. Nitrogen can be added to saccharified refined starch as a combination of stillage and an ammonium salt such as diamonnium phosphate. The saccharification and fermentation are summarized in Table III.

Table III.   Saccharification and Fermentation[a]

Concurrent	Consecutive
Dilute to 19 wt % solids; hold 1-2 hr; cool to 30°C and ferment	Hold to 95 + DE (36-72 hr); dilute to 19 wt % solids; cool to 30°C and ferment

[a]Liquefied starch: 60°C; pH 4-4.5; 0.22 ml/100 g DSB; Diazyme L-100.

LABORATORY-SCALE FERMENTATION

The atmospheric batch liquefaction and concurrent saccharification-fermentation processes are illustrated in the following laboratory-scale fermentation. The procedure is given in Table IV.

In choosing the solids concentrations to use in the fermentation, the fermentation capability of the yeast must be considered. The yeast used in the laboratory-scale fermentation can produce 8–10 g ethanol/100 ml before its viability is affected. The sample of ground whole corn used contained 10.2 wt % moisture. Thus, 100 g of ground corn slurry contained 19.9 g of dry solids. Using the average starch content of corn (71.5 wt % DB, Table I), the slurry contained 14.2 g starch, DB. The theoretical ethanol yield is 8.05 g, or 0.567 g ethanol/g starch. A 100-g slurry aliquot would contain 77.8 g water (neglecting the moisture in the corn). The filtrate of the fermentation would theoretically contain 9.38 wt % ethanol, or 9.24 g/100 ml, which is within the ethanol tolerance range of the yeast. For this yeast, a starch concentration giving greater than 10 g/100 ml would result in incomplete consumption of glucose.

In the laboratory study (Table IV), the temperature was raised to 90–95°C in 20 min. After 60 min at 90–95°C, no starch or dextrin color was evident with the iodine test. The reducing sugar level was 24.7 DE. The 2-hr saccharification period gave a reducing sugar concentration of 61 DE. A liquefied starch preparation with 14 DE reducing sugar would require 2 hr saccharification for 40 DE. Ethanol was determined by comparing the density of the filtered mash to the density of ethanol-water solutions prepared from reagent-grade ethanol.

Table IV. Ethanol Production

Liquefaction

    Corn slurry, 22.2 wt % ground whole corn, 500 mg/l calcium as $CaCl_2 \cdot 2H_2O$, NaOH to pH 6.4–6.5, 0.15 wt % DSB, Taka-Therm

    Hold at 60°C, 60 min, with stirring.

    Raise temperature and hold at 90–95°C, 60 min.

Saccharification

    Cool to 60°C, titrate to pH 4.2–4.5 with HCl.

    Add 0.22 ml/100 g DSB, Diazyme L-100.

    Hold at 60°C, 2 hr, cool to 30°C.

Fermentation

    Rehydrate dry yeast,[a] add to 0.3 g/100 ml.

    Ferment anaerobically at 30°C with slow agitation.

[a] Red Star® distillers active dry yeast, Universal Foods Corp., Milwaukee, WI.

The theoretical fermentation yield of 0.567 g ethanol/g starch was stated above. A molecule of water is added across each glycosidic bond in the hydrolysis of starch:

$$1.0 \text{ g starch} + H_2O \longrightarrow 1.11 \text{ g glucose} \qquad (1)$$

From Gay-Lussac's equation (Equation 2), 1.1 g glucose gives a stoichiometric yield of 0.567 g ethanol:

$$C_6H_{12}O_6 \longrightarrow 2 C_2H_5OH + 2 CO_2 \qquad (2)$$
$$\text{1.11 g} \qquad\qquad 0.567 \text{ g}$$

Converting to English units using the density of ethanol, 11.59 lb of starch results in 1 gal of 100% ethanol (1.39 kg starch for 1 liter ethanol). Taking a bushel of corn to be 56 lb (25.4 kg), containing 61 wt % starch (typically containing 14 wt % moisture), a bushel of corn would yield 2.95 gal (11.2 liters) of ethanol.

This theoretical yield does not take into account ethanol production loss due to carbohydrate used for yeast growth and for the formation of small amounts of nonethanol metabolic products by the yeast. A simplified pathway for conversion of glucose to ethanol is shown in Table V. This is the Embden-Meyerhof-Parnas scheme for glycolysis. Glycerol and lactic acid are formed in small amounts compared to ethanol synthesis, but do contribute to a less-than-stoichiometric formation of ethanol from glucose. Allowing for the growth of yeast cells and for the

Table V.   Ethanol Biosynthesis

Glucose
↓
↓
Glyceraldehyde -3-PO$_4$ → Glycerol
↓
↓
Pyruvic Acid → Lactic Acid
↓
↓
Ethanol

---

formation of fermentation by-products, the maximum fermentation efficiency is about 95% of the stoichiometric yield. The fermentation efficiency of the laboratory preparation is:

- yield: 8.0 wt %
- efficiency: 86%
- gallons/bushel: 2.5

assuming a mash of 22.2 g corn; 77.8 ml H$_2$O; 14.2 g starch; 8.05 g ethanol (9.38 wt % or 9.4 g/100 ml). Ethanol yield is commonly reported in gallons per bushel of corn. These values were calculated using the fermentation efficiency and the theoretical yield of 2.95 gal (11.2 liters) of ethanol per bushel of corn derived above.

PRODUCTION COST

The cost of producing one gallon ethanol (Table VI) is made up of substrate, capital and operating costs. The by-product distillers dry grains (DDG) is a significant credit that reduces the manufacturing cost. These factors can vary considerably, and the following assumptions were made for the example below. Corn is used as substrate; the facility would require proper grinding equipment. The price of corn has increased in the past year; the current cash market price was used. A fermentation yield of 2.5 gal ethanol/bu corn was assumed [19]. Plant size and location, fuel costs and plant design affect the capitol and operating costs. The values quoted [19] were derived for a two-million-gal/yr plant and would be lower for a 50 million-gal/yr plant, the typical size plants currently being considered. Enzyme cost is calculated from recommended use level and current price for industrial quantities.

Table VI.   Estimated Production Cost

Material	Unit Price
Corn [20]	$3.20/bu
Taka-Therm	$1.35/lb
Diazyme L-100	$2.63/l
At 2.5 gas ethanol/bu corn, production cost of ethanol	
Corn	$1.28/gal ($0.338/l)
Enzymes	$0.06/gal ($0.016/l)
Capital and operating costs[a] [19]	$0.65/gal ($0.172/l)
Total	$1.99/gal ($0.526/l)
Minus DDG[b] credit	$0.46/gal ($0.122/l)
Total production cost	$1.53/gal ($0.404/l)

[a]2 million gal/yr production.
[b]Distillers dry grains.

The current price of fuel-grade ethanol is $0.489/liter ($1.85/gal); thus an efficient operation is mandatory.

ACKNOWLEDGMENTS

The author wishes to thank Dr. R. E. Pyle for the use of his experimental results and Dr. J. J. Marshall for his helpful review of this manuscript.

REFERENCES

1.   Watson, S. A. "Manufacture of Corn and Milo Starches," in *Starch: Chemistry and Technology, Vol. 2,* R. L. Whistler and E. F. Paschall, Eds. (New York: Academic Press, 1967), pp. 15, 27.
2.   Kerr, R. W., Ed. *Chemistry and Industry of Starch,* 2nd ed. (New York: Academic Press, 1950), pp. 99, 225.
3.   Treadway, R. H. "Manufacture of Potato Starch," in *Starch: Chemistry and Technology, Vol. 2,* R. L. Whistler and E. F. Paschall, Eds. (New York: Academic Press, 1967), p. 90.
4.   Osman, E. M. "Starch in the Food Industry," in *Starch: Chemistry and Technology, Vol. 2,* R. L. Whistler and E. F. Paschall, Eds. (New York: Academic Press, 1950), p. 165.
5.   Miles Laboratories, Inc. "Taka-Therm® and Diazyme® L-100 Microbial Enzymes for Ethanol Production," Product Bulletin L-1161 (1980)
6.   Thoma, J. A. "The Oligo- and Megalosaccharides of Starch," in *Starch: Chemistry and Technology, Vol. 1,* R. L. Whistler and E. F. Paschall, Eds. (New York: Academic Press, 1965), p. 177.

7.   Vance, R. V., A. O. Rock and P. W. Carr. "Starch Liquefaction Process," U.S. Patent 3,654,081 (1972).

8.   Pyle, R. E. Personal communication (1980).

9.   Kooi, E. R., and F. C. Armbuster. "Production and Use of Dextrose," in *Starch: Chemistry and Technology, Vol. 2,* R. L. Whistler and E. F. Paschall, Eds. (New York: Academic Press, 1967), p. 553..

10.   Pazur, J. H., and T. Ando. "The Action of an Amyloglucosidase of *Aspergillus niger* on Starch and Malto-oligosaccharides, *J. Biol. Chem.* 234: 1966–1970 (1959).

11.   Lineback, D. R., I. J. Russell and C. Rasmussen. "Two Forms of the Glucoamylase of *Aspergillus niger," Arch. Biochem. Biophys.* 134:539–553 (1969).

12.   Marshall, J. J. "Some Aspects of Glucoamylase Action Relevant to the Enzymic Production of Dextrose from Starch," in *Food Process Engineering, Vol. 2. Enzyme Engineering in Food Processing,* P. L. Linko, Y. Mälkki, J. Olkku and J. Larinkari, Eds. (Barking, Essex, England: Applied Science Publishing, Ltd., in press).

13.   Holt, N. C., C. Bos and K. V. Rachlitz. "Method of Making Starch Hydrolysates by Enzymatic Hydrolysis," U.S. Patent 3,910,820 (1975).

14.   Denault, L. J., and P. R. Casey. "Process for Starch Liquefaction," U.S. Patent 3,378,462 (1968).

15.   Slott, S., and G. B. Madsen. "Procedure for Liquefying Starch," U.S. Patent 3,912,590 (1975)

16.   Blanchard, P. H. "Starch Thinning Process," U.S. Patent 4,062,728 (1977).

17.   Miles Laboratories, Inc. "Tenase® Jet Process for Starch Liquefaction," Product Bulletin L-1005 (1972).

18.   Reed, G., and H. J. Peppler. *Yeast Technology* (Westport, CT: The AVI Publishing Co., Inc., 1973), pp. 57, 212.

19.   "Small-Scale Fuel Alcohol Production," U.S. Dept. of Agriculture (1980).

20.   "Active Spot Commodities," *J. Commerce* (September 30, 1980), p. 6.

CHAPTER 16

THE NEW YORK UNIVERSITY
CONTINUOUS ACID HYDROLYSIS PROCESS:
HEMICELLULOSE UTILIZATION —
PRELIMINARY DATA AND ECONOMICS
FOR ETHANOL PRODUCTION

Barry Rugg, Peter Armstrong and Robert Stanton
Department of Applied Science
New York University
New York, New York

There has been considerable interest in the utilization of organic wastes for both material and energy recovery. Currently employed methods for dealing with these solid residues are inadequate for cost-effective utilization of their latent energy values; waste conversion via acid hydrolysis to pentose (xylose) and hexoses (glucose and mannose) followed by fermentation to ethanol offers an attractive alternative. Additionally, as is shown in Figure 1, the xylose, glucose and mannose from cellulosic wastes could be used as the basic raw material for the manufacture of many "petrochemicals" or more precisely, volume chemicals which are presently obtained from petroleum feedstocks.

The New York University (NYU) continuous acid hydrolysis process has been utilized for the conversion of crystalline α-cellulose to glucose. Under the rather severe conditions of high temperature required for this process, the amorphous hemicellulose fraction primarily composed of pentosans with some glucan and mannan in hardwoods is converted to sugars and decomposition products (primarily furfural from xylose). It has been the objective of our recent experiments to examine both the technical and economic feasibility of a continuous two-stage hydrolysis,

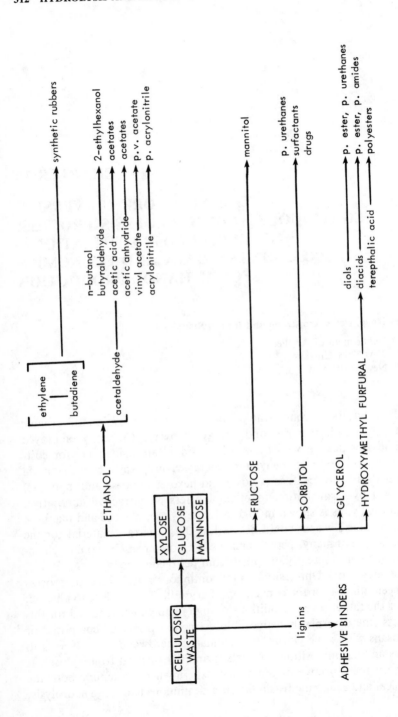

Figure 1.   Possible routes to petrochemicals from cellulosic wastes.

which would allow for more complete utilization of carbohydrate content. Conceptually, the process is shown in Figure 2. By using a mild prehydrolysis and extraction, it is possible to reclaim a greater portion of the hemicellulose fraction as xylose, glucose and mannose. Subsequently the remaining α-cellulose fraction is hydrolyzed continuously to glucose by the usual process.

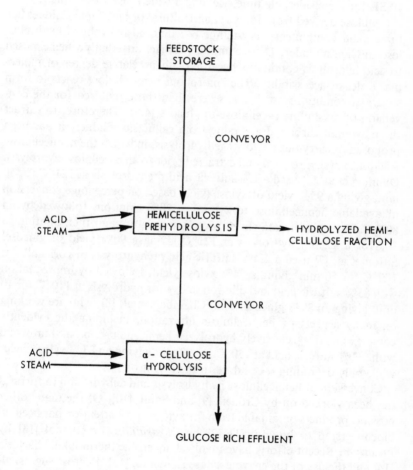

Figure 2.   Schematic of two-stage continuous acid hydrolysis.

HISTORICAL REVIEW

Acid hydrolysis of cellulose has been extensively studied for the better part of a century, particularly in connection with the manufacture of ethanol from wood wastes [1-3]. Attempts to commercialize this technology in Europe and the United States occurred only at wartime when petroleum was restricted. A significant effort has been ongoing in the USSR for a considerable time, yielding a wide range of products.

Cellulose derived from forestry, agricultural or municipal residues has three main components: crystalline α-cellulose, amorphous hemicellulose and lignin binder. These components react differently when exposed to acid hydrolysis conditions because of their relative degree of molecular order or accessibility. The amorphous hemicellulose reacts to form sugars at conditions much less severe than those required for the conversion of crystalline α-cellulose to glucose [4,5]. Therefore, to extract the maximum carbohydrate value from cellulosic residues, it has been proposed to carry out a two-stage hydrolysis, in which the hemicellulose fraction is prehydrolyzed and extracted prior to an α-cellulose hydrolysis. Dunning et al. [6] used a 1% sulfuric acid prehydrolysis at 121°C for 50 min, giving a 95% yield of xylose (yield based on percentage conversion of available hemicellulose to xylose) in 20% solution, followed by an 80% sulfuric acid treatment at room temperature and then dilution to 8% for a 45-min hydrolysis at 121°C giving a 90% yield of glucose. Sitton et al. [7] used a 4.4% sulfuric acid prehydrolysis of cornstalks at 100°C for 50 min giving a 94% xylose yield, followed by impregnation with 85% sulfuric acid and dilution to 8% for hydrolysis at 110°C for 10 min, giving an 89% glucose yield. Chambers et al. [8], who are working on many aspects of hemicellulose utilization, including the evidently complex kinetics, selectively hydrolyzed hemicellulose in oak hardwood with 0.2% sulfuric acid at 150°C for 90 min, giving an 83% xylose yield, while only degrading α-cellulose to the extent of 17%.

Conversion of hemicellulose by hydrolysis and dehydration to furfural has been worked on by Crönert [9] and Smuk [10]. Of the many other possible products obtainable from the xylose, most attention has been on bioconversion to butanol [11,12] by *Clostridium*, or ethanol [13] by *Fusarium*. Recent efforts have involved anaerobic thermophilic bacteria [14] and the use of the enzyme xylose isomerase [15]. Pentose sugars, although not normally fermented to ethanol by any of the large number of yeasts tried, have been used for growth of yeast as fodder [16,17], especially *Torula* yeast, which tolerates the inhibitors in acid hydrolsates.

Hydrolysis of α-cellulose has been reviewed by Harris [5]. Of direct relevance to the present work are the studies of Saeman [2] and later

Fagan et al. [18]. In general, economic and technical factors favor the use of dilute acid at high temperatures for short times to maximize the glucose yield. Fagan et al. obtained a maximum sugar yield of 55% with 0.4% acid at 230°C. The Fagan experiments were carried out with rather small samples (0.5 g) of ball-milled kraft paper and verified Porteous' predictions. Such kinetic studies are of considerable value in the development of improved process designs and economic data for waste cellulose to glucose or ethyl alcohol production facilities [2,18–21]. Grethlein [22] has recently proposed and built a plug-flow pipe reactor in which nearly isothermal conditions can be maintained. Verification of previously developed data based on a batch reactor is currently underway. The kinetic model indicates that high-temperature, short-time, dilute acid hydrolysis reactions favor the production of glucose versus its degradation; yields of 70–80% of the available glucose may theoretically be obtained under ideal conditions [22,23].

PREVIOUS WORK AT NYU

Experimental investigations on dilute acid hydrolysis of cellulose to glucose have been carried out at NYU over the past five years. The feedstock employed in the initial studies was newspaper pulp. This experimental work involved an evaluation of the cost-effectiveness of various pretreatments for enhancing the accessibility of the cellulose and the determination of the optimum reaction conditions for maximizing the sugar yields. Batch hydrolysis experiments were initially carried out with two differently sized stirred stainless steel autoclave reactors. The optimum reaction conditions were determined to be temperatures around 220–230°C and reaction times of less than 30 sec with about 1 wt% sulfuric acid. These results agree quite well with the results of the kinetic rate studies previously reported by Porteous and Fagan [18]. Over the past three years, studies at NYU have resulted in the design, costing, construction and operation of a continuous waste cellulose-to-glucose pilot plant with a nominal capacity of 1–2 ton/day dry feedstock.

The key to successful operation of a continuous acid hydrolysis process is the design of the hydrolysis reactor. The reactor must be capable of feeding, conveying and discharging cellulosic materials continuously while maintaining appropriate temperatures and associated pressures in a reaction zone. Because this hydrolysis requires exposure of the reactor components to dilute acids at high temperatures and pressures, materials of construction have to be resistant to corrosion, especially in the reaction zone. A Werner & Pfleiderer ZDSK 53 (53 mm) twin screw extruder

(Werner & Pfleiderer Corporation, Ramsey, NJ) was selected because of its capacity for conveying, mixing and extruding the required amounts of cellulosic feedstock. This machine allows accurate control of temperature, pressure, residence time, etc., within the previously established acid hydrolysis operating conditions while continuously feeding and discharging material. The equipment was obtained and installed at the Antonio Ferri Laboratories of NYU (Westbury, Long Island, NY), and considerable progress has been achieved in the development and characterization of reaction conditions. Conversions of 50–60% yield based on available α-cellulose have been reported [24–26]. Experiments have been run with diverse feedstocks such as paper pulp (10% solids) and hardwood sawdust (95% solids). It is anticipated that significant increases in yield will result with improved process control.

In recent experiments to improve carbohydrate utilization by continuous acid hydrolysis, the feedstock was a mixed hardwood sawdust. A representative analysis is shown in Table I. Analytical procedures for this complex system are being developed using high-pressure liquid chromatography (HPLC). Initial determinations of sugar yields from hemicellulose, however, utilized a dual-wavelength spectrophotometric technique with orcinol reagent [27]; this and similar early methods are, unfortunately, subject to significant interferences.

Initial prehydrolysis experiments for the hardwood sawdust were directed to the determination of reaction conditions (acid concentration, temperature and residence time) for satisfactory utilization of the hemicellulose fraction. Preliminary findings are presented in Table II. Subsequently, it is proposed to hydrolyze the residual α-cellulose fraction by previously described methods [24–26].

Table I.   Analysis of Mixed Hardwood Sawdust [4]

	wt%, dry basis
α-Cellulose (crystalline)	45
Hemicellulose Glucan	4
Mannan	4
Xylan	17.5
Others	8.5
Lignin	21
Total	100

Table II. Preliminary Data on the Effects of Acid
Concentration and Temperature on the Conversion of
Hemicellulose to Xylose in Hardwood Sawdust

Sulfuric Acid Conc (wt %)	Temperature (°C)	Conversion to Xylose (% of available xylan)
0.25	150	0
0.5	149	6
	164	12
1	153	14
	161	27
2	151	56
	160	75

ECONOMICS

Yield — Sugars and Ethanol

For the purpose of a preliminary analysis to determine the impact on ethanol yield of the proposed two-stage acid hydrolysis, comparative material flowcharts are presented in Figures 3 and 4. The single-stage hydrolysis (Figure 3) assumes a 60% conversion of α-cellulose to glucose and a 58% conversion of hemicellulose to glucose, mannose and xylose [21]. The balance of both the cellulose and hemicellulose fractions are either unreacted cellulose or decomposition products such as furfural or hydroxymethyl furfural. The two-stage hydrolysis (Figure 4) utilizes the same α-cellulose conversion with an 80% prehydrolysis conversion of hemicellulose. For both cases, bioconversion to ethanol is assumed to be 30% for pentose sugars and 45% for hexoses with 98% efficiency in recovery and 99% in distillation. Based on these assumptions, hemicellulose utilization would result in 236 liters of ethanol per dry metric ton of sawdust for the single-stage acid hydrolysis and 263 liter/metric ton for the two-stage continuous process (56.6 and 63.1 gal/ton, respectively).

Analysis

A comparative cost analysis of alcohol plants utilizing a single-stage or two-stage continuous acid hydrolysis process and consuming 1814 metric ton/day of hardwood sawdust is presented in Tables III–V. Table III is a

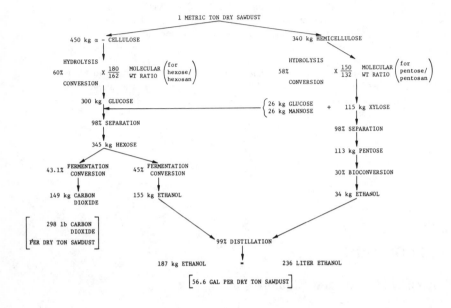

Figure 3.   Yields with single-stage hydrolysis.

production cost analysis which is subdivided into raw material costs, operating costs, overhead costs and by-product credits. The total adjusted manufacturing cost for the single-stage hydrolysis is $0.252/liter ($0.954/gal) for a plant producing 1.427 × 10⁸ liter/yr (37.7 Mgal/yr). The cost for the two-stage hydrolysis is $0.283/liter ($1.07/gal) for a 1.59 × 10⁸ liter/yr (42 Mgal/yr) plant.

This difference in cost is primarily due to the increased energy requirement in the two-stage hydrolysis. Table IV presents a capital cost analysis resulting in costs in 1980 dollars of $81.4 million for the single-stage hydrolysis plant and $94.3 million for the two-stage plant. The major difference here is due to the additional machines and engineering required for the prehydrolysis process.

Post–tax return calculations are presented in Table V for both single- and two-stage hydrolysis processes, with and without CO_2 by-product credit. The principal assumption here is that fuel grade ethanol can be sold at $0.423/liter ($1.60/gal) while competing with gasoline having a refinery price of $0.238/liter ($0.90/gal). This is arrived at by using two

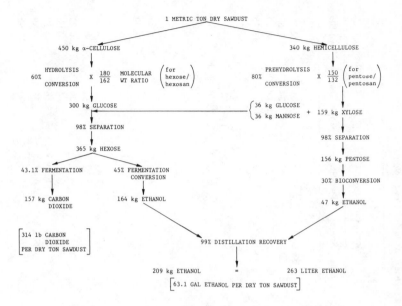

Figure 4. Yields with two-stage hydrolysis.

factors: (1) an allowable 0.79¢/liter (3¢/gal) premium due to 3 octane points increase when using 10% ethanol is equivalent to a 7.9 + ¢/liter (30¢/gal) subsidy; and (2) a 1.06¢/liter (4¢/gal) federal tax elimination on gasohol granted through 1992 is equivalent to a 40¢/gal ethanol subsidy. Payback times for these four cases vary from 4.4 yr (single-stage hydrolysis with CO_2 by-product) to 7.1 yr (two-stage hydrolysis without CO_2), giving credit for depreciation of equipment and a conservative investment tax credit spread over 10 yr.

CONCLUSIONS

Utilization of the hemicellulose fraction of hardwood sawdust has been analyzed for ethanol production with both single- and two-stage continuous acid hydrolysis. It is evident at this time that the increased

320   HYDROLYSIS AND EXTRACTION

Table III.   Production Cost Analysis (Basis: 1814 metric ton/day of Dry Sawdust)[a]

	$/metric ton Sawdust		¢/l Ethanol	
	Single-Stage	Two-Stage	Single-Stage	Two-Stage
Raw Material Costs				
Sawdust: $16.53/metric ton delivered wet, 50% H₂O	33.06	33.06	14.01	12.55
Acid: 98% H₂SO₄ at $66.12/metric ton[b]	0.80	1.61	0.34	0.61
Neutralizing Agent: Ca(OH)₂ at $33.06/metric ton[c]	0.25	0.51	0.11	0.19
Total Raw Material Costs	34.11	35.18	14.46	13.35
Operating Costs—Labor and Maintenance				
Labor—Single-Stage: 30 operators, 7/shift at $21,000; 6 foreman at $24,000; 1 supervisor at $28,000 = $802,000/yr	1.32		0.56	
Labor—Two-Stage: 35 operators, 6/shift at $21,000; 6 foremen at $24,000; 1 supervisor at $28,000 = $907,000/yr		1.50		0.57
Maintenance: material and labor at 6% of inside battery limits (see Table IV) one-stage, $55.8 million = $3.348 million/yr; two-stage, $66.8 million = $4.008 million/yr	5.53	6.63	2.34	2.52
Total Labor and Maintenance Costs	6.85	8.13	2.90	3.09
Operating Costs—Energy				
Hydrolysis: electricity at 4¢/kWh	9.92	17.63	4.20	6.69
steam generated from lignin	1.22	1.32	0.52	0.50
Fermentation: electricity at 4¢/kWh	0.13	0.13	0.06	0.05
cooling water at 5¢/3785 liters	0.10	0.11	0.04	0.04
Distillation: electricity at 4¢/kWh	7.45	11.30	3.16	4.29
steam at $4/454 kg	0.88	0.97	0.37	0.38
cooling water at 5¢/3785 liters				
Total Energy Costs	19.70	31.46	8.35	11.95
Total Operating Costs	26.55	39.59	11.25	15.04

Overhead Costs

Direct Overhead: 40% of labor & supervision (single-stage, $802,000 × 0.40 = $321,000/yr; two-stage, $907,000 × 0.40 = $363,000/yr)	0.53	0.60	0.22	0.23
General Plant Overhead: 65% of labor, supervision & maintenance (single-stage, $4.150 million × 0.65 = $2.698 million/yr; two-stage, $4.915 million × 0.65 = $3.195 million/yr)	4.46	5.29	1.89	2.01
Insurance and Property Taxes: 1.5% of total investment (single-stage, $81.4 million × 0.015 = $1.221 million/yr; two-stage, $94.3 million × 0.015 = $1.415 million/yr)	2.02	2.34	0.85	0.89
Total Overhead Costs	7.01	8.23	2.96	3.13
Total Manufacturing Cost	67.67	82.99	28.67	31.51
By-Product Credit: CO_2 at 5.5¢/kg	-8.19	-8.65	-3.47	-3.28
Total Adjusted Manufacturing Cost	59.48	74.34	25.20	28.23

[a] Plant summary and conversion factors. *One-stage:* production, 142.7 million liter/yr of fuel-grade ethanol, plus 90,700 metric ton/yr of CO_2 by-product; conversion, 60% κ-cellulose to glucose, 58% hemicellulose to sugars. *Two-stage:* production, 159 million liter/yr of fuel-grade ethanol, plus 93,400 metric ton/yr of CO_2 by-product; conversion, 80% hemicellulose to sugars; no CO_2 from pentose. *Both:* throughput, 1814 metric ton/day (604,062 metric ton/yr) of sawdust; conversion, 98% recovery of sugars, 8.6% glucose solution, 45% glucose to ethanol, 30% pentose to ethanol, 43% glucose to CO_2; capacity scaling factor, $(63/57)^{0.7} = 1.072$.

[b] 1.22 and 2.44% of sawdust dry weight for single- and two-stages, respectively.

[c] 1.24 and 2.48% of sawdust dry weight for single- and two-stages, respectively.

Table IV. Capital Cost Analysis (Basis: 1814 metric ton/day of Sawdust Feedstock)

	Installed Cost (million $)	
	Single-Stage[a]	Two-Stage[b]
Materials Handling and Hydrolysis		
Bulk unloading and storage for wood	1.8	1.8
Sawdust handling, grinding and feeding	1.2	1.2
Acid unloading, storage and feeding (two-stage = (63/0.7) × one-stage = 1.374 × one-stage)	1.0	1.4
Water metering	0.2	0.2
Extruder reactors	17.5	17.5
Prehydrolysis reactors		6.5
Engineering	2.0	2.7
Installation of extruders	1.8	2.4
Installation of rest of system	2.5	3.4
Subtotal	28.0	37.1
Separation and Continuous Fermentation	6.4	6.9
Distillation	9.0	9.6
Piping, Refrigeration, Steam and Controls	6.4	6.9
CO_2 Recovery Plant	6.0	6.3
Total Inside Battery Limits	55.8	66.8
Outside Battery Limits	25.6	27.5
Total (site, building, cooling tower, waste disposal/ cleanup, etc.)	81.4	94.3

[a]Fuel-grade alcohol produced: 142.7 million liter/yr. CO_2 by-product produced: 90,700 metric ton/yr.
[b]Fuel-grade alcohol produced 159 million liter/yr. CO_2 by-product produced: 94.4 million metric ton/yr.

output of 1.59×10^8 liter/yr (42.0 million gal/yr) for the two-stage process as compared to $1.427 \times 10_8$ liter/yr (37.7 million gal/yr) for the single-stage system, from the same amount of wood, is not justified, due to both increased capital costs and increased production costs. Future studies on cellulosic waste utilization should therefore concentrate on optimization of the single-stage process for maximum conversion of α-cellulose and hemicellulose, and minimum degradation of the resulting sugars. Little work has been done to date on the hydrolysis of hemicellulose at temperatures as high as those employed here. Further studies are also needed on the relative merits of the various possibilities for converting the sugars produced to alcohol, namely, separation of pentoses from hexoses prior to conversion (no method appears economical yet); successive conversion (requiring removal of ethanol as formed to avoid

Table V. Post-tax Return Calculations—Capital Cost Summary ($ million)

	Single-Stage[a]		Two-Stage[b]	
	Without CO_2	With CO_2	Without CO_2	With CO_2
Inside Battery Limits	49.8	55.8	60.5	66.8
Outside Battery Limits	25.6	25.6	27.5	27.5
Post-tax Return—With CO_2				
1.427×10^8 liter/yr × (42.3 − 25.2) × 55%		13.42		
1.59×10^8 liter/yr × (42.3 − 28.3) × 55%				12.24
Tax credit for depreciation				
[(55.8 million × 10%/yr) + (25.6 million × 2%/yr)] × 45% tax		2.74		
[(66.8 million × 10%/yr) + (27.5 million × 2%/yr)] × 45% tax				3.25
Investment tax credit spread over 10 years				
(10% × 81.4)/10		0.81		
(10% × 94.3)/10				0.94
Total return		16.97/yr		16.43/yr
Post-tax Return—Without CO_2				
142.7 million liter/yr × (42.3 − 28.8) × 55%	10.58			
142.7 million liter/yr × (42.3 − 31.5) × 55%			9.47	
Tax credit for depreciation				
[(49.8 million × 10%/yr) + (25.6 million × 2%/yr)] × 45%	2.47			
[(60.5 million × 10%/yr) + (27.5 million × 2%/yr)] × 45%			2.97	
Investment tax credit spread over 10 years				
(10% × 75.4 million)/10	0.75			
(10% × 88.0 million)/10			0.88	
Total return	13.80/yr		13.32/yr	

[a] Alcohol production, 142.7 million liter/yr.
[b] Alcohol production, 159 million liter/yr.

inhibition of microbes); and simultaneous conversion of pentoses and hexoses (requiring further development of organisms, probably thermophilic bacteria, as well as removal of the product ethanol).

ACKNOWLEDGEMENTS

The authors would like to thank Kuan Ming Ang and Adam Dreiblatt for their efforts in carrying out the experimental and analytical aspects of the study. We would also like to thank Mr. Charles Rogers of the U.S. Environmental Protection Agency for his support of these studies under EPA Grant No. R-805239-030. Finally we would like to acknowledge the faithful and skillful work of Helen Jones in the preparation of the manuscript.

REFERENCES

1. Faith, W. L. "Development of the Scholler Process in the United States," *Ind. Eng. Chem.* 27:9-11 (1945).
2. Saeman, J. "Kinetics of Wood Saccharification," *Ind. Eng. Chem.* 37:43-52 (1945).
3. Forest Products Laboratories Report R 1475 (1945).
4. Rydholm, S. A. *Pulping Processes* (New York: Interscience, 1965).
5. Harris, J. F. "Acid Hydrolysis and Dehydration Reactions for Utilizing Plant Carbohydrates," *Appl. Polym. Symp.* (28):131-144 (1975).
6. Dunning, J. W., and E. C. Lathrop. "The Saccharification of Agricultural Residues," *Ind. Eng. Chem.* 37:24-29 (1945).
7. Sitton, O. C., G. L. Foutch, N. L. Book and J. L. Gaddy. "Ethanol From Agricultural Residues," *Chem. Eng. Prog.* (December 1979), pp. 52-57.
8. Chambers, R. P. et al. "Xylose Recovery from Hemicellulose by Selective Acid Hydrolysis," Proceedings of the International Solar Energy Society Meeting M-111-C-8, 50-53 (1979).
9. Crönert, H., and D. Loeper. "New Industrial Paths in the Continuous Production of Furfural," *Escher Wyss (Corp.) News* (Switzerland) (1969/2-1970/1), pp. 69-77.
10. Smuk, J. M. "Engineering Studies on the Production of Furfural from Aqueous Xylose Solutions," PhD Thesis, University of Wisconsin (1960).
11. Sjolander, N. O. et al. "Butyl Alcohol Fermentation of Wood Sugar," *Ind. Eng. Chem.* 30:1251-1255 (1938).
12. Underkofler, L. A., L. M. Christensen and E. I. Fulmer. "Butyl-Acetonic Fermentation of Xylose and Other Sugars," *Ind. Eng. Chem.* 28:350-354 (1936).
13. Hajny, A. J. "Biological Utilization of Wood for Production of Chemicals and Foodstuffs, " Forest Products Research Paper FPL 385, Madison, WI (1981), pp. 7-12.

14. Ljungdahl, L. G., and J. Wiegel. "Ethanol Fermentation Using Anaerobic Thermophilic Bacteria," Proceedings of the 27th IUPAC Congress, Helsinki, Finland (1979).
15. Gong, C. S., L. F. Chen, M. C. Flickinger, L. C. Chiang and A. T. Tsao. "Fermentation of D-Xylose to Ethanol by Yeasts through Isomerization," Appl. Environ. Microbiol. (submitted).
16. Peterson, W. H., J. F. Snell and W. C. Frazier. "Fodder Yeast from Wood Sugar," Ind. Eng. Chem. 37:30–35 (1945).
17. Harris, E. E. et al. "Fermentation of Wood Hydrolyzates by Torula utilis," Ind. Eng. Chem. 40:1216–1220 (1968).
18. Fagan, R. D., H. E. Grethlein, A. O. Converse and A. Porteous. "Kinetics of the Acid Hydrolysis of Cellulose Found in Paper Refuse," Environ. Sci. Technol. 5:545–547 (1971).
19. Converse, A. O. et al. "A Laboratory Study and Economic Analysis for the Acid Hydrolysis of Cellulose in Refuse to Sugar and Its Fermentation to Alcohol," Final Report under PHS Grant No. UL-00597-02. Thayer School of Engineering, Dartmouth College, Hanover, NH (1971).
20. Fagan, R. D., A. O. Converse and H. E. Grethlein. "The Economic Analysis of the Acid Hydrolysis of Refuse," Thayer School of Engineering, Dartmouth College, Hanover, NH (1970).
21. Grethlein, H. E. "Comparison of the Economics of Acid and Enzymatic Hydrolysis of Newsprint," Biotechnol. Bioeng. XX:503–535 (1978).
22. Thompson, D. R., and H. E. Grethlein. "Design and Evaluation of a Plug Flow Reactor for Acid Hydrolysis of Cellulose," Ind. Eng. Chem. Prod. Res. Dev. 18(3):165–169 (1979).
23. Grethlein, H. E. Thayer School of Engineering, Dartmouth College, Hanover, NH Personal communication (1980).
24. Rugg, B. A., and W. Brenner. "Development of Continuous Acid Hydrolysis Process for the Utilization of Waste Cellulose," in Proceedings of the 5th Annual Research Symposium; Municipal Solid Waste; Resource Recovery, EPA-600/9-79-0236 (1979).
25. Rugg, B. A., and W. Brenner. "Continuous Acid Hydrolysis of Waste Cellulose for Ethanol Production," in Proceedings of the 7th Energy Technology Conference and Exposition, (Washington, DC Government Institutes, Inc., 1980).
26. Rugg, B. A., P. N. Armstrong and R. P. Stanton. "Preliminary Results and Economics of the New York University Process: Continuous Acid Hydrolysis of Cellulose Producing Glucose for Fermentation," Devel. Ind. Microbiol. Vol. 22 (1981).
27. Brown, A. H. "Determination of Pentose in the Presence of Large Quantities of Glucose," Arch. Biochem. II:269–278 (1946).

CHAPTER 17

CHEMICAL FEEDSTOCKS FROM WOOD: AQUEOUS ALCOHOL AND PHENOL TREATMENT

S. M. Hansen and G. C. April

Department of Chemical Engineering
The University of Alabama
Tuscaloosa, Alabama

The current world petroleum crisis has caused renewed interest in alternative and renewable sources of fuel and feedstock chemicals. One such alternative receiving attention involves the separation of wood into its components, cellulose, hemicellulose and lignin, followed by conversion to chemical feedstocks. Treatment of wood with an aqueous organic solvent has been found to be effective in removing lignin and hemicellulose from the cellulosic fibers.

Early work utilized aqueous solutions of organic alcohols to remove lignin and to produce wood pulp [1]. It was determined that aqueous *n*-butanol was the most effective solvent for producing a well-pulped residue and for the removal of lignin. It was found that delignification occurred on solvent extraction in two distinct pseudo-first order stages [2,3]. Delignification was considerably faster during the first part of the batch cook. More recently, continuous alcohol pulping using aqueous ethanol was reported to be economically feasible if the by-products (lignin and hemicellulose hydrolysis products) can be upgraded and sold [4]. A pay-out time of 3.2 yr with a 22% return on investment was calculated for a plant producing 136 metric ton/yr of dry pulp.

This chapter reviews recent work on the delignification of southern yellow pine using aqueous alcohol and phenol. The emphasis of this work is on the rate of lignin removed and the extent of pulp recovery. The use of the data to make preliminary process calculations regarding technical and economic feasibility is also addressed.

METHODS

The batch delignification experiments were conducted in a stirred bomb (shown schematically in Figure 1). The batch reactor is a 300-ml Parr bomb with a mechanical agitator (1725 rpm) and an electrically heated jacket. A Love temperature controller (Model 69-1) is available to maintain the temperature to $\pm 2.5°C$ using an iron-constantan thermocouple as a temperature sensor. The procedure for the batch delignification experiments is to load the bomb with 10 g of wood meal and 150 ml of solvent mixture. The electical heater and agitator are then turned on, and after some heatup period (30–45 min), the system is held at the desired temperature level for the required period. After this period, the heater is removed and the bomb is allowed to cool to room temperature. The residual pulp is separated from the solvent mixture by filtration, and the solvent volume is determined. The wood meal is weighed, dried and reweighed, and the material balances for wood and solvent are calcu-

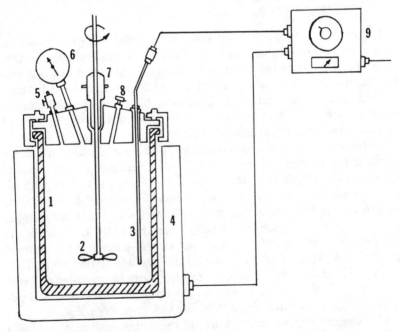

Figure 1. Schematic diagram of Parr bomb unit used for aqueous-organic solvent delignification of wood. (1) Parr minireactor (300 ml); (2) 4-cm-diameter 3-blade propeller; (3) thermocouple (Iron-Constanan); (4) heating mantle; (5) pressure release valve; (6) pressure gauge; (7) cooling jacket for agitator shaft; (8) rupture disk assembly; (9) temperature controller (Love).

lated. The residual pulp can be analyzed for the amount of lignin remaining by standard TAPPI methods. The amount of carbohydrate residue can be determined by difference or independently using TAPPI methods.

The semibatch reactor system is shown schematically in Figure 2. Water and an organic solvent are placed in the reservoirs and are pumped at preselected solvent-to-water ratios using variable-speed pumps. For a solvent system that is a single phase at room temperature, the liquids can be mixed together at the desired ratio and pumped through the system using a single pump. The reactor volume is approximately 200 ml. The system is heated using an electrical furnace and the temperature inside the reactor is determined using an iron-constantan thermocouple. The system temperature is controlled manually using a powerstat. The system pressure is controlled by adjustment of a pressure relief valve in the exit line. The effluent is cooled to room temperature in the condenser and collected in 100-ml samples for analysis. The procedure for the semibatch runs is to load the reactor with 10 g of wood chips. After filling the system with solvent using the pumps, the heater is turned on. When the operating temperature is reached (40–50 min heatup time), the pumps are

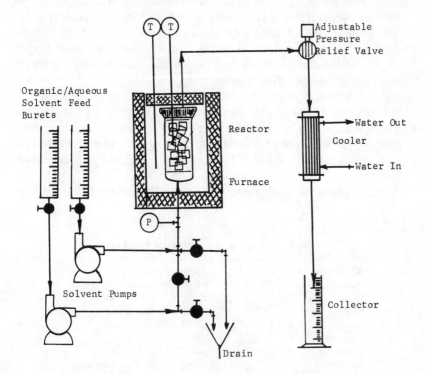

Figure 2.   Schematic diagram of the semibatch delignification system.

again turned on and the system is held at this temperature for the required period. The effluent solvent is collected in 100-ml samples for analysis. At the end of the run, the heater is turned off, but the solvent pumps are left on until the vessel temperature is reduced to a level at which the delignification process rates are negligible (130°C). The same analytical procedures used in the batch procedures are applied in evaluating the residual pulp from the semibatch runs. In addition, the solvent samples from the semibatch runs are analyzed for lignin and carbohydrate content. This analysis provides information about the reaction rate curves for the lignin and carbohydrate components and pinpoints any shift in the relative magnitudes of the removal of these components.

RESULTS AND DISCUSSION

Batch Delignification Studies

A majority of the work reported in the literature has been performed using batch equipment. Initial screening studies at The University of Alabama likewise followed this mode of investigation. A summary of the various conditions investigated is shown in Table I. The results of the batch studies are presented in Table II and graphically illustrated in Figure 3 as the fractional residual pulp versus time-at-temperature. A rapid decrease in the residual pulp during the early stages of the cook is observed. Figure 3 shows the characteristic two-stage process that has been reported as two distinctive, first-order expressions [5,6]. For the systems investigated in this study, the rate constants for each step were calculated to be between 5.0×10^{-3} and 2.0×10^{-2} min^{-1} for the rapid initial step, and between 4.0×10^{-4} and 4.0×10^{-3} min^{-1} for the slower second step.

Figure 4 shows the lignin removed vs time. The percent lignin removed

Table I.   Summary of Experimental Conditions

Wood Type	Southern yellow pine, sweet gum
Wood Size	Meal 0.2–1.65 mm
Solvent Type	Ethanol, butanol, phenol
Solvent/Water	50/50 by volume
Solvent/Wood	15 ml solvent/g wood
Temperature	175–205°C
Pressure	1140–2180 kPa
Additives	Alum, AlCl$_3$, anthraquinone

Table II. Summary of Batch Results

	Solvent	Temperature (°C)	Pressure (kPa)	Time (hr)	% Residual Pulp	% Lignin Removed
Southern Yellow Pine	n-butanol	175	1,140	0.0	85.3	15.0
				0.5	74.1	36.0
				1.0	71.8	36.0
				4.0	67.5	22.0
		205	2,170	0.0	65.3	27.2
				2.0	60.7	36.0
				4.0	56.0	35.0
	Phenol	205	1,490	0.0	58.0	35.9
				0.5	43.0	83.1
				1.0	41.1	84.7
				1.5	39.5	87.0
				2.0	38.0	89.3
				4.0	36.2	89.7
Sweet Gum	n-butanol	175	1,140	0.00	90.8	17.2
				0.67	69.1	45.3
		205	2,170	0.0	69.2	42.2
				0.5	56.5	60.9
				1.0	50.4	78.6
				1.5	47.8	83.6
				2.0	45.7	88.8
				4.0	43.1	85.7

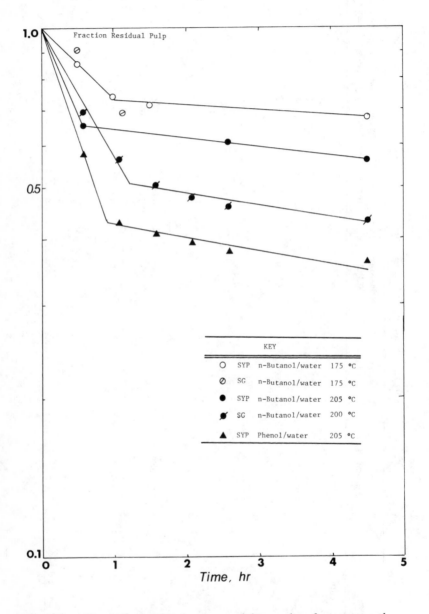

Figure 3. Fraction of residual pulp remaining vs time for aqueous alcohol delignification of sweet gum (SG) and southern yellow pine (SYP).

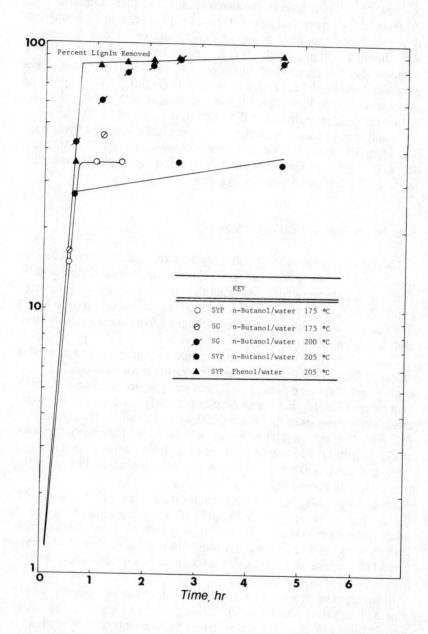

Figure 4.   Percent lignin removed vs time for aqueous alcohol delignification of sweet gum (SG) and southern yellow pine (SYP).

reaches a maximum level in the batch studies and then decreases. This decrease in the amount of lignin removed is likely due to repolymerization and precipitation of removed lignin, and/or to the decrease in solubility of lignin in the solvent as it cools to room temperature. Figure 4 shows that the removal of lignin occurs in several first-order stages paralleling the results noted for residual pulp (Figure 3). Corresponding rate constants for lignin are in good agreement with those values reported for residual pulp ($2.0–6.0 \times 10^{-2}$ min^{-1} and 3.0×10^{-4} to 1.0×10^{-3}) indicating a common mechanism during initial stages of hydrolysis. The reported values for the rate constants are between 1.5×10^{-2} and 3.0×10^{-2} min^{-1} using spruce and poplar wood for times corresponding to the initial first-order removal presented here [2,3].

Semibatch Delignification Studies

To eliminate problems of solubility and repolymerization/redeposition of lignin, and to simulate more closely commercial delignification processing, data were collected using the semibatch apparatus shown in Figure 2. In this scheme, the solvent phase is continuously passed over a fixed bed of woodchips. The results obtained from these runs are summarized in Table III.

These values show that for delignification to proceed at an appreciable rate with southern yellow pine, the temperature must be above 175°C. It is also seen that more lignin is removed using aqueous ethanol as a solvent (over n-BuOH), but the amount of residual pulp decreases due to increases in hydrolysis of the carbohydrates (runs 2 and 8). The results likewise show that anthraquinone has no beneficial effect in aqueous ethanol at 175°C (run 3). The use of AlCl$_3$ as an additive causes a severe loss of the carbohydrates. This loss makes AlCl$_3$ an undesirable additive for the production of wood pulp.

In aqueous n-BuOH, solvent delignification at 175°C, alum (Al$_2$ (SO$_4$)$_3$) had no beneficial effect (run 9). The use of anthraquinone as an additive causes the percent lignin removal to increase from 20.0 to 37.9% with a corresponding decrease in residual pulp from 95.5 to 74.6% (runs 8 and 11). The use of AlCl$_3$ again leads to high lignin removals with rapid hydrolysis of the carbohydrates.

Recovery data from a 4-hr semibatch extraction of southern yellow pine with aqueous ethanol at 205°C are shown in Figure 5. These data were collected by sampling the effluent at 100-ml intervals. The samples were stripped of solvent using a rotary evaporator followed by analysis of the residue for lignin content. The corresponding rate of removal of the lignin was described by a first-order model producing a rate constant

Table III. Semi-Batch Results

Run No.	Solvent	Wood	Temp, °C	Time (hr)	Flow (cm³/min)	% Residual Pulp	% Lignin Removed	Additive[a]
1	EtOH	SYP	175	2.50	7.20	91.1	3.7	
2	EtOH	SG	175	0.67	7.20	74.0	47.3	
3	EtOH	SG	175	0.67	7.20	73.2	48.9	Anthraquinone
4	EtOH	SG	175	0.67	7.20	30.6	97.9	AlCl$_3$
5	EtOH	SYP	205	1.33	3.20	52.1	59.4	
6	EtOH	SYP	205	2.50	7.20	65.4	38.9	KHC$_8$H$_4$O$_4$
7	EtOH	SYP	205	4.00	7.20	47.0	62.2	
8	BuOH	SG	175	0.67	7.20	95.5	20.0	
9	BuOH	SG	175	0.67	7.20	96.8	15.3	Alum
10	BuOH	SG	175	0.67	7.20	43.2	87.2	AlCl$_3$
11	BuOH	SG	175	0.67	7.20	74.6	37.9	Anthraquinone
12	BuOH	SG	200	0.67	7.20	67.8	45.2	
13	BuOH	SYP	205	4.00	5.43	65.0	43.7	
14	BuOH	SYP	205	4.00	6.30	69.3	43.6	

[a] Additive concentration is 0.005 wt %.

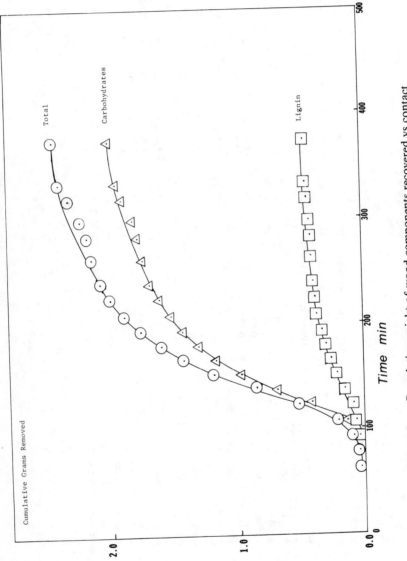

Figure 5.   Cumulative weight of wood components recovered vs contact time in the semibatch system using aqueous ethanol at 205°C.

of 1.5×10^{-2} min^{-1}, which is within the range of values calculated for batch treatment.

Using the above information and data on the physical properties of the solvents, a preliminary mass balance can be made to help evaluate the treatment method. Commercial applications of various process schemes have been proposed for organic solvent delignification including aqueous ethanol extraction [4]. Figure 6 shows a variation of that process modified for n-butanol–water treatment. It is comprised of four sections: extraction, solvent separation, solvent recovery and product recovery. A fifth, pretreatment of the wood, may also be included as a major section if hemicellulose recovery is considered as an option. The tabulated values are based on the recovery data and extraction efficiencies generated in the experimental investigation for n-butanol–water treatment. In general, 1000 kg of wood is needed to produce 650 kg of pulp (25% lignin), 150 kg of dry extracted lignin and 200 kg of hemicellulose hydrolysis products. The pulp could be processed further for use as a paper stock or could be hydrolyzed to glucose. Processes exist for converting glucose to ethanol or n-butanol, making either solvent process totally independent of petroleum sources.

Approximately 1% of the solvent would be retained by the pulp and 1% would be removed with the lignin slurry. Approximately 80–90% of this could be readily removed by steam stripping to give an overall alcohol loss of 25 kg of n-BuOH/1000 kg of wood processed. It is also estimated that 275 kcal/kg of wood processed/extractor would be required assuming a 20% energy loss. This compares quite favorably with the reported value of 268 kcal/kg wood using a battery of nine extractors [4].

CONCLUSIONS

As reported in the literature [2,3,6], batch delignification occurs in two distinct steps which can be described by first order kinetics. The data in the more rapid first step can be described using rate constants between 5.0×10^{-3} and 2.0×10^{-2} min^{-1} for the residual pulp and 2.0×10^{-2} and 6.0×10^{-2} min^{-1} for the lignin removed over the range of conditions investigated. The constants for the second step fall between $4.0 \times 10^{+4}$ and 4.0×10^{-4} and 4.0×10^{-3} min^{-1} and 3.0×10^{-4} and 1.0×10^{-3} min^{-1} for residual pulp and lignin removed, respectively.

Ethanol delignification data for the semibatch delignification can be fit to a single first-order model giving a rate constant of 1.5×10^{-2} min^{-1} for total wood removal to 205°C. This corresponds favorably with the rate constants derived from the first stage batch treatment method.

The effect of the additive tested depended on the solvent system. Aluminum chloride was an unsatisfactory additive in the ethanol and

Table IV.   Tabular Legend for Figure 6

Stream	1	2	3	4	5	6	7	8	9	10
nBuOH (kg)	–	10840	10740	90	9990	750	9910	80	750	–
H_2O (kg)	–	13470	13370	110	2720	10640	2720	–	10450	200
Cellulose (kg)	400	–	–	400	–	–	–	–	–	–
Hemicellulose(kg)	300	–	200	100	–	200	–	–	–	200
Lignin (kg)	300	–	150	150	150	–	–	150	–	–
Total Mass (kg)	1000	24310	24400	850	12860	11590	12630	230	11200	400
Temp (°C)	25	205	205	205	40	40	117	117	100	100
Q kcal x 10^6	0.005	-8.89	–	–	0.518	1.064	3.411	0.005	7.144	0.014

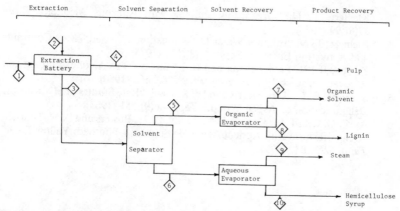

Figure 6.   Process schematic for the aqueous *n*-butanol delignification
of southern yellow pine.

n-butanol systems because of the high loss of residual pulp. The residual
pulp values from the aluminum chloride runs were 30.6 and 43.2% in
ethanol and butanol, respectively, versus 74.0 and 95.5% in the nonaddi-
tive runs at the same conditions. Anthraquinone showed no beneficial
effect in the removal of lignin in ethanol (48.9% vs 47.3% lignin
removed in the conventional run) but showed an increase in the lignin removed in the *n*-butanol system (37.9% vs 20.0% lignin removed in the
conventional run).

A continuous process for the *n*-butanol pulping of wood was shown.
The major process steps include extraction, solvent separation, solvent
recovery and product recovery. This process is comparable to the continuous ethanol process with the exception of the solvent separation step.

ACKNOWLEDGMENTS

The authors would like to thank the National Science Foundation
(Contract No. AER-78-02651) and the School of Mines and Energy
Development, The University of Alabama (Project 102) for their support
of this research effort.

REFERENCES

1.   Aronovsky, S. I., and R. A. Gortner. "The Cooking Process in Pulping
Wood with Alcohols and Other Organic Reagents," *Ind. Eng. Chem.*
28(11):1270 (1936).

2. Kleinert, T. N. "Organosolv Pulping with Aqueous Alcohol," *Tappi* 57(8):99 (1974).
3. Kleinert, T. N. "Ethanol-Water Delignification of Wood – Rate Constants and Activation Energy," *Tappi* 58(8):170 (1975).
4. Katzen, R., R. Frederickson and B. F. Brush. "The Alcohol Pulping and Recovery Process," *Chem. Eng. Prog.* 76(2):62 (1980).
5. Springer, E. L., J. E. Harris and W. K. Neill. "Rate Studies of Hydrotropic Delignification of Aspen Wood," *Tappi* 46(9):551 (1963).
6. April, G. C., M. M. Kamal, J. A. Reddy, G. H. Bowers and S. M. Hansen. "Delignification with Aqueous-Organic Solvents – Southern Yellow Pine," *Tappi* 62(5):83 (1979).

SECTION 4

FERMENTATION ETHANOL

CHAPTER 18

FUEL ALCOHOL PRODUCTION
FROM WASTE MATERIALS

B. A. Friend and K. M. Shahani
Department of Food Science and Technology
University of Nebraska
Lincoln, Nebraska

Whey is the liquid separted from the curds during cheese making [1]. Because cheese production has increased to nearly 1.9 billion kg per year, approximately 17 billion kg of whey is generated annually in the United States. Although considerable advances have been made recently in developing novel approaches for the utilization of surplus whey, more than half of the U.S. total is thrown away or dumped into the sewer (Table I). As shown for Nebraska, only 100 of the 300 million kg of whey produced per year is utilized, while 200 million kg is discarded. In 1979 a Nebraska dairy plant produced 0.8 million kg of cottage cheese and discarded approximately 4.5 million kg of whey into the sanitary sewer. For this, a city sewage charge of $72,000, or 1.6¢/kg of whey was levied.

Approximately 85% of the whey produced in the United States is sweet, and is obtained from the manufacture of hard cheeses such as cheddar, mozzarella and swiss. The remainder, a by-product of cottage cheese, is called acid whey because of its high acid content. As shown in Table II, liquid whey contains 93% water and 7% solids, of which 1% is protein, 5% lactose and 1% fat and salts. Whey protein has a good balance of amino acids and is easily digestible. The lactose in whey is fermentable and the remainder of the nutrients are valuable as animal feed additives.

Dumping huge quantities of whey down the drain increases pollution and/or imposes a huge biochemical oxygen demand (BOD) on wastewater treatment facilities, is costly to the dairy industry and constitutes a

343

Table I.   Annual Production of Cheese Whey

	Fluid Whey Produced (10^9 kg)	Fluid Whey Discarded (10^9 kg)
Nebraska	0.3	0.2
Ohio	0.5	0.3
Wisconsin	5.9	4.4
Other States	10.5	5.5
U.S. Total	17.2	10.4

Table II.   Proximate Composition of Whole Whey,
Sweet Whey and Sweet Whey Permeate[a]

Constituent	Acid Whey [1] (%)	Sweet Whey [1] (%)	Sweet Whey Permeate [2] (%)
Water	93.7	93.5	94.3
Total Solids	6.5	6.4	5.7
Protein	0.75	0.8	0.03
Lactose	4.9	4.85	4.9
Fat	0.04	0.5	<0.01
Ash	0.8	0.5	0.5
Lactic Acid	0.40	0.05	
pH	4.4–4.6	5.9–6.3	

[a]Sweet whey which has been subjected to ultrafiltration to remove protein.

significant loss of a potential food and energy source. Fermentation of the whey lactose to ethanol for use in gasohol (10% alcohol/90% gasoline fuel blend) would eliminate a costly environmental pollutant while extending our limited petroleum reserves.

Fermentation of Whey

As early as 1941, Browne [3] produced ethanol from small batches of cheese whey and noted, "Organoleptic tests indicated that the alcohol produced from whey was of high quality but a detailed examination of the product has not as yet been made." Later studies showed that *Torula cremoris* was the most efficient fermenter of the lactose in whey [4]. Approximately 91% of the theoretical alcohol was produced under laboratory conditions, while only 84% was produced under semiplant conditions.

Bernstein and co-workers [5,6] developed a closed fermentation system for the production of single-cell protein and/or ethanol. *Saccharomyces fragilis* was grown continuously on either acid or sweet whey previously supplemented with growth factors. By spiking the fermentation broth with additional lactose in the form of dried whey, these workers were able to attain 9% alcohol or a conversion efficiency of 90%.

O'Leary and co-workers [7,8] determined the feasibility of hydrolyzing the lactose in whey prior to fermentation. This process allows nonlactose fermenters such as *S. cerevisiae* to be used for alcohol production in whey. Fermentation was found to proceed more slowly, however, in prehydrolyzed acid whey as compared to normal whey [7]. Investigation revealed that the slower rate was the result of diauxic fermentation of glucose and galactose by the lactose fermenter *Kluyveromyces fragilis*.

When acid whey concentrates were prehydrolyzed and fermented with an alcohol-tolerant strain of *S. cerevisiae*, the alcohol yields were increased [8]. These workers noted, however, that the process was wasteful, since galactose, which accounted for almost half of the available sugar, was not fermented. They suggested that a mixed culture of the alcohol tolerant *S. cerevisiae* and a galactose-utilizing organism could improve the commercial feasibility of the process.

A review of the production of alcohols from whey was published by Mann [9]. He noted that extensive investigations on whey fermentation processes and products have been completed in the Netherlands. The studies indicated that whey alcohol was the most feasible product, especially should the price of lactose decrease.

Danish workers [10,11] have developed a continuous two-stage system for the fermentation of whey permeate by *K. fragilis*. Based on a fermentation efficiency of 80%, it was estimated that 42 liters of permeate would be required to produce 1 liter of 100% alcohol. A plant utilizing 500,000 liters of permeate per day could produce the alcohol at a cost of 25-26¢, which is competitive with the traditional fermentation process using molasses. The whey alcohol was found to be comparable in quality to molasses alcohol and could be used directly for industry or rectified for use in alcoholic beverages.

Kosikowski [12] reported that continuous distillation procedures are being used industrially in Ireland and Australia for alcohol obtained from cheese whey. Nevertheless, the relatively low yield of alcohol from whey (2-2.5% alcohol) is considered a major disadvantage. Since cheese whey contains only 4-5% fermentable carbohydrate as compared to 10-18% carbohydrate in other substrates such as molasses, plants which utilize whey incur higher costs for fermentation and distillation equipment and processes.

To make whey fermentation more economical, the use of whey concentrates has been investigated. Gawel and Kosikowski [13] showed that a lactose-fermenting *Kluyveromyces* yeast could be adapted to fermenting the 24% lactose in whey concentrates to ethanol. Up to 12 vol% alcohol was obtained from concentrated, demineralized whey permeate; however, a fermentation time of two weeks was required [13,14].

Burgess and Kelly [15] used whey containing 5-15% lactose and the organisms *K. fragilis* and *Candida pseudotropicalis*. They obtained 6-9% alcohol within 30-36 hr. Laham-Guillaume et al. [16] also obtained 8% alcohol with the same two organisms in concentrated whey. Similary, a recent British patent [17] describes the production of ethanol in concentrated, demineralized permeate with *C. pseudotropicalis*.

This chapter summarizes the production of fuel alcohol from cheese whey alone and in combination with other raw materials.

METHODS

Culture Preparation

Kluyveromyces fragilis and *Saccharomyces cerevisiae* were the organisms used in this study. Stock cultures were maintained on YM Agar slants (Difco) at 4°C. The stock slant was washed with 5 ml of sterile distilled water and the inoculum was aseptically transferred to 500 ml of YM broth (Difco). The YM broth contained, per 1000 ml of distilled water, 20 g yeast extract, 3 g malt extract, 5 g peptone and 50 g carbohydrate. The carbohydrate for *K. fragilis* was lactose and for *S. cervisiae* was glucose. The medium was sterilized at 121°C and 117 kPa for 15 min prior to inoculation. The broth was shaken at 28-30 for 2 days. A 200-ml sample of medium was then centrifuged, washed with 0.01 M KH$_2$PO$_4$/ K$_2$HPO$_4$ buffer and resuspended in sterile water for inoculation. Counts were determined in a Levy counting chamber.

Fermentation Substrates

Sweet whey permeate, prepared by ultrafiltration of mozzarella cheese whey, was obtained from Dodge Dairy, Dodge, NE. Whole acid whey was from Fairmont Foods, Lincoln, NE and whole sweet whey from the University of Nebraska Dairy Plant, Lincoln, NE. Whey samples were frozen and stored at −20°C until used. Elevator run, commercial grain corn from Crete Mills, Crete, NE, was ground to 14 mesh and stored at room temperature. Two potato varieties were obtained from the University of Nebraska Experiment Station, Scottsbluff, NE.

The conversion of whey and whey:grain used α-amylase (Maxamyl 1200) and glucoamylase (G-zyme) from the Enzyme Development Corporation, NY. Similar enzymes (Takatherm and Diazyme L-100) from Miles Laboratories, Elkhart, IN were used for the potato fermentation.

Fermentation of Sweet Whey Permeate

Aliquots (100 ml) of permeate were placed in 250-ml Erlenmeyer flasks, steamed for 15 min, cooled to room temperature and then inoculated with a suspension of *K. fragilis* or a combination of *K. fragilis* and *S. cerevisiae* at a cell concentration of 10^8/ml. A fermentation lock containing distilled water was placed on the flask to ensure anaerobiosis. Incubation was at 28–30°C, a compromise between the optimum growth temperature of *Kluyveromyces* and *Saccharomyces*. At designated time intervals, 5-ml aliquots were aseptically removed and stored at −20°C for later analysis.

Whey:grain Fermentation

The general fermentation scheme is outlined in Figure 1. Corn (100g) was blended with 200 ml of sweet whey permeate and 200 ml of water. Control mashes were also prepared using the standard 100 g of corn/400 ml of water. After blending, the raw mash was autoclaved at 121°C and 117 kPa for 30 min to sterilize the mash and gelatinize the starch. The mash was then cooled to 65°C, and the pH was adjusted to 5.5. A 1-ml volume of each enzyme was blended into the mash, the mixture cooled to 40°C and then inoculated with the appropriate yeast suspension. The mixture was transferred to flasks and incubated as described for the whey fermentation. At each sampling time, appropriate flasks were removed from the incubator, sealed and frozen for later analysis.

For the reduced grain studies, 80 rather than 100 g of corn was blended with 420 ml of undiluted whey permeate and the fermentation was carried out as in the whey:grain process.

Potato Fermentation

A 500-g sample of potatoes was shredded in an Acme juicerator and the volume was made up to 1250 ml with distilled water. Fermentation

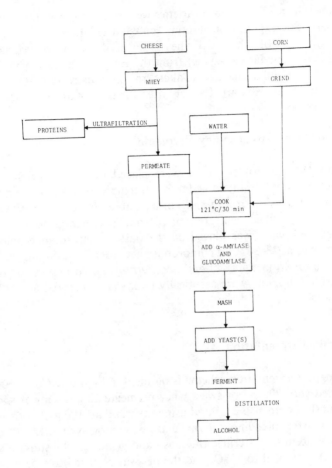

Figure 1. Schematic of combining whey and grain to produce industrial alcohol.

was based on the procedure for grain outlined by Stark et al. [18]. The mash was placed in a large beaker, and the pH was adjusted to 5.6. The α-amylase was added and the mixture heated for 20 min in a water bath. The sample was held in a boiling water bath under constant stirring for 1 hr, then steamed for 30 min, cooled for glucoamylase addition and held for 15 min at 60°C. The pH was then readjusted to 4.5 and the volume was returned to 1250 ml. A 200-ml aliquot was placed in each Erlenmeyer flask and inoculated with yeast at a final titer of $2-5 \times 10^6$ cell/ml. Incubation was at 30°C for 68–72 hr.

Chemical Analyses

A rapid reducing sugar assay by Teles et al. [19] was used to monitor lactose during the fermentation of the whey permeate. Glucose was determined enzymatically with glucose oxidase and peroxidase [20]. Initially, alcohol production was estimated with a distillation-oxidation procedure developed by Aull and McCord [21]. Later samples were distilled in a Kjeldahl unit, and the amount of alcohol in the distillate determined with an immersion refractometer as described by Stark [22].

RESULTS AND DISCUSSION

To optimize the economics of the fermentation process, the highly nutritious and relatively expensive whey proteins were recovered by ultrafiltration and the resulting permeate was used for alcohol production. As shown in Table III, there were no differences in the utilization of lactose or the production of ethanol when either *K. fragilis* or the mixed culture was used. Both groups produced 2% alcohol in 24 hr. Theoretically, 180 g of lactose would be expected to yield 92 g of ethanol and 88 g of carbon dioxide. The 5% lactose in whey would yield approximately 2.5% ethanol, assuming 100% efficiency. The 2% alcohol obtained in our study represents 80% of the theoretical yield and is in agreement with previous reports [5-7,10,11]. Assuming 75% efficiency, the 10.4-billion-kg whey surplus in the United States could produce 260 million liters of ethanol per year. The gasohol potential, therefore, is 2.6 billion liters. According to our estimates (Table IV), the cost would be approximately 28¢/l. These figures, however, are based on the fermentation of dry whey powder, and do not take into account the lack of feasibility of dis-

Table III.   Fermentation Efficiency of Selected Organisms
in Sweet Whey Permeate[a]

Culture	% Residual Lactose		% Ethanol
	12 hr	24 hr	24 hr
K. fragilis	0.21[b]	0.10	2.0
K. fragilis + S. cerevisiae	0.25	0.12	2.0

[a]The whey permeate contained 5.1% lactose initially.
[b]Each value represents an average of two separate trials.

Table IV.   Costs[a] ($) for Producing 1000 Liters of Ethanol
from Dried Whey Powder

Whey costs	289
By-product Credit	-224
Raw Material Cost	65
Direct Costs	
Conversion	105
Loan Interest	34
Indirect Costs[b]	
Depreciation	34
Taxes	21
20% Return	21
Total Costs	$280/1000 liters of ethanol or 28¢/liter

[a] Based on 2400 kg of dried whey at 12¢/kg.
[b] Figures adapted from Scheller [23].

tilling a dilute (2% alcohol) ferment. Low-energy distillation processes are required to make the fermentation of liquid whey more economically feasible. Therefore, a more energy-efficient process involving fermentation of whey and other raw materials was conceived in our laboratory. As shown in Figure 1, the whey permeate replaces all or part of the water required in the preparation of the grain mash. Residual whey solids become part of the distiller's dried grain, and the process requires no equipment modification other than the addition of a whey handling facility at the alcohol plant.

For an initial study, the permeate replaced half of the water in the mash, and samples were inoculated with K. fragilis, S. cervisiae and a mixed culture of K. fragilis and S. cerevisiae. In addition, a control mash prepared with water (no whey) and inoculated with S. cerevisiae was also run to determine the effect of the addition of the fermentable sugars in whey. As shown in Figure 2, by 36 hr, the mixed culture produced the highest level of alcohol, 9.7%; K. fragilis produced 9.4%, S. cerevisiae produced 9.3% and the S. cerevisiae control produced 9.1%. From 36 to 60 hr, the same order of alcohol production was maintained and by 60 hr, all had produced from 9.6 to 10.3% alcohol. The fermentation of grain does not appear to be inhibited by the whey permeate; in fact, production of alcohol is enhanced by the addition of whey at the mashing stage. Since the whey provides significant levels of fermentable sugars, it seemed reasonable to reduce the amount of corn and still obtain sufficient alcohol production. A reduced grain mash was, therefore, prepared with 20% less ground corn than normal and undiluted sweet whey

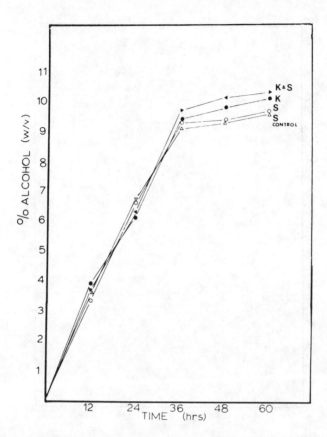

Figure 2.   Alcohol production in a grain system in which half of the water
is replaced by undiluted sweet whey permeate. Each data point represents an average of four trials using duplicate samples. Samples were
inoculated with *K. fragilis* (K), *S. cerevisiae* (S) and a combination of *K.
fragilis* and *S. cerevisiae* (K + S). Mash inoculated with *S. cerevisiae* (S
control) served as the control.

permeate substituted for the water. As shown in Figure 3, the *K. fragilis*
produced 10.9% alcohol by 36 hr and 12.2% by 60 hr. That is in contrast
to 10.1% alcohol produced in the whey:grain system (Figure 2). The total
replacement of water by whey permeate in a reduced grain system appears to be a feasible process. To ensure flexibility in an industrial process, several types of whey were tested. Substitution of whole sweet whey
or whole acid whey for the whey permeate did not appear to affect
alcohol production; all three types of whey produced 12% alcohol by 60
hr.

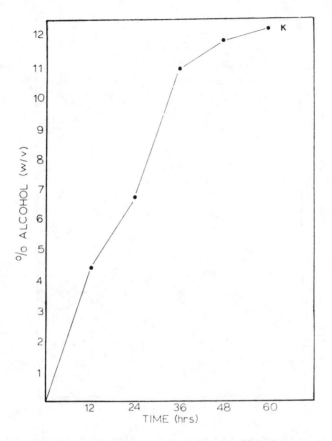

Figure 3.   Alcohol production in a grain system in which the water is replaced by undiluted sweet whey permeate and the concentration of grain
is 20% less than normally used. Each data point represents an average
of four separate trials using duplicate samples.

A feasibility study is currently underway to utilize the Nebraska
whey:grain fermentation process at an 80-million-liter alcohol plant in
Wisconsin. Such a plant would require 160 million kg of corn and 360
million kg of whey per year based on a 15% reduction in the corn requirement. The substitution of whey would amount to a savings of more
than 28 million kg corn per year for a total of nearly 3 million dollars or
3.5¢/l. of alcohol.

At the present, studies to incorporate whey into fermentation processes using other surplus raw materials are in progress. Assuming 90%
conversion of sugars, only 283 gal of alcohol can be produced per acre of
corn. That is in comparison to 300 gal/acre of sweet sorghum, 316 for
potatoes, 495 for sugar beets and 650 for sugar cane. As shown in Table

V, both potatoes and sugar beets are produced in significant quantities in Nebraska.

In preliminary studies, both Norchip and Bounty potatoes yielded 3.3% alcohol or 1 gal/100 lb when fermented with *K. fragilis*. That represented a conversion efficiency of 69%, assuming 29 gal/ton as the theoretical yield (Table V). Block et al. [24] reported efficiencies of 52–88% when fermenting saccharized potato peel waste. Fermentation of whey:potato mixtures with *K. fragilis* and a mixed culture of *K. fragilis* and *S. cerevisiae* is currently being evaluated.

Table V. Annual Supply of Raw Materials in Nebraska and Their Potential Yield of Alcohol

Raw Material	Supply (Million Units)	Conversion Factor	Alcohol Yield (million gal)
Grain	1000 bu	2.6 gal/bu	2600
Whey Solids	0.019 ton	84 gal/ton	1.6
Potato	180 lb	1.4 gal/100 lb	2.5
Sugar Beets	1.5 ton	27 gal/ton	40.5

CONCLUSIONS

Fuel alcohol has been produced from cheese whey using a "trained" lactose fermenting *Kluyveromyces* yeast. The 10.4-billion-kg/yr whey surplus in the United States could produce 260 million liters of alcohol at a cost of 28¢/l. This figure is based on fermentation of dried whey powder and does not include concentration costs.

Whey can also be fermented in conjuction with grain, potatoes or other raw materials. Whey has been successfully substituted for water at the grain mashing stage. Since whey provides fermentable carbohydrate, the grain was reduced 20% with no loss in alcohol production. Studies with the whey:reduced grain system showed that *K. fragilis* produced 10.9% alcohol within 36 hr and 12.2% alcohol in 60 hr when a high-level inoculum was used. A 15% decrease in corn would be expected to decrease the cost of alcohol production by 3.5¢/l.

ACKNOWLEDGMENTS

This work was supported by the Nebraska Water Resources Center Grant No. A-040-NEB and Nebraska Agricultural Products Utilization Commission.

REFERENCES

1. Kosikowski, F. *Cheese and Fermented Milk Foods* (Ann Arbor, MI: Edwards Brothers, Inc., 1977).
2. Cotin, D. "The Utilization of Permeates from UF of Whey and Skim Milk," paper presented at Int. Dairy Fed. Mtg., September 1979.
3. Browne, H. H. "Ethyl Alcohol from Fermentation of Lactose in Whey," *News Ed. Amer. Chem. Soc.* 19:1272, 1276 (1941).
4. Rogosa, M., H. H. Browne and E. O. Whittier. "Ethyl Alcohol from Whey," *J. Dairy Sci.* 30:263-269 (1947).
5. Tzeng, C. H., D. Sisson and S. Bernstein. "Protein Production from Cheese Whey Fermentation," US EPA 600/2-76-224 (1975).
6. Bernstein, S., C. H. Tzeng and D. Sisson. "The Commercial Fermentation of Cheese Whey for the Production of Protein and for Alcohol," in *Proceedings of Biotechnology and Bioengineering Symposium No. 7* (New York: John Wiley and Sons, Inc., 1977), pp. 1-9.
7. O'Leary, V. S., R. Green, B. C. Sullivan and V. H. Holsinger. "Alcohol Production by Selected Yeast Strains in Lactase-Hydrolyzed Acid Whey," *Biotechnol. Bioeng.* 19:1019-1035 (1977).
8. O'Leary, V. S., C. Sutton, M. Bencivengo, B. Sullivan and V. H. Holsinger. "Influence of Lactose Hydrolysis and Solids Concentration on Alcohol Production by Yeast in Acid Whey Ultrafiltrate," *Biotechnol. Bioeng.* 19:1689-1702 (1977).
9. Mann, E. J. "Alcohols from Whey," *Dairy Ind. Int.* 45(3):47-48 (1980).
10. Reesen, L. "Alcohol Production from Whey," *Dairy Ind. Int.* 43:9, 16 (1978).
11. Reesen, L., and R. Strube. "Complete Utilization of Whey for Alcohol and Methane Production," *Process Biochem.* 13:21-22, 24 (1978).
12. Kosikowski, F. V. "Whey Utilization and Whey Products," *J. Dairy Sci.* 62:1149-1160 (1979).
13. Gawel, J., and F. V. Kosikowski. "Improving Alcohol Fermentation in Concentrated Ultrafiltration Permeates of Cottage Cheese Whey," *J. Food Sci.* 43:1717-1719 (1978).
14. Kosikowski, F. V., and W. W. Zorek. "Whey Wine from Concentrates of Reconstituted Acid Whey Powder," *J. Dairy Sci.* 60:1982-1986 (1977).
15. Burgess, K. J., and J. Kelly. "Alcohol Production by Yeast in Concentrated Ultrafiltration Permeate from Cheddar Cheese Whey," *Irish J. Food. Sci. Technol.* 1:107-115 (1977).
16. Laham-Guillaume, M., G. Moulin and P. Galzy, "Selection de souches de levures envue de la production d'alcool sur lactoserum," *LeLait* 59:489-496 (1979).
17. Philliskirk, G. British Patent 1,524,618 (1978).
18. Stark, W. H., S. L. Adams, R. E. Scalfand and P. Kolachov. "Laboratory Cooking, Mashing and Fermentation Procedures," *Ind. Eng. Chem.* 15:443-446 (1943).
19. Teles, F. F. F., C. K. Young and J. W. Stull. "A Method for Rapid Determination of Lactose," *J. Dairy Sci.* 61:506-508 (1978).
20. Barton, R. R. "A Specific Method for Quantitative Determination of Glucose," *Anal. Biochem.* 14:258-260 (1966).

21. Aull, J. C., and W. M. McCord. "Simple Apparatus and Procedure for the Determination of Blood Alcohol," *Am. J. Clin. Pathol.* 43:315-319 (1964).
22. Stark, W. H. "Alcohol Fermentation of Grain," in *Industrial Fermentations, Vol. I,* L. A. Underkoffler and R. J. Hickey. Eds. (New York: Chemical Publishing Co., Inc., 1954).
23. Scheller, W. A. "The Use of Ethanol-Gasoline Mixtures for Automotive Fuel," in *Symposium Papers—Clean Fuels From Biomass and Wastes* sponsored by IGT, Orlando, FL, January 25-28, 1977, pp. 185-200.
24. Bloch, F., G. E. Brown and D. F. Farkas. "Utilization of Alkaline Potato Waste by Fermentation, Amylase Production by *Aspergillus foetidus* NRRL 337, and Alcoholic Fermentation", *Am. Potato J.* 50:357-364 (1973).

CHAPTER 19

ECONOMIC OUTLOOK FOR THE PRODUCTION
OF ETHANOL FROM FORAGE PLANT MATERIALS

A. R. Moreira and J. C. Linden
Department of Agricultural and Chemical Engineering

D. H. Smith
Department of Agronomy
Colorado State University
Fort Collins, Colorado

R. H. Villet
Solar Energy Research Institute
Golden, Colorado

The potential of solar biotechnology is immense, not only for liquid fuels, but also for the range of petrochemical substitutes that can be produced by fermentation [1]. Since fermentation based on easily fermentable substrates such as sugar and starch is established, these materials are being used to produce ethanol for gasohol in the near term. However, the feedstock cost represents a large fraction (more than 50%) of the cost of producing ethanol. If grain prices were to rise dramatically, the final product cost of ethanol would soar. An alternative and relatively cheap substrate is lignocellulose. The processing technology, however, is not fully developed as yet. Lignocellulose is not readily converted because of cellulose crystallinity and also since lignin shields cellulose and hemicellulose from attack by enzymes. The only biological process which has been operated successfully at greater than the bench scale is based on municipal solid waste. In the Emert process [2], ethanol (190 proof) has been produced at 284 liter/day (75 gal/day) from about 1 metric ton/day of waste. The development of alternative processing technology using thermophilic anaerobes, for converting lignocellulose directly to ethanol is being pursued [3,4]. Most cost analyses predict an ethanol production cost well above $0.40/1 [5,6].

In herbaceous plant materials, cell walls are composed of cellulose, lignin, hemicellulose and minor amounts of gums, pectins and other compounds. The major barrier to efficient hydrolysis of cellulose, either by acid or with enzymes, are complexes of lignin and hemicellulose with cellulose. While covalent bonds between these components have been demonstrated [7], limitation of hydrolysis is thought to be primarily due to sheathing of cellulose microfibrils with the lignin-hemicellulose matrix [8]. Access of the hydrolysis catalyst and reactants to the glucosyl linkages is retarded until lignin is removed. Because of the high cost of reducing lignocellulosic complexes to hydrolyzable form, it would seem reasonable to utilize sources of cellulose with minimal lignin content. During the growth and development of plant cells, lignification occurs at a stage after cellulose biosynthesis [9]. This fact suggests that immature vegetable parts of plants may be a source of readily available cellulose.

The possibility of using sorghum fiber for biomass and for papermaking pulp has already prompted numerous agronomic and chemical studies [10-12]. Sweet sorghum is attracting interest in this respect in all agriculturally productive regions of the United States; high-sucrose hybrids suitable even for the northern states are now available. Potential for utilizing sucrose invert sugar and starch as substrates for ethanolic fermentation and for utilizing the fiber as a source of fuel energy, or alternatively, of synthetic gas is promising but is hampered by the relatively poor storability of harvested cane [13]. The practice of ensiling forage materials has interesting potential as a means of storage of the fiber feedstock for alcohol production schemes. During ensiling, the organic acids produced from soluble sugars by the *Lactobacillus* and *Streptococcus* bacteria may cause hemicellulose-lignin sheathing to break down. As a result, the accessibility of water to cellulose for hydration and of enzymes for hydrolysis is reportedly improved [14].

The experimental basis for the economic study described in this chapter consisted of obtaining samples of selected herbaceous plant species and subjecting them to enzymatic hydrolysis. The results of this work have been previously reported [15]. Our objective is to provide a preliminary economic assessment of the alcohol fermentation potential of these species based on projected yields and laboratory results.

METHODOLOGY FOR ECONOMIC ASSESSMENT

Ethanol production costs were obtained for a process similar to the Natick process [6]. A simplified diagram of the processing operations is shown in Figure 1. The process consists of mild mechanical size-reduction of the biomass, cellulase production, enzymatic hydrolysis of the lignocellulosic materials, filtration of the undigested solids and production of 95% ethanol using conventional yeast fermentation and distillation technology. Enzyme hydrolysis is assumed to occur over a 48-hr period at an enzyme load of 10 IU/g of substrate and without enzyme recycle.

While the laboratory hydrolysis data reported in this paper was obtained at an enzyme load of 86.7 IU/g of substrate, it was found that hydrolysis performed at an enzyme load of 8.7 IU/g of substrate over a period of 48 hr gave 95% of the original values. It is thus felt that the hydrolysis conditions used for the plant design will be representative of the laboratory data.

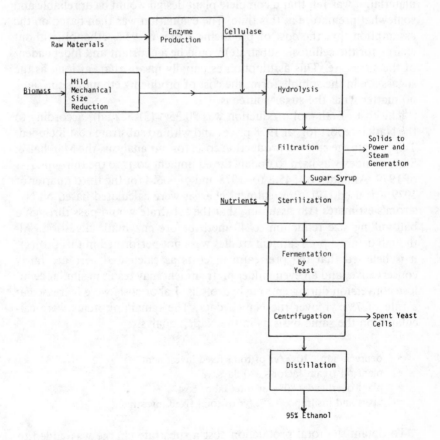

Figure 1. Simplified process flow diagram for ethanol production from vegetative forage crops.

Forage biomass culturing and harvesting costs were charged according to Saterson et al. [16] at the following levels:

- alfalfa, $26.78/metric ton
- sudangrass, $17.75/metric ton
- sorghum (any type), $22.71/metric ton

where the sudangrass cost was estimated assuming an average forage yield of 22.15 metric ton/ha [16] and the same harvesting costs as for sorghum. Figure 2 shows two of the species selected for this analysis.

A preliminary economic evaluation ($\pm 25\%$) was then performed using the Natick information [6]. Since the sole experimental data available were on the 24-hr sugar yield from the enzymatic hydrolysis of the forage material, it was felt that a complete plant design would be unreliable and somewhat premature at this time. The evaluation was then based on the assumption that the cost of producing 1 liter of 95% ethanol (without charge for the cellulosic substrate) would be a constant and independent of the substrate. This assumption essentially means that, as long as the sugars are in the soluble form, the cost of producing ethanol is the same no matter what the sugar source is.

The cost of ethanol production was $0.35/l ($1.32/gal), according to the Natick report [6], at 1978 prices and with no substrate cost included. To generate the ethanol production costs for our analysis, the Marshall & Stevens index was used to update the equipment costs to the third quarter of 1979. An index of 545.3 for 1978 and of 606.4 for the third quarter of 1979 was used [17]. Pretreatment charges were calculated based on Nystrom's estimates [18] assuming that the substrate would pass through a ball-milling size reduction to 40 mesh before enzymatic digestion. Although detailed pretreatment studies were not performed in this project, it is believed that the pretreatment costs as calculated here are fairly conservative, and an even milder pretreatment may result in similar cellulose conversion during enzyme hydrolysis. Labor costs were increased at a rate of 7%/yr over the Natick data. The remaining items were calculated on the same basis as in the Natick analysis:

- depreciation - 10%/yr of total fixed investment
- plant onstream factor - 330 days/yr
- plant overhead - 80% of total labor cost
- taxes and insurance - 2%/yr of total fixed investment

To obtain the total production cost a substrate charge was added to this cost as calculated according to the following formula:

Figure 2. Field-grown sudangrass (top) and forage sorghum (bottom), representing two of the species used in these studies (provided by DeKalb AgResearch, Inc.).

$$\frac{\text{substrate charge}}{(\$/1 \text{ of } 95\% \text{ ethanol})} = \frac{\underset{(\$/\text{metric ton})}{\text{forage crop cost}}}{\underset{(\text{kg/metric ton})}{\text{glucose yield}} \times \underset{(\text{kg/kg})}{\text{ethanol conversion}} \times (1 \text{ liter ethanol}/0.789 \text{ kg})}$$

The main limitation of this economic analysis lies in the fact that a 10% glucose syrup after hydrolysis as assumed in the Natick study may not be possible for all the forage materials included in this work using an enzyme load of 10 IU/g of substrate. This would make a concentration step necessary in some cases; however, since no data were available on the maximum substrate charge possible on the hydrolyzer, no calculations were made in this study for this purpose.

RESULTS AND DISCUSSION

Experimental

Lignin content is related directly to plant maturity. The conversion of the cellulose component of forage crops to glucose by enzymatic hydrolysis is related inversely to the lignin content. Generally, hydrolysis of cellulose from young plant tissues is superior to that from mature tissues. In Tables I and II and the following paragraphs are presented examples of these findings from studies on alfalfa, sudangrass, sorghum silage and brown-midrib sorghum mutants.

Mature alfalfa tissue contains proportionally more lignin than does younger tissue. The percent conversion of cellulose proportionally varies from 41% for the most mature tissue to 84% for the youngest parts of the plant. Fermentable sugar yields from the most easily hydrolyzed top segment of the plants are, however, less than those from the mature bottom segment because of the higher cellulose content of the bottom fraction.

Studies on whole plant samples of half-grown and mature sorghum supported the stated relationships between maturity, lignin content and cellulose hydrolysis. As an example, mature sorghum with 6.5% lignin gave 31% of the theoretical conversion of cellulose while vegetative material with 3.1% lignin gave 47% conversion. Mature sorghum, but not vegetative sorghum, contains considerable fermentable sugars which are extractable from leaves and stalks. The differences were compensating and resulted in similar glucose yields after cellulolytic hydrolysis of mature and of vegetative sorghums.

Ensiling would provide a means of storage of vegetative feedstock and

Table I.   Enzymatic Hydrolysis Products and Theoretical
Conversion of Cellulose to Glucose from Forage Crops at
Various Stages of Maturity

	Total Glucose	Extractable Glucose	Net Hydrolysis[a]	Cellulose Conversion[b]
	(mg glucose/g dry substrate-day)			(%)
Dekalb FS-25A + Sorghum				
Vegetative	155	0	155	47
Mature	151	57	94	31
Silage	188	0	188	71
Frontier 214 Sorghum				
Vegetative	103	0	103	34
Silage	175	0	175	68
Sudangrass, Vegetative	204	64	140	56
Brown-Midrib Mutants of Sorghum (vegetative, field-grown)				
bmr_6	215	61	154	75
bmr_{12}	251	80	171	77
bmr_{16}	236	84	152	68
bmr_{17}	257	74	183	89
bmr_{18}	288	69	159	70
Alfalfa (first cutting, vegetative)				
Top	NA			
Next-to-top	NA			
Next-to-bottom	NA			
Bottom	128	5	123	43
Alfalfa (second cutting, vegetative)				
Top	89	0	89	84
Next-to-top	113	1	112	77
Next-to-bottom	131	1	130	55
Bottom	148	4	144	41

[a] By difference.
[b] Values obtained by dividing net hydrolysis by respective cellulose contents from Table II
and multiplying by 100.

a biological process to improve the conversion of constituent cellulose.
The hydrolysis of the silage of the same sorghum variety described above
resulted in 71% theoretical cellulose conversion as compared to that
from the mature sorghum equal to 31%. Since the lignin content of the
silage was equal to that of the mature material, changes in the fiber structure resulting from ensiling apparently improve accessibility of enzymes
to the fibers. Hydrolysis of the cellulose in silage may be enhanced by the
action of organic acids (resulting pH ~3.8–4.5 in well-ensiled material)

Table II.   Fiber Composition of Forage Sorghum Varieties
(%, dry weight basis)[a]

	Cell-Soluble Material	Acid Detergent Fiber	Hemicellulose	Cellulose	Lignin
Dekalb FS-25A + Sorghum					
Vegetative	40.6	38.9	20.5	33.0	3.1
Mature	37.6	39.0	23.3	30.4	6.5
Silage	38.7	37.0	24.3	26.3	6.7
Frontier 214 Sorghum					
Vegetative	44.2	38.8	17.0	30.3	3.9
Silage	45.1	35.3	19.5	25.9	4.5
Sudangrass, Vegetative	45.6	29.7	24.7	25.2	3.1
Brown Midrib Mutants of Sorghum (vegetative)					
bmr$_6$	51.0	26.5	22.5	20.5	4.4
bmr$_{12}$	48.3	25.4	26.3	22.3	1.9
bmr$_{16}$	51.9	26.9	21.2	22.0	2.5
bmr$_{17}$	50.1	24.2	25.7	20.5	2.2
bmr$_{18}$	51.7	26.4	21.9	22.7	1.9
Alfalfa (first cutting)					
Top	68.0	26.9	5.1	18.6	7.8
Next-to-top	54.9	39.6	5.5	23.8	12.9
Next-to-bottom	49.0	45.4	5.6	26.7	13.6
Bottom	39.4	46.1	14.5	28.9	15.9
Alfalfa (second cutting)					
Top	83.5	15.8	0.7	10.5	4.8
Next-to-top	73.0	25.6	1.4	14.8	8.0
Next-to-bottom	56.8	39.1	4.2	23.7	10.3
Bottom	43.7	50.1	6.2	34.9	13.8

[a]Analysis by permanganate oxidation procedure of Goering and Van Soest [19].

on lignocellulosic structures over time. During enzymatic hydrolysis, the loss of the glucose product to the acid-forming *Lactobacillus* and *Streptococcus* bacteria was prevented by addition of 0.01% (w/v) of agricultural grade tetracycline hydrochloride. This level of antibiotic did not inhibit the fermentation of the hydrolyzed sugars by *Saccharomyces cerevisiae*.

Unlike sorghum, sudangrass in vegetative growth contained considerable amounts of sugars that were extractable from leaves and stalks. Cellulolytic hydrolysis added to the extractable 6.4% glucose and yielded a total of 20.4% fermentable sugar on a dry weight basis. This material contained 3.1% lignin, and the cellulose was converted to 56% of theoretical.

Conversions of cellulose averaging 75% of theoretical were obtained from brown-midrib sorghum mutant lines. The average lignin content of these materials was 2.6%. The literature described mature brown-midrib mutants as having lignin content 61 percent lower than isogenic normal lines [20]. These mutants in vegetative growth contained 7.4% extractable glucose and on hydrolysis yielded a total of 23.7% glucose on a dry weight basis.

Economics

The results obtained by a detailed analysis of the bioconversion process of the various forage materials are shown in Tables III to VIII. Table III shows that the total fixed investment for a 95-million-liter/yr ethanol plant is estimated at about $59 million, or about $0.62/l of installed capacity, which is considered a reasonable figure by most of the researchers working in this area. Startup and working capital estimates bring the total capital investment to about $74 million.

Table IV presents a breakdown of the ethanol production costs from the forage crops without a substrate charge. The processing costs are estimated at $0.33/l, well below the $0.35–0.45/l range reported by other researchers [5,6]. Enzyme production is the major factor in the ethanol cost (47% of the total), followed by fermentation and distillation (26%), hydrolysis (15%) and pretreatment (12%). This finding stresses once more the need for strong research efforts in the area of cellulase production.

Table III. Estimated Capital Investment for a
95-million-liter/yr Ethanol Plant ($1,000s,
Third Quarter 1979)

	Pretreatment	Enzyme Production	Hydrolysis	Ethanol Production	Total
Major Equipment	1,320	17,243	13,350	15,186	47,099
Offsite Investment	462	1,869	108	4,242	6,681
General Service Facilities	179	1,911	1,346	1,943	5,379
Total Fixed Investment	1,961	21,023	14,804	21,371	59,159
Startup (8.5% TFI)					5,029
Working Capital (16.5% TFI)					9,761
Total Capital Investment					73,949

Table IV.   Cost (¢/1) Analysis, Ethanol from Cellulose[a]

	Pretreatment	Enzyme Production	Hydrolysis	Ethanol Production	Total
Total Material	1.40	8.88	0.35	0.51	11.14
Total Utilities	2.10	1.57	1.20	3.16	8.03
Total Direct Labor	0.20	1.37	0.82	1.30	3.69
Total Direct Cost	3.60	11.82	2.37	4.97	22.76
Plant Overhead	0.16	1.10	0.66	1.04	2.96
Tax and Insurance	0.04	0.44	0.31	0.45	1.24
Depreciation	0.21	2.22	1.56	2.26	6.25
Factory Cost	4.01	15.58	4.90	8.72	33.21
Percent Total Cost	12	47	15	26	100

[a] Basis: 95% ethanol, no substrate charge.

Estimates for the ethanol yield from the forage crops included in this study are shown in Table V. These estimates are based on a 45% ethanol yield from glucose during anaerobic fermentation. As expected, sudangrass and the brown-midrib mutants of sorghum show the highest potential with, respectively, 2583 and 2338 liters of ethanol/ha-yr. The ensiled sorghum materials show the second best possibility with an ethanol yield close to 1900 liter/ha-yr. Vegetative Frontier 214 sorghum and vegetative alfalfa rank at the bottom with respectively 1016 and 903 liters/ha-yr.

The estimated total production costs are shown in Table VI. These costs show that vegetative sudangrass and brown-midrib mutants of sorghum are the most promising substrates with the ensiled sorghum crops being the second best. Total ethanol production costs are now at least $0.48/l, with alfalfa and Frontier 214 sorghum reaching $0.72/l of 95% ethanol. A breakdown of the total production costs presented in Table VI can be seen in Table VII. It can be observed that substrate costs represent the major fraction of the total cost, ranging from a minimum of 31% to a maximum of 54%. Enzyme costs rank second, ranging from 22 to 33%, followed by fermentation and distillation costs, which vary from 12 to 18% of the total. Hydrolysis and pretreatment costs represent the minor fraction, varying from 5 to 10% each of the total production costs.

Table VIII shows the estimated total ethanol production costs for a fermentation yield of 50% (weight of ethanol/weight of glucose). As expected, a decrease in the production costs relative to those in Table VI is observed, reflecting the smaller quantity of forage raw materials required for the same ethanol production rate. The decrease averages about 3¢/l and reflects the high cost of the raw materials and the need for efficient substrate conversion at all stages of the process.

Table V.   Estimated Ethanol Yields from Several Forage Materials[a]

Raw Materials	Total Glucose Yield (kg glucose/ dry metric ton-day)	Substrate Yield (metric ton/ha-yr)	Ethanol Yield (liter/ha-yr)
Dekalb FS-25A + Sorghum			
Vegetative	155	17.3	1530
Mature	151	17.3	1490
Silage	188	17.3	1855
Frontier 214 Sorghum			
Vegetative	103	17.3	1016
Silage	175	17.3	1727
Sudangrass, Vegetative	204	22.2	2583
Brown-Midrib Mutants of Sorghum, Vegetative (average)	237	17.3	2338
Alfalfa, Vegetative (average)	120	13.2	903

[a] Basis: ethanol yield during glucose fermentation = 45% on a weight basis.

Table VI.   Estimated Total Ethanol Production Costs from Several Forage Materials[a]

Raw Materials	Substrate Cost ($/metric ton)	Substrate Charge to Ethanol Cost ($/l 95% Ethanol)	Total Ethanol Production Cost ($/l 95% Ethanol)
Dekalb FS-25A + Sorghum			
Vegetative	22.71	0.26	0.59
Mature	22.71	0.26	0.59
Silage	22.71	0.21	0.54
Frontier 214			
Vegetative	22.71	0.39	0.72
Silage	22.71	0.23	0.56
Sudangrass, Vegetative	17.75	0.15	0.48
Brown-Midrib Mutants of Sorghum, Vegetative (average)	22.71	0.17	0.50
Alfalfa, Vegetative (average)	26.78	0.39	0.72

[a] Ethanol processing costs = 33.21¢/l (from Table IV). Ethanol yield during glucose fermentation = 45% on a weight basis.

Table VII. Relative Cost Factor Analysis of
Ethanol Production Costs from Several Forage Materials[a]

Raw Materials	Substrate	Pretreatment	Enzyme Production	Hydrolysis	Ethanol Production
Dekalb FS-25A + Sorghum					
Vegetative	44	7	26	8	15
Mature	44	7	26	8	15
Silage	39	7	29	9	16
Frontier 214 Sorghum					
Vegetative	54	5	22	7	12
Silage	41	7	28	9	15
Sudangrass, Vegetative	31	8	33	10	18
Brown-Midrib Mutants of Sorghum, Vegetative (average)	34	8	31	10	17
Alfalfa, Vegetative (average)	54	5	22	7	12

[a] Figures given are % of total cost.

Table VIII.   Estimated Total Ethanol Production Costs from
Several Forage Materials[a]

Raw Materials	Substrate Charge to Ethanol Cost ($/l 95% EtOH)	Total Ethanol Production Cost ($/l 95% EtOH)
Dekalb FS-25A + Sorghum		
Vegetative	0.23	0.56
Mature	0.24	0.57
Silage	0.19	0.52
Frontier 214 Sorghum		
Vegetative	0.35	0.68
Silage	0.20	0.53
Sudangrass, Vegetative	0.14	0.47
Brown Midrib Mutants of Sorghum, Vegetative (average)	0.15	0.47
Alfalfa, Vegetative (average)	0.35	0.68

[a]Ethanol processing costs = 33.21¢/l (from Table IV). Ethanol yield during glucose fermentation = 50% on a weight basis.

CONCLUSIONS

The production of ethanol by fermentation of the glucose obtained via enzymatic hydrolysis of the vegetative forage crops considered in this study requires further research and development before economic feasibility can be attained. The total production costs ranges from $0.48/l for vegetative sudangrass to $0.72/l for vegetative alfalfa. These high costs are not totally unexpected, since the forage crops considered here have a high cash value. It should be noted that the costs obtained in this study do not account for the use of reducing sugars other than glucose and do not include any by-product credits. If these credits were included, the costs reported in this study could be lowered by as much as 14¢/l. Since only a mild pretreatment is required for the vegetative forage materials, processing costs are at least about 10% lower than other published processing costs [6]. This represents a considerable advantage of vegetative forage crops over other lignocellulosic materials.

Substrate costs constituted, in most instances, the major fraction of the total production costs, varying from 31 to 54%. In view of this, an efficient substrate conversion must be obtained at all stages of the process. Enzyme production costs were also very important, ranging from 22–33% of the total cost; this indicates the need for continued research on cellulase production technology. The total capital investment for a 95-million-liter/yr ethanol plant was found to be about $74 million. This represents a fixed capital investment of about $0.62/l ethanol capacity. To reduce substrate costs, one might either look at less expensive means of culturing and harvesting the crops or coupling to other operations. Examples could be coupling alfalfa hydrolysis to a soluble protein extraction operation or harvesting sorghum grain and stalks simultaneously but separately. Alternatively, one may obtain other substrates whose culture is indigenous to a growing area. Such unconventional plants may have the same processing costs, yet may be obtained for 0-3¢/l of ethanol.

These studies were definitive in showing how hydrolysis and endogenous sugar levels influence the yield of fermentable sugar. This yield is also proportional to the biomass yield. Saterson et al. [16] in work supported under a Department of Energy contract to A. D. Little Coporation and Jackson [21] at Battelle Columbus Laboratories screened herbaceous plants for potential biomass production in 10 regions of the contiguous United States. Many were plants whose culture was indigenous to a growing area. Some were unconventional as food and forage crops, but were good candidates in terms of their projected biomass production potential. Crops appropriate for the Great Plains included 14 species of grasses and legumes and 9 species of unconventional crops and/or weeds. The comparative analysis of Heichel of cultural energy requirements placed such crops high with respect to total energy yield [22]. Sweet sorghum rated highest in that study, but in terms of practical energy recovery, cane storage and juice expression present major difficulties at present [13]. Future crops for alcohol fermentation may include other traditional food crops, certain weeds, syrup sorghum, Jerusalem artichoke and forage grasses. The latter are adapted to a wider range of growing conditions than other crops and are the more productive under adverse conditions. Since they are grown primarily for plant material they are more likely to produce significant yields of biomass than other crops. Warm-season grasses possess the more efficient photosynthesis route, permit multiple cuttings which maintain the plant at a high rate of photosynthesis for a large part of the growing season, have low water requirements, and their culture requires less energy than other crops. The use of such crops as raw materials may bring the cost of fermentation ethanol down to the economically viable range.

The high cost of feedstock is a major barrier to the conversion of biomass to alcohol fuels [4]. To reduce substrate costs, one must optimize the efficiency of either production or conversion. Production costs are reduced when yields are increased, when means of culturing and harvesting are the most energy efficient in terms of cultivation, irrigation and fertilization, and when the harvesting costs can be discounted, as with the simultaneous collection of grain and straw. Conversion costs are reduced when the biomass requires no pretreatment to obtain high percentages of cellulose hydrolysis, when a significant proportion of the plant dry matter is soluble fermentable sugar, and when the fermentation system can utilize both cellulose and hemicellulose hydrolysis products. For these reasons, it is important to study simultaneously the agronomic and biochemical aspects of a potential biological conversion feedstock as a production-conversion system [1]. An advantage gained by the production of great quantities per unit area of biomass is offset if the cellulose is resistant to hydrolysis. On the other hand, materials containing relatively little lignin can be hydrolyzed very efficiently and would be very attractive as feedstock if biomass yields were reasonable. The balance between the potential for production and conversion must be known in a controlled comparative experimental setting.

SUMMARY

In this research project, we have tested vegetative alfalfa, vegetative sudangrass, and vegetative, mature and ensiled sorghum species as possible feedstocks for ethanol production. Results were presented for the yield of sugars via cellulose hydrolysis of these materials and for the projected alcohol production costs for a 95-million-liter/yr plant. These costs ranged from $0.48/l for vegetative sudangrass to $0.72/l for vegetative alfalfa. Substrate costs comprised the major fraction of the total cost. This leads to the conclusion that feasible process economics depend on options such as use of unconventional crops, stillage protein credit, cohydrolysis of starch in immature grain component and sharing of feedstock production cost with mature grain harvest.

REFERENCES

1. Villet, R. H. *Biotechnology for Producing Fuels and Chemicals from Biomass: Recommendations for R&D, Vol. 1,* SERI/TR-332-360 (Golden, CO: Solar Energy Research Institute, 1979).
2. Emert, G. H., and R. Katzen. "Chemicals from Biomass by Improved Enzyme Technology," in *Biomass as a Nonfossil Fuel Source,* D. L. Klass, Ed. (Washington, DC: American Chemical Society, 1981), pp. 213–225.

3. Wang, D. I. C., I. Biocic, H.-Y. Fang and S.-D. Wang, "Direct Microbiological Conversion of Cellulosic Biomass to Ethanol," in *Proceedings of the Third Annual Biomass Energy Systems Conference* (Golden, CO: U.S. Department of Energy, 1979), p. 61.

4. Pye, E. K., and A. E. Humphrey. "Production of Liquid Fuels from Cellulosic Biomass," in *Proceedings of the Third Annual Biomass Energy Systems Conference* (Golden, CO: U.S. Department of Energy, 1979), p. 69.

5. Wilke, C. R., R. D. Yang, A. F. Sciamanna and R. P. Freitas. "Raw Materials Evaluation and Process Development Studies for Conversion of Biomass to Sugars and Ethanol," paper presented at the Second Annual Symposium on Fuels from Biomass, Troy, NY, June 20–22, 1978.

6. Spano, L., A. Allen, T. Tassinari, M. Mandels and D. Y. Ru. "Reassessment of Economics of Cellulase Process Technology: For Production of Ethanol from Cellulose," paper presented at the Second Annual Symposium on Fuels from Biomass, Troy, NY, June 20–22, 1978.

7. Morrison, I. M. "Structural Investigations on the Lignin-Carbohydrate Complexes of *Lolium perenne*," *Biochem. J.* 139:197–204 (1974).

8. Polein, J., and B. Bezuck. "Investigation of Enzymatic Hydrolysis of Lignified Cellulosic Materials," *Wood Sci. Technol.* 11:275–290 (1977).

9. Esau, K. *Plant Anatomy*, 2nd ed. (New York: John Wiley & Sons, Inc., 1953).

10. White, G. A., T. F. Clark, J. P. Craigmiles, R. L. Mitchell, R. G. Robinson, E. L. Whiteley and K. H. Lessman. "Agronomic and Chemical Evaluation of Selected Sorghums as Sources of Pulp," *Econ. Bot.* 28:136–144 (1974).

11. Crookston, R. K., C. A. Fox, D. S. Hill and D. N. Moss. "Agronomic Cropping for Maximum Biomass Production," *Agron. J.* 70:899–902 (1978).

12. Schmidt, A. R., R. D. Goodrich, R. M. Jordan, G. C. Marten and J. C. Meiske. "Relationships among Agronomic Characteristics of Corn and Sorghum Cultivars and Silage Quality," *Agron. J.* 68:403–406 (1976).

13. Lipinsky, E. S., S. Kresovitch, T. A. McClure and W. T. Lawhon. "Fuels from Sugar Crops," report to U.S. DOE, TID-28191 (1978).

14. Leatherwood, J. M., R. E. Mochrie, E. J. Stone and W. E. Thomas. "Cellulose Degradation by Enzymes Added to Ensiled Forages," *J. Dairy Sci.* 46:124–127 (1963).

15. Linden, J. C., A. R. Moreira, D. H. Smith, W. S. Hedrick and R. H. Villet. "Enzymatic Hydrolysis of the Lignocellulosic Component from Vegetative Forage Crops," *Biotechnol. Bioeng. Symp.* (10):199–212 (1980).

16. Saterson, K. A., and M. K. Luppold. "Herbaceous Screening Program—Phase I," DOE Contract ET-78-C-02-5035-A000 (1979).

17. "Marshall & Stevens Equipment Cost Index," *Chem. Eng.* 86(24):7 (1979).

18. Nystrom, J. "Discussions of Pretreatment to Enhance Enzymatic and Microbiological Attack of Cellulosic Materials," *Biotechnol. Bioeng. Symp.* (5):221–224 (1975).

19. Goering, H. K., and P. S. Van Soest. "Handbook No. 379," Agricultural Research Service, U.S. Department of Agriculture, Washington, DC (1970).

20. Fritz, J. O., R. P. Cantrell, V. L. Lechtenberg, J. C. Axtel and J. M. Hertel. "Fiber Composition and IVDMD of Three bmr Mutants in Grain

Grass Type Sorghums," *Abstr. Crop. Sci. Soc. Am. Ann. Mtg.* (1979), p. 128.
21.   Jackson, D. R. "Common Weeds and Grasses May Prove to Be Valuable Sources of Energy," *Biomass Dig.* 1(11):2 (1979).
22.   Heichel, G. H. "Energetics of Producing Agricultural Sources of Cellulose," *Biotechnol. Bioeng. Symp.* (5):43–47 (1975).

CHAPTER 20

PILOT-SCALE CONVERSION
OF CELLULOSE TO ETHANOL

D. K. Becker, P. J. Blotkamp and G. H. Emert
Biomass Research Center
University of Arkansas
Fayetteville, Arkansas

The importance of piloting a complete process for the conversion of cellulose to ethanol was recognized by this laboratory in 1974. The complexity of combining the material handling of bulky slurries such as air-classified municipal solid waste (MSW) and pulp mill waste (PMW) with the aseptic operation of an enzyme production facility posed a unique set of problems which could not adequately be addressed on a laboratory scale. To address these problems, it was believed that the design of a pilot plant should include the flexibility of handling feedstocks of widely varying composition and moisture content. Operation of a pilot plant would allow the identification and testing of equipment for the preparation and transfer of slurries, sterilization, and liquid solid separation.

Conversion of cellulose to chemicals is a research area that many groups have participated in and approached in different ways. A brief review of the groups and processes that have contributed to the cellulose to chemicals technology is appropriate before outlining the objectives of our research.

The Natick Process is based on enzymatic hydrolysis of a cellulosic material such as urban waste, wheat straw or poplar to yield a sugar solution that is subsequently fermented to ethanol [1–5]. The hydrolysis step is separate from the fermentation step. The cellulase enzyme complex used in the hydrolysis step is produced by growing the fungus *Trichoderma reesei* on some cellulosic material in an aerobic fermentation in the presence of appropriate nutrients. The *Trichoderma* cellulases are supplemented with β-glucosidase from *Aspergillus phoenicis*. The cellulosic substrate used in the hydrolysis step (saccharification) is pretreated in some manner to increase the susceptibility of the cellulose to the cellulase enzymes. Pretreatment involves a reduction in particle size, decrease in crystallinity, and a decrease in the degree of polymerization. The reported glucose concentration after hydrolysis is in the range of 100-120 g/l. This yields 40-50 g/l of ethanol when fermented. This work was carried out in vessels as large as 400 liter (287 liter working volume) prepilot-plant fermenters.

The SRI-Wilke process is based on the enzymatic hydrolysis of cellulosic material like newsprint and agricultural wastes such as corn stover [5-8]. This process incorporates a proposed enzyme recycle step that passes the liquid from the hydrolysis step through fresh unreacted substrate to bind enzyme that is in the saccharification effluent. Before hydrolysis, the cellulosic material is shredded and hammer-milled to reduce particle size and the agricultural residue is treated with dilute sulfuric acid to remove the hemicellulose for SCP production, before further hydrolysis is carried out. The resulting sugar solution is concentrated before being fermented with *Saccharomyces cerevisiae*. Conventional distillation is used to recover that ethanol. However, a modification of this is to perform the fermentation under vacuum and condense the vapors followed by further distillation to achieve 95-100% ethanol. Residual solids could be burned to help generate steam and electricity required by the plant. The yeast recovered from the fermentation could be a by-product stream.

The process under development at Purdue involves acid hydrolysis to pretreat cellulosic material including agricultural residues and urban waste [5,9-12]. The first step involves the removal of the hemicellulose by using dilute sulfuric acid. The remaining cellulose and lignin are subjected to 70% sulfuric acid to solubilize the cellulose and lignin, which are precipitated using methanol. The solids are removed and the methanol-sulfuric acid solution is separated by distillation. The precipitated solid fraction is then hydrolyzed using cellulase enzymes with fermentation following hydrolysis.

The University of Pennsylvania-G.E. process uses poplar trees for the proposed feedstock which are chipped and then exposed to a hot, alkaline, aqueous butanol solution to delignify the material [5,13,14]. Cellulose is recovered as solids from this step while the hemicellulose is recovered from the aqueous phase. The xylans can be hydrolyzed using xylanases produced from *Thermomonospora* sp. (a thermoactinomycete that also produces cellulases). The lignin recovered would be in the form of a butanol-lignin slurry which could possibly be substituted for diesel fuel. The cellulose would be hydrolyzed using the cellulases produced from *Thermomonospora*. Three methods of cellulose utilization are proposed. The first is the use of the extracellular enzyme and immobilized cells to provide β-glucosidase activity to saccharify the cellulose followed by fermentation. The next method is a simultaneous saccharification and cellobiose fermentation using *Clostridium thermocellum* at 60°C under partial vacuum. The third method is simultaneous saccharification and fermentation at 30°C using *Clostridium acetobutylicum* to produce butanol. Butanol recovery would be by solvent extraction.

MIT has proposed a process using *Clostridium thermocellum*, a thermophilic anaerobe, for the utilization of cellulosic material [15]. The proposed process would use *C. thermocellum* to simultaneously produce cellulase, hydrolyze the cellulose and ferment the hydrolysate to ethanol in one reactor. A variation of this process is to use a mixed culture to produce other chemicals such as acetic acid, acetone, and butanol.

Auburn University has developed a process primarily for the utilization of hemicellulose [16]. The hemicellulose is hydrolyzed by a percolation reactor. By using dilute sulfuric acid at temperatures ranging from 120 to 160°C, the selective recovery of hemicellulose sugars (pentoses) is almost quantitative. Fermentation to butanediol or butanol is proposed after hydrolysis. Cellulose hydrolysis is also accomplished in a percolation reactor. This process essentially reproduces the Madison process and the current Soviet process [17]. Process improvements include a continuous, high yield, short residence time, multiple-percolation reactor system. The average glucose concentration is about 80 g/l using Solka-Floc as the substrate.

The University of Arkansas process involves the biochemical conversion of cellulose to ethanol and other chemicals. The enzyme system used to hydrolyze the cellulose is produced using *Trichoderma reesei* QM 9414. The crude *T. reesei* culture is used in the hydrolysis step with no purification or concentration steps. The proposed feedstock (cellulosic material) used in the process consists of 75% MSW and 25% PMW. The particular feedstock blend can change depending on the location of the plant and feedstock availability. These feedstocks were chosen on the basis of evaluations carried out on more than 200 potential feedstock samples representing agricultural residues, various waste sludges and MSW. The materials chosen exhibit characteristics that are necessary for use in a process of this type. The greatest quantity of cellulosic material available in a relatively small area is MSW, and PMW in selected areas. A city with a population of approximately 450,000 could supply a commercial scale plant with approximately 1000 ton/day of MSW. Agricultural residues pose a collection problem that increases the cost of this type of feedstock along with the consideration of more intensive pretreatment. The proposed pretreatment of the feedstock is mechanical where it is warranted and is designed to increase the surface area enabling higher rates of hydrolysis to be obtained [18]. Another factor enabling higher rates of hydrolysis is the use of a simultaneous saccharification and fermentation (SSF) step developed in this laboratory [19]. The addition of yeast to the saccharification initially prevents any build-up of glucose that would cause feedback inhibition of the cellulases. Cellobiose does not create a problem in the SSF step because the enzymes produced by *T. reesei* in the U of A process contain ample β-glucosidase to convert all the cellobiose to glucose making supplemental enzymes unnecessary. The solid residue can be used to generate steam and electricity for the plant while the liquid effluent from the process can be concentrated and used as an animal feed supplement.

The economic feasibility of a capital intensive process such as the cellulose-to-ethanol process requires that the use of highly specialized exotic equipment be kept to a minimum. As a result of this, low cost chemical reactors were evaluated as fermentation vessels. The vessels first tested as "off the shelf items" could then be modified as necessary to accommodate the individual requirements of each set of fermentation conditions. In this way paramenters such as agitation, aeration, temperature, pH control and sterility could be evaluated and adjusted as needed. Using these criteria the biochemical conversion of cellulose to ethanol was scaled-up approximately 100 fold from 10-liter laboratory fermenters to 1000-liter vessels in a pilot facility capable of processing 1 ton per day of cellulosic feedstock.

METHODS AND MATERIALS

Three strains of yeast were used during the pilot investigations of SSF (19,20). These were (1) *Saccharomyces cerevisiae* ATCC 4132, obtained from the American Type Culture Collection, Rockville, MD; (2) *Candida brassicae* IFO 1664, obtained from the Institute for Fermentation, Osaka, Japan; and (3) a strain of *Saccharomyces* obtained from Budweiser, Joplin, MO. Stock cultures were stored on Difco YM agar slants at 4°C. Seed cultures of each yeast were prepared by the addition of a portion of a stock culture into a shake flask containing a medium shown in Table I. Chemicals used in media formulations were mostly technical or reagent grade; however, in the past year many of the compounds used were fertilizer or food grade.

Table I.   Yeast Growth Medium (Flask)

	g/l
D-Glucose	20.0
Yeast Extract	5.0
Malt Extract	5.0
Bacto-peptone	5.0

Shake flasks were inoculated at 28°C for 18 hr. The shake-flask culture was used to inoculate a 130-liter fermenter made by Fermentation Design, Inc., containing 100 liters of the medium in Table II. This culture was incubated for 18 hr at 30°C, pH 5.0, with an agitation speed of 120 rpm. The yeast seed culture was harvested into sterilized 15-gal aluminum barrels prior to use in SSF. If the yeast was not used immediately, the barrels were stored in a cold room at 4°C for no longer than 48 hr.

The mold *Trichoderma reesei* QM 9414 was obtained from ATCC. This organism was grown on potato dextrose agar at 29°C until use. *T. reesei* seed cultures were prepared by inoculating shake flasks with a portion of a spore plate. The culture medium used in the shake flasks is shown in Table III. The one-liter shake flasks were scaled up stepwise to 100-liter fermenters. Physical parameters controlled in the fermenters were aeration at 0.5 vol/vol-min and agitation speed at 300 rpm (100-liter fermenter). The seed cultures were incubated for 24 hr and then harvested aseptically into 15-gal aluminum barrels to be transported to the pilot facility where it was used as inoculum for enzyme production.

A 10% v/v inoculum was used for intiation of the cellulase induction state in both batch and continuous phases of enzyme production [21]. The medium used in enzyme production is described in Table IV. Avicel PH 105, comparable to MSW in inducing cellulase enzymes, was chosen as a model substrate because of its ease of handling and uniformity.

Table II.   Yeast Growth Medium (Fermenter)

	g/l
D-Glucose	20.5
$(NH_4)_2SO_4$	1.5
$MgSO_4 \cdot 7H_2O$	0.11
$CaCl_2$	0.06
Cornsteep Liquor	7.5

Table III.   *T. reesi* Growth Medium

	g/L
D-Glucose	20.0
KH₂PO₄	2.0
(NH₄)₂HPO₄	1.23
MgSO₄•7H₂O	1.0
CaCl₂	3.0
FeSO₄	0.05
AnSO₄	0.014
MnSO₄	0.016
CoCo₂	0.04
(NH₄)₂SO₄	2.62
(NH₄)₂CO	1.7
Cornsteep	7.5

Table IV.   *T. reesei* Enzyme Production Medium

	g/l
Cellulose (Avicel 105)	20.0
KH₂PO₄	2.0
(NH₄)₂HPO₄	1.23
MgSO₄•7H₂O	1.0
CaCl₂	3.0
FeSO₄	0.05
ZnSO₄	0.014
MnSO₄	0.016
CoCl₂	0.04
(NH₄)₂SO₄	2.62
(NH₄)₂CO	1.72
Cornsteep Liquor	7.5
Tween 80	0.2%

Avicel PH 105 was obtained from American Viscose Co., Division of FMC, Marcus Hook, PA. The length of incubation of the culture depended on the mode of enzyme production being used. Batch enzyme production lasted 96–120 hr whereas continuous enzyme production has a residence time of 50 hr (D = 0.02). Batch SSF were run for 24 hr unless experimental design dictated otherwise. Semicontinuous SSF were run for 96–120 hr with the residence time varying from 24 to 48 hr. Three major types of feedstocks were used: purified cellulose (Solka floc.), PMW (digester rejects, primary sludges and digester fines), and MSW.

None of the feedstocks received any type of pretreatment before use in the SSF. However, MSW was at times pasteurized depending on experimental conditions. The MSW used in the SSF had been shredded so that it would pass a 4-in. screen and then air classified prior to arrival at the pilot plant.

Assays for measurement of enzyme activity and protein concentration were conducted as described by Blotkamp et al. [19]. Glucose measurements were made with the use of a Yellow Springs Instrument Company Model 23A glucose analyzer. Total reducing sugars were measured by the dinitrosalicylic acid method [22]. Ethanol was analyzed using a Perkin-Elmer Model 3920 B gas chromatograph or a Hewlett-Packard Model 5730 A gas chromatograph equipped with flame ionization detectors, an electronic integrator and a 6-ft column of Porapak Q. Isothermal analysis was performed at 150°C.

Yeast populations were monitored by using dilution plating. Cellulose concentration of samples used in SSF was determined by a modified version of the Van Soest procedures [23,24]. Moisture determinations were performed on an Ohaus moisture balance.

EQUIPMENT

The vessels used for enzyme production and SSF were of 1250 liter capacity manufacture by Pfaudler (L/D = 0.78). Four of the five vessels were capable of aseptic operation. The vessels were constructed of stainless steel with carbon steel jackets. The vessels were fully jacketed for adequate temperature control and sterilization.

All process piping was stainless steel with welded connections except where piping entered the vessel. Flanged fittings with Teflon® gaskets were used at these points. No pumps were used as a precaution against contamination. The liquids and slurries were moved with pressure (sterile air or steam) or gravity. The agitator shafts were equipped with double mechanical seals filled with oil. Enzyme production vessels used two flat blade impellers, each having four blades (Di/Dv = 0.456). Agitation speed was 120 rpm and aeration was 0.5 vol/vol-min, at which the $k_L a$ was 84 hr^{-1} vs 330 hr^{-1} on the lab scale (with water).

The baffle tray stripping column was constructed from 9-in i.d. glass pipe with trays made of Monel to resist corrosion. Associated process lines on the stripper were stainless steel. Pumps were used on the beer feed lines on the stripping column and recirculation loops to maintain solids in suspension.

A brief process flow diagram is presented in Figure 1. After the

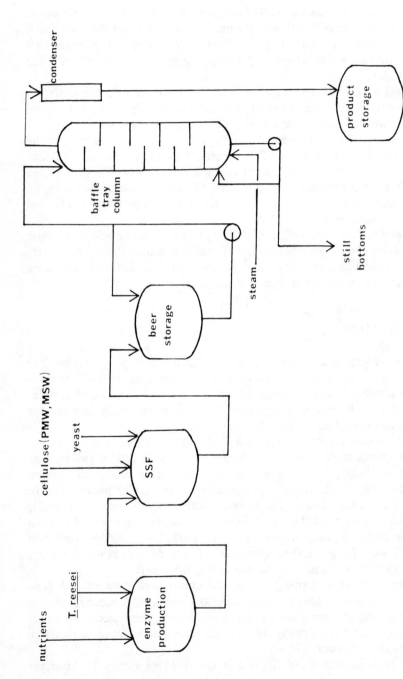

Figure 1.   Process flow diagram of pilot plant.

enzyme production vessels were filled with nutrients and sterilized, the seed inoculum was transferred aseptically from the aluminum barrels to the vessels using nitrogen to pressurize the barrels. From this point, enzyme production could be run in either a batch or continuous mode. When enzyme was ready to be harvested, a portion of the whole culture enzyme broth was transferred to the SSF vessel into which the cellulosic feedstock (PMW or MSW) would be added along with the yeast. The SSF could be run in either batch or semicontinuous modes in which one half of material was transferred out every one half residence time. As the SSF was harvested, the resulting beer slurry was moved to the beer storage tank where it could be pumped into the stripper column for ethanol recovery.

RESULTS

Enzyme Production

Performance of batch enzyme production can be typified by the data presented in Figures 2 and 3. Relatively high levels of protein and β-glu-

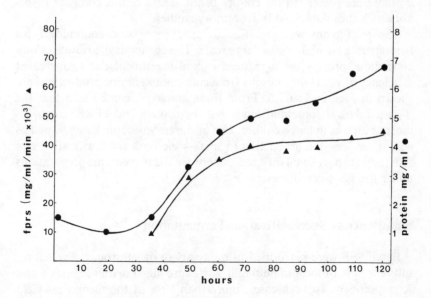

Figure 2. Batch enzyme production FPRS activity and protein concentration.

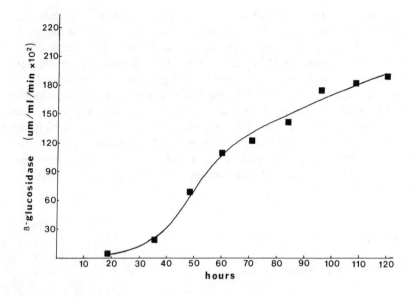

Figure 3.   Batch enzyme production β-glucosidase activity.

cosidase are present in the culture broth. These results compare favorably with those obtained in laboratory studies.

The pilot plant was modified to produce enzyme continuously to demonstrate feasibility on a large scale. The economical advantages of a continuous process lie in reduced capital investment due to increased efficiency of vessel use. Results from continuous enzyme productions are shown in Figures 4 and 5. From these graphs, it can be seen that the β-glucosidase is somewhat lower but the protein and FPRS remain almost as high as in batch culture. Use of the enzyme from batch as well as continuous enzyme production in small-scale flask saccharification and SSF indicate only small differences between the two enzyme preparations under the same conditions.

Simultaneous Saccharification-Fermentation

Batch SSF were performed using a variety of substrates. Typical results for Solka-Floc and pulp mill wastes are illustrated in Figures 6 and 7, respectively. In both cases, more than 50% of the theoretical yield from cellulose to ethanol was achieved. Batch SSF were run with cellulose concentration ranging from 5 to 15%.

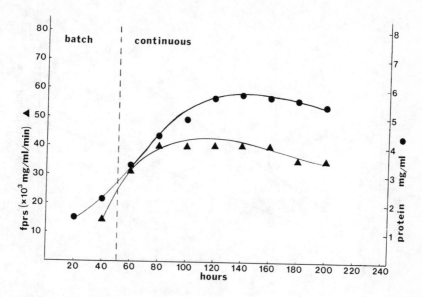

Figure 4. Continuous enzyme production FPRS activity and protein concentration.

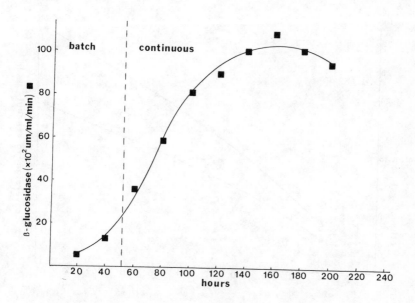

Figure 5. Continuous enzyme production β-glucosidase activity.

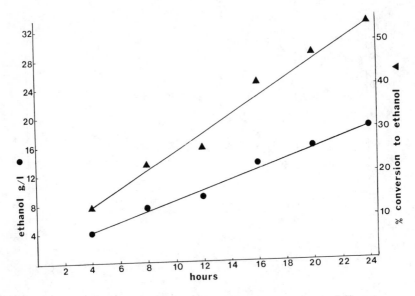

Figure 6. Batch SSF using Solka-Floc. Ethanol production and % conversion to ethanol (6% cellulose).

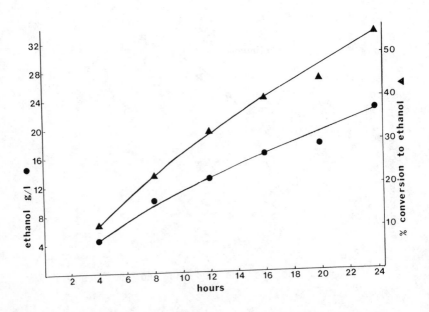

Figure 7. Batch SSF using/PMW. Ethanol production and % conversion to ethanol (7% cellulose).

Semicontinuous SSF utilized PMW and MSW as primary feedstocks. Ethanol production can be seen in Figure 8. Both MSW and PMW showed the same trend (Figure 9) concerning ethanol yield, base utilization for pH control and bacterial contaminant population. The presence of contaminants and increased base usage indicate the production of other acidic products. Lab-scale continuous SSF operation has proven to be significantly better than batch SSF per unit time.

Stripping Operations

After the SSF were completed, the resultant beer slurry was pressured to the beer storage tank (Figure 1). From the beer storage tank, the slurry was pumped to the top of the baffle tray column while steam was injected into the bottom of the column [25]. As the beer slurry cascaded down the column, the hot vapor from below contacted the descending liquid and effected the stripping of the ethanol from the beer feed. The column was designed to handle beer slurries with solids content as high as 10% and deliver a product stream of approximately 25 wt % ethanol from a feed containing 2.0–3.5% ethanol. The still bottoms ethanol concentration remained as low as 0.04%. In a large-scale plant the product from the slurry stripper will be rectified further to yield 95–100% industrial or motor grade ethanol as necessary.

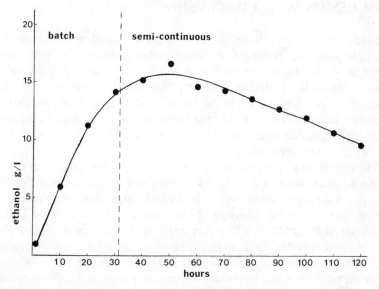

Figure 8. Semicontinuous SSF using PMW or MSW. Ethanol production (8% cellulose).

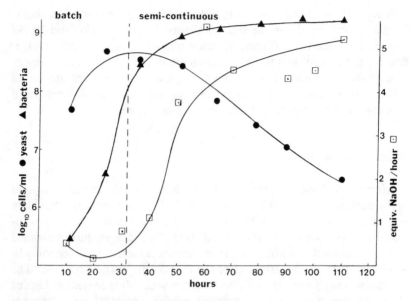

Figure 9.   Semicontinuous SSF using PMW or MSW. Yeast cell count, bacterial cell count and base utilization.

DISCUSSION AND CONCLUSION

Many pieces of equipment used for materials handling were tested in the pilot plant. An example is a 750-gal pulper which worked with some wood products but not very well with MSW because of the plastics and metal cans in the material. A rotary vacuum filter was used for dewatering some slurries but for the majority of feedstocks it was not acceptable. For these reasons, the feedstocks used at the pilot plant, as outlined in this chapter, received no pretreatment and were used in the process just as they were received.

The operation of the pilot plant in both a batch and continuous mode using potential industrial feedstocks demonstrated the enzymatic cellulose-to-ethanol technology on a substantially larger scale than had previously been reported. The size of the plant enabled the use of bulky materials, such as MSW, which was difficult on a laboratory scale. The results from the pilot plant enzyme production compared very favorably with the laboratory results; however, in the case of the SSF, the data from the pilot plant and the laboratory are only comparable for approximately the first 24 hr after which the pilot plant results lagged behind.

For example, on batch SSF that ran longer than 24 hr at the pilot plant, the percent conversion to ethanol did not continue to rise as in the laboratory. With pulp mill wastes in laboratory studies, SSF of 85–90% of theoretical conversion to ethanol was achieved in 48 hr compared to 55–60% conversion at the pilot plant. The reasons for the difference in results can be explained in part by the lack of adequate environmental controls such as temperature and pH due to poor heat and mass transfer in the high solids slurry of the SSFs. Contamination was also a problem in SSF that ran for extended periods as evidenced by the increase in base utilization for pH control and the concomitant decrease in ethanol yields (Figures 8 and 9). This is readily explained by the use of unpasteurized feedstock in the pilot plant.

The data gathered from the operation of the pilot plant were used for extensive economic analysis of the cellulose-to-ethanol technology [26,27]. The results of this analysis along with the problem areas mentioned above indicate further scale-up of the process from the 1-ton/day to a 50-ton/day facility should be carried out to identify specific equipment to be used on a commercial scale and execute process modifications toward enhancing the economic viability of the technology.

NOMENCLATURE

a = area
D = dilution rate
Di = impeller diameter
Dv = vessel diameter
k_L = mass transfer coefficient
L = vessel length

REFERENCES

1.   Nystrom, J. M., and R. K. Andren. "Pilot Plant Conversion of Cellulose to Glucose," *Process Biochem.* (December 1976).
2.   Spano, L. A., J. Medeiros and M. Mandels. "Enzymatic Hydrolysis of Cellulosic Wastes to Glucose," *Resource Recov. Conserv.* (1):279–294 (1976).
3.   Reese, E. T. "History of the Cellulase Program at the U.S. Army Natick Development Center," *Biotechnol. Bioeng. Symp.* (6):9–20 (1976).
4.   Nystrom, J. M., and A. L. Allen. "Pilot Scale Investigations and Economics of Cellulase Production," *Biotechnol. Bioeng. Symp.* (6): 55–74 (1976).

5. *Energy From Biological Processes: Volume II — Technical and Environmental Analysis,* (Washington, DC: Office of Technology Assessment, 1980), pp. 167–172.
6. Wilke, C. R., R. D. Yang and U. V. Stockar. "Preliminary Cost Analysis for Enzymatic Hydrolysis of Newsprint," *Biotechnol. Bioeng. Symp.* (6):155–175 (1976).
7. Wilke, C. R., G. R. Cysewski, R. D. Yang and U. V. Stockar. "Utilization of Cellulosic Materials through Enzymatic Hydrolysis, II. Preliminary Assessment of an Integrated Processing Scheme," *Biotechnol. Bioeng.* XVIII(4):1315–1323 (1976).
8. Wilke, C. R. U. V. Stockar and R. D. Yang. "Process Design Basis for Enzymatic Hydrolysis of Newsprint," in *AICHE Symp. Ser.* 72 (158): 104–114.
9. Tsao, G. T., M. Ladisch, C. Ladisch, T. A. Hsu, B. Dale and T. Chou. "Fermentation Substrates from Cellulosic Materials: Production of Fermentable Sugars from Cellulosic Materials," in *Annual Reports on Fermentation Processes, Vol. 2,* D. Perlman, Ed. (New York: Academic Press, Inc., 1978), p. 1.
10. Flickinger, M. C., and G. T. Tsao. "Fermentation Substrates from Cellulosic Materials: Fermentation Products from Cellulosic Materials," in *Annual Reports on Fermentation Processes, Vol. 2,* D. Perlman, Ed. (New York: Academic Press, Inc., 1978), p. 23.
11. Ladisch, M., C. M. Ladisch and G. T. Tsao. "Cellulose to Sugars: New Path Gives Quantitative Yield," *Science* 201(23):743–745 (1978).
12. Ladisch, M. R. "Fermentable Sugars from Cellulosic Residues," *Process Biochem.* 14(1):21–25 (1979).
13. Humphrey, A. E. "Economical Factors in the Assessment of Various Cellulosic Substances as Chemical and Energy Resources," *Biotechnol. Bioeng. Symp.* (5):49–65 (1975).
14. Pye, E. K., A. E. Humphrey, J. R. Forro, E. Nolan and J. K. Alexander. "The Biological Production of Liquid Fuels from Biomass," SERI Contract No. RFP-R20-9129-1 (1980).
15. Cooney, C. L., D. I. C. Wang, S. D. Wang, J. Gordan and M. Jiminez. "Simultaneous Cellulose Hydrolysis and Ethanol Production by a Cellulolytic Anaerobic Bacterium," *Biotechnol. Bioeng. Symp.* (8):103–114.
16. Chambers, R. P., Y. Y. Lee and T. McCaskey. "Liquid Fuel and Chemical Production from Cellulosic Biomass — Hemicellulose Recovery Pentose Utilization," Contract DE AS0579ET23051 (1980).
17. Harris, E. E., and E. Beglinger. "Madison Wood Sugar Process," *Ind. Eng. Chem.* 38(9):pp. 890–895 (1946).
18. Horton, G. L., D. B. Rivers and G. H. Emert. "Preparation of Cellulosics for Enzymatic Conversion," *Ind. Eng. Chem. Prod. Res. Dev.,* 19(3): 422–429 (1980).
19. Blotkamp, P. J., M. Takagi, M. S. Pemberton and G. H. Emert. "Enzymatic Hydrolysis of Cellulose and Simultaneous Fermentation to Alcohol," in *AIChE Symp. Ser.* 74(181):85–90, (1978).
20. Takagi, M., S. Suzuki and W. F. Gauss. "Manufacture of Alcohol from Cellulosic Materials Using Plural Ferments," U.S. Patent 3,990,994 (November 9, 1976).

21. Gracheck, S. J., K. E. Giddings, L. C. Woodford and G. H. Emert. "Continuous Enzyme Production as Used in the Conversion of Lignocellulosics to Ethanol," Energy Futures, Am. Soc. Agri. Eng. Energy Symposium, (submitted).
22. Miller, G. L. "Use of Dinitrosalicylic Acid Reagent for Determination of Reducing Sugars," *Anal. Chem.* 31(3) (1959).
23. Von Soest, P. J., and R. H. Wine. *J. Assoc. Off. Agric. Chem.* 51(4):780–785 (1968).
24. Updergraff, D. M. *Anal. Biochem.* 32:420–424 (1969).
25. Katzen, R., V. B. Diebold, G. D. Moon, Jr., W. A. Rogers and K. A. LeMesurier. "A Self-Descaling Distillation Toner," *Chem. Eng. Prog.* 64(1) 1968.
26. Emert, G. H., and R. Katzen. "Gulf's Cellulose to Ethanol Process," *Chemtech* 10(10):610–614 (1980).
27. Emert, G. H., and R. Katzen. "Economic Update of the Bioconversion of Cellulose to Ethanol," *Chem. Eng. Prog.,* 76(9):47–52 (1980).

CHAPTER 21

LOW-ENERGY DISTILLATION SYSTEMS

R. Katzen, W. R. Ackley, G. D. Moon, Jr.,
J. R. Messick, B. F. Brush and K. F. Kaupisch
Raphael Katzen Associates International, Inc.
Cincinnati, Ohio

Over the past decades, we have developed a series of highly efficient alcohol distillation systems for recovery of various grades of ethyl alcohol from synthetic and fermentation feedstocks. For each of these systems, the prime goal is minimization of energy consumption.

The RKAII distillation system for production of high-quality spirits or industrial-grade alcohol, uses a four-tower distillation train. The product is first-quality neutral spirits at 96° G.L. (192° U.S. proof) ethanol. When the crude ethanol feed is obtained by the synthetic process, e.g., direct hydration of ethylene, only three towers are required. For motor fuel-grade alcohol, where a high-quality product is not necessary, simpler abbreviated systems are used to reduce investment and operating costs. The distillation system is shown in Figure 1 [1,2]. The process has successfully been operated commercially with four different fermentation feedstocks: molasses, grain (corn or milo), corn wet milling middlings and sulfite waste liquor. In addition, it has been operated with an ethylene-based synthetic crude.

Production of 96° G.L. Hydrous Alcohol

Beer from the fermenters, containing approximately 6-8 wt % alcohol and 8-10% total solids (suspended and dissolved) is preheated to near saturation temperature and fed to the beer still. An overhead condensed

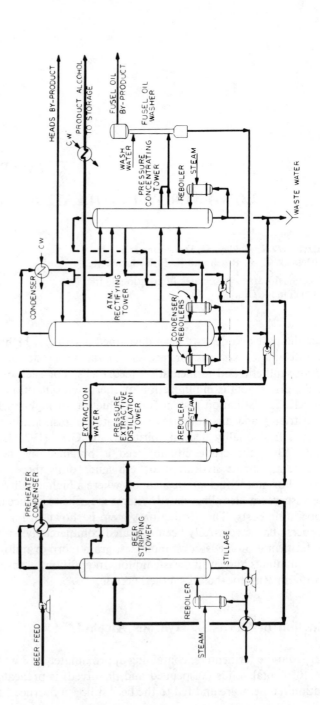

Figure 1.   High-grade hydrous alcohol system.

product, at 75–85° G.L. (150–170° U.S. proof) is taken to the high wines drum, and the bottoms liquid (stillage), containing not more than 0.02 wt % alcohol, is treated further for animal feed production. The high wines distillate from the beer still is mixed with recycled alcohol from the concentrating tower and the combined stream is fed to the extractive distillation tower. The extractive tower is designed to separate substantially all impurities (aldehydes, esters and higher alcohols) from the ethanol. The extraction technique relies on the volatility inversion of the higher alcohols with respect to ethanol in solutions containing high concentrations of water. The net result is that a substantially pure ethanol/water mixture is removed from the bottom of the extractive tower while the impurities are taken overhead.

Dilute alcohol from the base of the extractive tower is stripped and concentrated to product strength in the rectifying tower. A heads purge is taken from the overhead condensate. Product ethanol at 96° G.L. (192° U.S. proof) is withdrawn near the top of the rectifying tower, and a water stream containing trace amounts of alcohol is discharged from the base. Heads and side draw fusel oil purges are fed to the concentrating tower to prevent any buildup of impurities in the rectifying tower. The overhead stream from the extractive tower also is fed to the concentrating tower. Heads and fusel oil are concentrated in this tower and removed from the system, with the recovered alcohol being recycled to the extractive tower. Steam economy is achieved by multistage preheating of beer feed, and by operating the extractive and concentrating towers at higher pressures. The overheads from these pressure towers supply thermal energy to the reboilers of the rectifying tower. Such pressure cascading results in a 30–50% reduction in virgin steam.

The key features of the RKAII high quality alcohol distillation system are:

1. Extractive distillation accomplishes a higher degree of impurity removal than is possible in more conventional systems. Product ethanol contains only 20–30 ppm of total impurities.
2. The use of pressure cascading permits substantial heat recovery and reuse. In the system described above and in Figure 1, the extraction tower and concentrating tower are operated at a pressure higher than the rectifying tower. The overhead vapors from these pressure towers supply thermal energy to the rectifying tower reboilers. By operating in this manner, steam usage is kept to a minimum. Commercial facilities using this pressure cascading technique show steam usages of only 3.0–4.2 kg of steam per liter (25–35 lb/gal) of 96° G.L. (192° U.S. proof) ethanol compared to 6.0 kg/l in earlier conventional systems.
3. Substantially all (95 to 98%) of the ethanol in the crude feed is recovered as first-grade product.

4.   Design of highly efficient tower trays permit high turndown capability. These trays are designed to be self-descaling in the stripping section of the beer towers.
5.   A highly advanced control system, developed through years of experience, provides for sustained stable operation; only part-time attention of an operator is required. Product quality is maintained with less than 30 ppm of total impurities. Permanganate time is in excess of 60 min.

PRODUCTION OF ANHYDROUS (99.5–99.98° G.L.) ALCOHOL

Anhydrous alcohol is produced by azeotropic distillation. A high–grade product of 99.98° G.L. (199.96° U.S. proof) concentration is produced for use in food and pharmaceutical aerosol preparations. A product of 99.5° G.L. (199° U.S. proof) concentration is produced for blending with gasoline for motor fuel. The RKAII two-tower dehydrating system design (Figure 2) has been installed and successfully operated in seven different alcohol plants in North America and the Caribbean.

The 96° G.L. (192° U.S. proof) product is withdrawn from the side of the rectifier in the hydrous distillation process. The hydrous alcohol is fed to an atmospheric dehydrating tower. Removal of water from the feed is achieved by use of benzene, heptane, cyclohexane or other suitable entraining agents. A ternary azeotrope is taken overhead from the dehydrating tower. The overhead vapors are condensed and the two liquid phases are separated in a decanter. The entrainer-rich phase is refluxed to the dehydrating tower. A reboiler is used to supply vapor to this tower with heat supplied by either low pressure steam, recovered flash vapor, or hot effluent and condensate streams from the hydrous alcohol unit. The aqueous phase from the decanter is fed to a stripper. The entraining agent is recovered, along with alcohol, in the overhead vapor. Water is removed from the bottom of the stripper. Direct steam may be used in this stripper. The bottoms stream from the azeotropic dehydrating tower is the anhydrous alcohol product.

Design know-how consists of optimizing the balance between capital costs and utility consumption, with stable control. Specific features which contribute to overall process efficiency and reliability to the RKAII anhydrous alcohol distillation system are:

1.   use of a common condenser and decanter for the dehydration and stripping towers to reduce capital costs;
2.   design of highly efficient tower trays for high turndown capability;

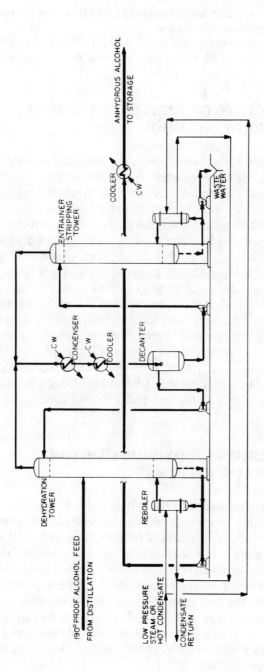

Figure 2. High-grade anhydrous alcohol system.

3. low consumption of entraining agent (less than 0.1 kg/1000 liters of anhydrous alcohol); and
4. low consumption of steam (1–1.5 kg/l or 8.3–12.5 lb/gal of anhydrous alcohol), or equivalent hot condensate or waste streams.

PRODUCTION OF ANHYDROUS MOTOR FUEL-GRADE ALCOHOL

For motor fuel–grade alcohol, the beer feed is preheated in a multi-stage heat exchange sequence. A pressure stripper-rectifier (Figure 3) is used to separate the beer feed into an overhead fraction of about 95° G.L. (190° U.S. proof) alcohol and a bottoms stream containing less than 0.02 wt % alcohol. Side draws are made to remove fusel oils. These oils are recovered by water washing, and reblended as a component of the motor fuel–grade alcohol. In addition, a pasteurizing section is used to concentrate low–boiling impurities. These are removed by taking a small heads draw which is burned in the plant reboiler. Dehydration of the hydrous product is accomplished in two additional towers. Energy is supplied to the reboilers of the two towers in the dehydration step by condensing the overhead vapors from the pressure stripper-rectifier. By operating the beer stripper-rectifier at a higher pressure [3] than the two-tower dehydration system, very low total steam consumption can be achieved. The steam usage is 1.8–2.4 kg/l (15–20 lb/gal) of 99.5° G.L. (199° U.S. proof) motor fuel–grade alcohol product [4].

PRODUCTION OF HYDROUS MOTOR FUEL-GRADE ALCOHOL

For a product to be utilized in neat alcohol engines (no gasoline in the blend), further steam economy can be achieved when only 85–95° G.L. (170–190° U.S. proof) alcohol product is desired. This is accomplished by splitting the stripping-rectifying duties between two towers (Figure 4) [5]. The first stripper-rectifier tower is operated at a pressure higher than the second tower and receives 50–60% of the beer feed. The overhead vapors from the first tower are used to boil vapor in the second tower.

The steam usage is 1.2–1.5 kg/l (10–12 lb/gal) of 85–96° G.L. (170–192° U.S. proof) motor fuel–grade alcohol (on a 100% ethanol basis). Along with steam economy, cooling water requirements are reduced proportionally.

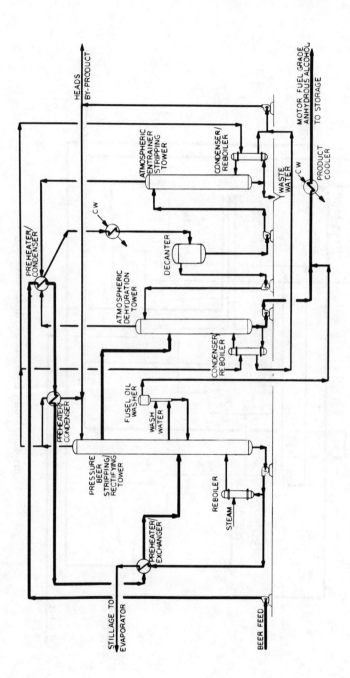

Figure 3.   Motor fuel–grade anhydrous alcohol system.

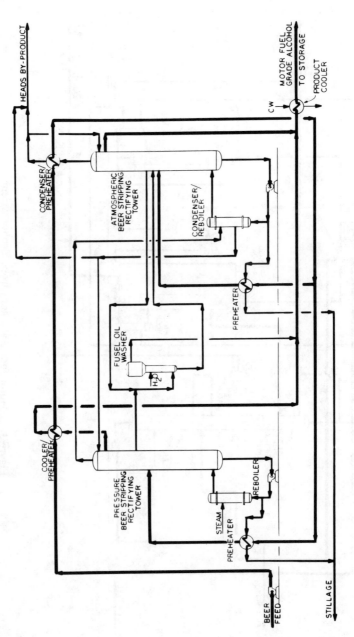

Figure 4.   Motor fuel-grade hydrous alcohol system.

Table I.   Low-Energy Distillation Systems: Summary of Investment and Utilities, 190 × 10⁶ liter/yr (50 × 10⁶ gal/yr)

	High-Grade (96° G.L.) Industrial Alcohol	Anhydrous (100° G.L.) Industrial Alcohol	Anhydrous (99.5° G.L.) Motor Fuel Alcohol	Hydrous Motor Fuel-Grade Alcohol
Figure	1	2	3	4
Alcohol Product, U.S. Proof	192	200	199	190
Distillation Unit Investment, $ million (1980)	7.3	2.8	5.1	3.4
Steam Use, kg/1 (lb/gal)	4.1 (34)	1.4 (11.7)	2.2 (18)	1.2 (10)
Cooling Water, metric ton/hr	1866	934	1311	482
Electric Power, kW	289	31	133	177

SUMMARY

By energy reuse, pressure cascading and waste heat recovery, the expenditure of energy in the distillation of ethanol can be reduced greatly. Such energy savings have been demonstrated industrially in three systems described in this chapter. For high-grade industrial ethanol production, a steam consumption of 3.0–4.2 kg/l (25–35 lb/gal) 100° G.L. alcohol is realized. For motor fuel–grade anhydrous alcohol, the steam consumption is 1.8–2.4 kg/l (15–20 lb/gal) 99.5° G. L. alcohol, and for hydrous motor fuel–grade alcohol, the steam consumption is 1.2–1.4 kg/l (10–12 lb/gal) 96° G.L. alcohol.

An investment summary for the typical low-energy distillation systems shown in Figures 1 to 4 for production of 190×10^6 liter/yr (50×10^6 gal/yr) of alcohol is given in Table I. Also shown are the steam, cooling water and electric energy requirements for each system.

REFERENCES

1. "Extractive Distillation of C-1 and C-3 Alcohols and Subsequent Distillation of Purge Streams," U.S. Patent 3,445,345 (1969).
2. "Improvement in Alcohol Distillation Process," U.S. Patent 3,990,952 (1976).
3. "Distillation System for Motor Fuel Grade Anhydrous Alcohol," U.S. Patent 4,217,178 (1980).
4. Raphael Katzen Associates. "Grain Motor Fuel Alcohol. Technical and Economic Assessment Study," U.S. DOE Contract No. EJ-78-C-6639 (1979).
5. "Novel Energy Efficient Process for Production of 170 to 190° Proof Alcohol," U.S. Patent (applied).

SECTION 5

NATURAL AND THERMAL LIQUEFACTION

CHAPTER 22

NATURAL PRODUCTION OF HIGH-ENERGY LIQUID FUELS FROM PLANTS

E. K. Nemethy, J. W. Otvos and M. Calvin

Laboratory of Chemical Biodynamics
University of California
Berkeley, California

The growing of green plants as a renewable energy source is attracting increasing interest [1-4]. The concept of "energy farms" involves the purposeful cultivation of selected species either for use as a solid fuel (wood), or for other energy-rich products. In the latter case, the product is a derivative of the total biomass, and after it is separated from the cellulosic plant material, it can be used directly as diesel fuel in some cases, or it can be converted to a convenient liquid fuel, such as gasoline. This approach, the cultivation of plants which already produce hydrocarbon-like compounds is attractive, since the conversion of this type of plant extract to a high-quality liquid fuel is expected to be energy-efficient, because the material is already in a reduced state.

A large number of plant species are capable of reducing CO_2 beyond carbohydrates to isoprenoids or other hydrocarbon-like compounds. The use of *Euphorbias*, for example, has been proposed for hydrocarbon production [1-4]. Recently, Buchanan and co-workers have screened several hundred plants for their oil and rubber content [5,6]. Among these, *Euphorbia lathyris* was identified as one of the few potential hydrocarbon-producing crops [7].

Euphorbia lathyris is a herbaceous member of the family *Euphorbiaceae*. This family of plants consists of approximately 2000 species, ranging from small herbs and succulents to large trees. Perhaps its best known member is the rubber tree *Hevea brasiliensis,* whose white, milky latex is the source of natural rubber. Many other *Euphorbia* species also produce a milky latex which may contain polyisoprenes, and is usually rich in low-molecular-weight reduced isoprenoids. *E. lathyris* is also a latex-bearing plant which grows wild in California, consequently we have started to investigate this particular Euphorbia as an "energy farm" candidate.

Figure 1 shows a close-up view of *E. lathyris*. The plants grow with multiple branching in the field to a height of 1.5 m. In the greenhouse, however, *E. lathyris* can attain a height of 2.5 m. The best planting time for *E. lathyris* in California is early spring, for a late fall (November–December) harvest. In the Southwest *E. lathyris* can be planted in October and harvested in May or June. For seed production, *E. lathyris* must be treated as a biennial, i.e., sown in one calendar year and harvested in the next.

The first effort to cultivate this plant started in 1977–1978, when test plots from wild seeds were established at the South Coast Field Station of the University of California in Santa Ana. Figure 2 shows a stand of *E. lathyris* in a typical test plot after a nine-month growing season. At this time the plants attained a typical dry weight of 200 g, and grew to a height of approximately 1 m. Planting density was at 1-ft centers, and the plots were irrigated receiving a total of 0.5 m of water, 0.25 m of which was natural rainfall. Preliminary results from these experiments indicate that a biomass yield of 22 dry metric ton/ha-yr may be achieved with *E. lathyris*. Further agronomic experiments are under way to determine the optimal growing conditions for this plant.

METHODS

E. lathyris exudes a milky latex when cut. However, this plant is not amenable to continuous tapping like some other Euphorbs. To obtain the reduced photosynthetic material, the entire plant is extracted after drying at 70°C to 4% moisture content. The reduced organic material is not uniformly distributed throughout the plant; the leaves contain twice as much as the stems per unit weight. Therefore, for uniform sampling, the dried plant is ground in a Wiley Mill to a 2-mm particle size and is subsequently thoroughly mixed. A portion of the plant material is then extracted in a soxhlet apparatus with boiling solvent for 8 hr. Different solvents can be used to extract various plant constituents. One scheme which yields cleanly separated fractions and reproducible results is shown in Figure 3.

Acetone can also be used as the initial solvent instead of heptane. However, acetone brings down a variable amount of carbohydrates

Figure 1.   *Euphorbia lathyris.*

which precipitate out of solution. These can be filtered off, leaving behind a pure acetone-soluble portion, which is 8% of the dry weight of the plant. This is equivalent to the sum of fractions I and II of the extraction scheme shown in Figure 3.

RESULTS AND DISCUSSION

The heptane extract of *E. lathyris* (fraction I of Figure 3) has a low oxygen content and a heating value of 42 MJ/kg, which is comparable to

Figure 2.   Plot of *E. lathyris* in Santa Ana, CA.

that of crude oil (44 MJ/kg). These qualities indicate a potential for use as fuel or chemical feedstock material. Therefore we have investigated the chemical composition of this fraction in some detail. Since the amount of the methanol fraction is quite substantial we have also identified the major components of this fraction.

Terpenoids (Heptane Extract)

The heptane extract is a complex mixture, which can be separated into crude fractions by absorption chromatography on silica gel. The characteristics of the resulting fractions are shown in Table I. We have examined each of these column fractions further by gas chromatography and have obtained structural information on the major components by combined gas chromatography/mass spectroscopy (GC/MS). Molecular formulae were obtained by high resolution mass spectroscopy [8].

The data from the GC/MS analyses indicate that more than 100 individual components comprise the heptane extract. About 50 of these are major ones; these we have either identified or classified. The major part

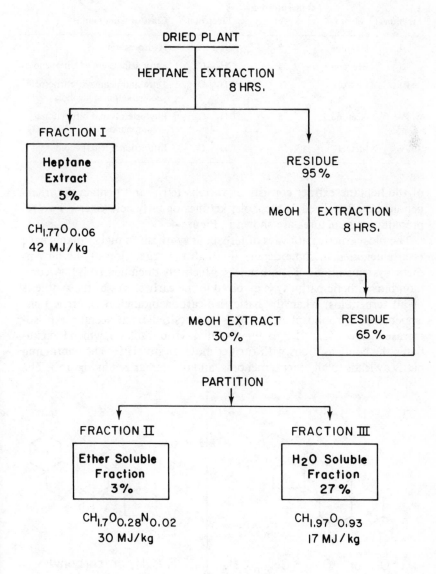

Figure 3.   Scheme for extraction.

Table I.   Silica Gel Column Fractions of the Heptane Extract

Fraction	Eluent	Concentration (%)	Elemental	Class of Compounds
I	Heptane	7	CH_2	Hydrocarbon
II	Benzene	33	$CH_{1.72}O_{0.03}$	Fatty acid esters of triterpenoids
III	EtOAc	41	$CH_{1.67}O_{0.03}$	Tetra and pentacyclic triterpenoids; ketones; alcohols
IV	Acetone	5	$CH_{1.63}O_{0.15}$	Phytosterols and bifunctional compounds
V	MeOH	15	$CH_{1.69}O_{0.22}$	Bifunctional triterpenoids

of the heptane extract consists of various tetra- and pentacyclic triterpenoid functionalized as alcohols, ketones or fatty acid esters. Two representative structures are shown in Figure 4.

The biosynthetic pathways of terpenoid synthesis in higher plants have been elucidated in some detail, although the regulation of the biosynthetic systems involved is much less effectively documented [9]. A common biosynthetic pathway is involved in the early stages of the synthesis of all terpenoids, when the basic five-carbon biological isoprene unit, isopentenyl pyrophosphate (IPP) is synthesized from acetate. An isomerase then converts IPP to dimethylallylpyrophosphate, which then initiates the condensations with further molecules of IPP. The continuing pathway leads to all other terpenoids, and is summarized in Figure 5. The

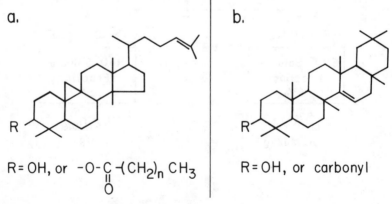

a.

b.

R = OH, or $-O-\underset{\underset{O}{\|}}{C}(CH_2)_n CH_3$

R = OH, or carbonyl

Figure 4.   Representative structures for tetra- and pentacyclic triterpenoids found in *Euphorbia lathyris*. The carbon skeletons are those of (a) cycloartenol and (b) taraxerol.

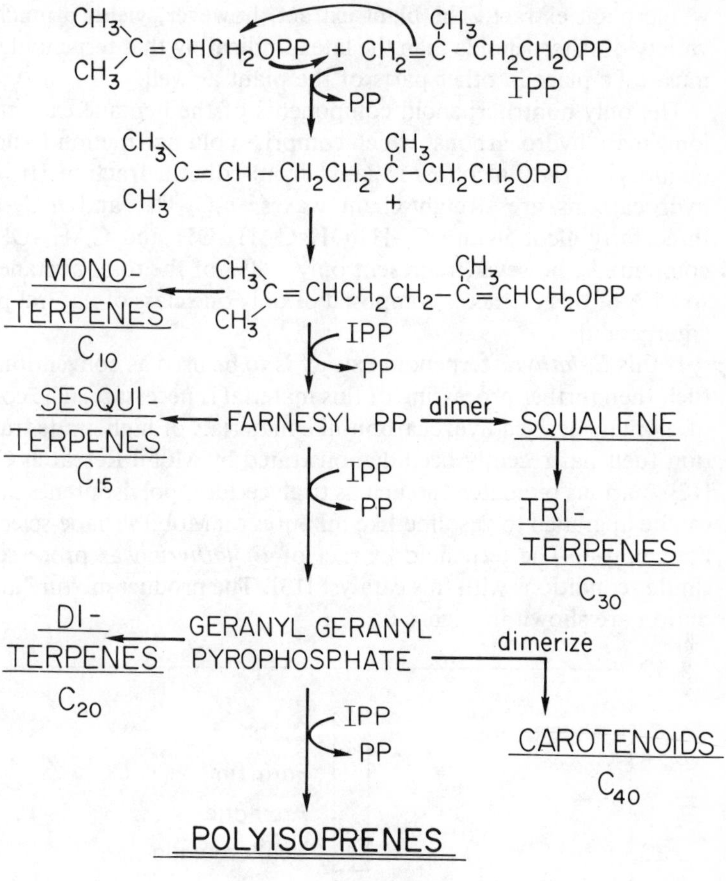

Figure 5.   Pathways of terpenoid biosynthesis.

triterpenoids arise via the enzyme mediated cyclization of squalene 1, 2-oxide followed by rearrangement sequences to yield a large array of interrelated C_{30} compounds.

In *E. lathyris,* terpenoid biosynthesis is evidently shunted almost exclusively via this pathway, since no major amounts of any other class of terpenoids have been detected. (Some irritant and toxic diterpenoids are present in the latex of *E. lathyris* in very low concentration, approx. 0.01% [10].) The major terpenoid components of the latex itself have been identified as five triterpenoids [11]; all of the latex components, with the exception of euphol (a minor one), could also be detected in the whole plant extract. The plant extract, however, yields a much greater variety of triterpenoids than the latex, indicating that terpenoid synthesis must take place in other parts of the plant as well.

The only nontriterpenoid components of the heptane extract are two long chain hydrocarbons, which comprise column fraction I and a small quantity of fatty alcohols isolated from column fraction III. The two hydrocarbons are straight-chain waxes: n-$C_{31}H_{64}$ and n-$C_{33}H_{68}$; the three fatty alcohols are $C_{27}H_{53}OH$ $C_{28}H_{57}OH$ and $C_{29}H_{57}OH$. These compounds, however, represent only ~8% of the total heptane extract, so 85% of this extract is composed of only one class of natural products: triterpenoids.

If this $E.$ $lathyris$ terpenoid extract is to be used as conventional liquid fuel, then further processing of this material is necessary. The conversion of biomass derived hydrocarbon-like materials of high grade transportation fuels has recently been demonstrated by Mobil Research Company [12]. Various biomaterials such as triglycerides, polyisoprenes and waxes can be upgraded to gasoline-like mixtures on Mobil's shape-selective zeolite catalyst. The terpenoid extract of $E.$ $lathyris$ was processed under similar conditions with this catalyst [13]. The product mixture and distribution are shown in Figure 6.

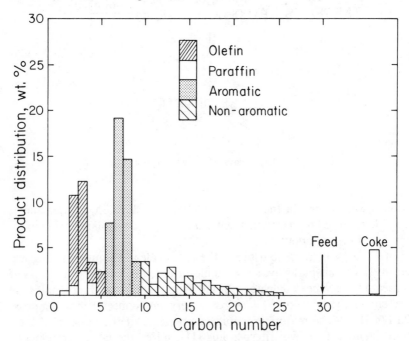

Figure 6. Catalytic conversion of $E.$ $lathyris$ terpenoids.

The products obtained from the conversion of *E. lathyris* terpenoids are seen to simulate a gasoline-type mixture; furthermore they are rich in compounds which are premium raw materials for the chemical industry.

Carbohydrates (Methanol Extract)

As the data in Figure 3 indicate, a substantial amount (30%) of the dried plant weight can be extracted with methanol. The empirical formula of the water soluble portion of this extract is indicative of carbohydrates. Since simple hexoses can be directly fermented to ethanol, a useful liquid fuel, we have determined the carbohydrate content of *E. lathyris* and identified the specific sugars. The results of gel-permeation chromatography of fraction III (Biogel-P-2) indicated that there are no poly- or even oligosaccharides present in this fraction. The carbohydrate-containing fractions were identified by the Molish test, and were further characterized by two-dimensional paper chromatography and high-pressure liquid chromatography (HPLC). In both of these systems, only four simple sugars were detected: sucrose, glucose, galactose and fructose. The HPLC trace of the total sugar fraction as well as the relative amounts of the individual components are shown in Figure 7.

We have determined that the entire crude carbohydrate (fraction III, Figure 3) extract is fermentable to ethanol without further purification. Since there is certainly no specific yeast available for *Euphorbia* sugar fermentation, we have tried several different types: brewers' yeast, bakers' yeast and two yeasts used in the wine industry: Champagne and Montrachet. The best results were obtained with the Montrachet type, a yeast which is very tolerant to phenolic impurities. The fermentations were carried out in 200 ml of an approximately 10% sugar solution to which 130 mg of commercial wine yeasts nutrients were added. The temperature was maintained at 23°C for 90 hr. Under these conditions, an 80% fermentation efficiency was obtained, yielding 8.4 ml of ethanol from 25 g of the crude water extract. This alcohol yield corresponds to 66% fermentable sugar content of the sample based on the established fermentation efficiency, and is in excellent agreement with the chromatographic data shown in Figure 7. We can therefore obtain not only hydrocarbons from *E. lathyris* but a substantial quantity of ethanol as well.

Processing and Energy Balance

Since *E. lathyris* and other hydrocarbon-producing crops are new species from the point of view of cultivation, their agronomic character-

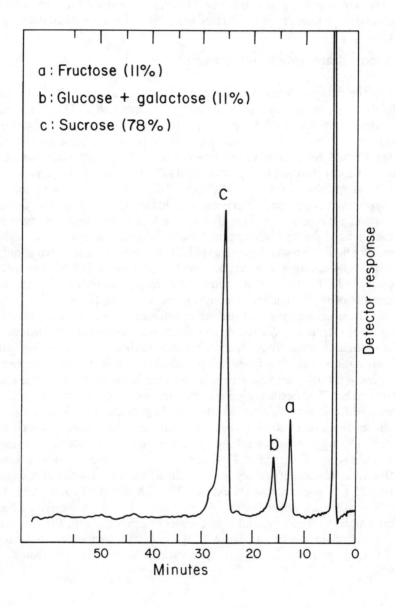

Figure 7. HPLC trace of sugar fraction. Altex NH_2 column, Mobile phase: $CH_3CN:H_2O_2$, 80:20.

istics, requirements and yield potentials are not yet well known. Consequently, any conceptual economic or technical evaluation will contain uncertainties. A recent study of SRI International on the feasibility of growing *E. lathyris* for energy use identified the major uncertainties of the feedstock cost and supply [14]. This conceptual process study is based on solvent extraction, an existing technology in the seed oil industry. The carbohydrates are recovered by extraction with water. The overall scheme and energy balance are shown in Figure 8. As seen in Figure 8, the cellulosic plant residue (bagasse) is used to generate the energy required for solvent extraction and recovery. According to this model, a considerable quantity of bagasse is left over after recovery of the useful products. If an estimate of the required energy input for cultivation is included in this model, the entire process still remains energy positive [15].

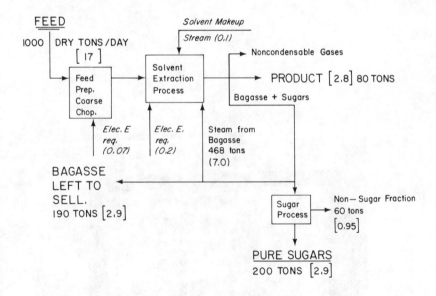

Figure 8.   Conceptual processing sequence to recover terpenoids and sugars from *E. lathyris*. Numbers in brackets indicate energy (TJ).

Plant Selection

To determine whether the seed source has any effect on the terpenoid content of *E. lathyris,* we have first investigated (in collaboration with R. Sachs, Dept. of Environmental Horticulture, University of California-

Davis) the two cultivars native to California: the northern and the southern varieties. These two ecotypes are shown in Figure 9. One hundred individual plants of each of these ecotypes were grown in the green house to approximately 100 g fresh weight (4-month growing season). The plant samples were then dried and ground in the usual manner and extracted with refluxing heptane to determine whether there are any differences in terpenoid content between the two seed samples. The frequency distributions for the northern and southern seed source plants are shown in Figures 10 and 11. In both cases, the average percent extractables are lower than the 5% obtained from mature field-grown *E. lathyris;* this is probably due to the shorter growing period as well as the reduced photosynthetic activity in the greenhouse environment. However, as the data indicate, the difference between the average percent extractables for the northern and southern sets is compatible with zero.

Figure 9. Northern (right) and southern (left) *E. lathyris.*

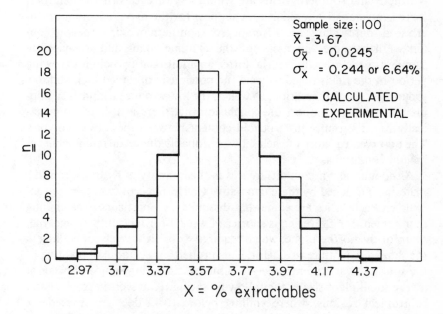

Figure 10. Frequency distribution for northern *E. lathyris*. Number of samples vs wt % extractables.

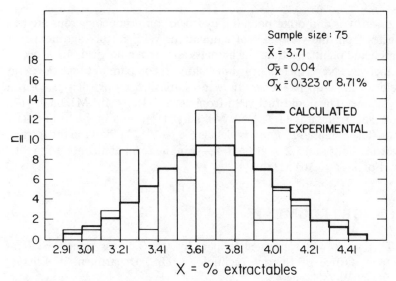

Figure 11. Frequency distribution for southern *E. lathyris*. Number of samples vs wt % extractables.

Northern and southern yields are within 1% of each other. Therefore, there is no statistically significant support for the hypothesis that northern seed source plants are different from southern ones in terpenoid content. Furthermore, an assay of the extreme high- and low-yielding samples of the greenhouse trial (after an additional two-month growing period in the field) failed to yield the previously observed high and low groupings. This observation, as well as the low variance within each population, tends to support the conclusion that there is little or no genetically based variance in terpenoid content between these two ecotypes. The observed variance of each set was probably due to different environmental conditions.

Although there are no differences in the quantity of hydrocarbon-like materials produced by these two native California ecotypes, their agronomic characteristics are somewhat dissimilar. In particular, the chilling requirement for flowering is shorter (4 weeks) for the southern ecotype, than for the northern one, which requires six to eight weeks of chilling at the same temperature (8°C)[16]. This property precludes fall planting of the southern variety in temperate climates, since flowering will occur at a very small plant size in the following spring. However, in tropical and subtropical regions, where chilling below 10°C does not exceed six weeks, year-round plantings are feasible.

CONCLUSION

E. lathyris and other potential hydrocarbon-producing crops are new species from the viewpoint of cultivation. With further agronomic research and plant selection, the biomass and terpenoid yields are expected to increase. Nevertheless, it is interesting to compare in terms of energy yield a new crop like *E. lathyris* to established crops such as corn or sugarcane. The liquid fuel yield from corn is 4.2×10^4 MJ/ha-yr [17]; from sugarcane it is 11.7×10^4 MJ/ha-yr [15], both in the form of ethanol. The potential *E. lathyris* yield is 6.5×10^4 MJ/ha-yr in the form of hydrocarbons and 5.2×10^4 MJ/ha-yr in the form of alcohol, for a total yield of 11.7×10^4 MJ/ha-yr.

ACKNOWLEDGMENTS

This work was supported in part by the Biomass Energy Systems Branch, Office of Energy Technology, U.S. Department of Energy, Contract W-7405-eng-48.

REFERENCES

1. Calvin, M. "Petroleum Plantations," in *Solar Energy: Chemical Conversion and Storage,* R. R. Hatula and R. B. Kings, Eds (Clifton, NJ: The Humana Press, 1979), p. 1.
2. Calvin, M. "Petroleum Plantations for Fuel and Materials," *Bioscience* 29:533–537 (1979).
3. Johnson, J. D., and C. Hinman. "Oils and Rubber from Arid Land Plants," *Science* 208:460–464 (1980).
4. Coffey, S. G., and G. M. Halloran. "Higher Plants as Possible Sources of Petroleum Substitutes," *Search* 10(12):423–428 (1979).
5. Buchanan, R. A., I. M. Cull, F. H. Otey and C. R. Russell. "Hydrocarbon and Rubber Producing Crops," *Econ. Bot.* 32:131–145 (1978).
6. Buchanan, R. A., I. M. Cull, F. H. Otey and C. R. Russell. "Hydrocarbon and Rubber Producing Crops," *Econ. Bot.* 32:146–163 (1978).
7. Buchanan, R. A., F. H. Otey, C. R. Russell and I. M. Cull. "Whole Plant Oils, Potential New Industrial Materials," *J. Am. Oil Chem. Soc.* 55:657–662 (1978).
8. Nemethy, E. K., J. W. Otvos and M. Calvin. "Analysis of Extractables from One Euphrobia," *J. Am. Oil Chem. Soc.* 56:957–960 (1979).
9. Banthorpe, D. V., and B. V. Charlwood. "The Isoprenoids," in *Secondary Plant Products,* E. A. Bell and B. V. Charlwood, Eds. (Heidelberg: Springer-Verlag, 1980).
10. Adolf, W. and E. Hecker. "On the Active Principles of the Spurge Family III. Skin Irritants and Cocarcinogenic Factors from the Caper Spurge," *Z. Krebsforsch.* 84:325–344 (1975).
11. Nielsen, P. E., H. Nishimura, J. W. Otvos and M. Calvin. "Plant Crops as a Source of Fuel and Hydrocarbon-like Materials," *Science* 198:942–944 (1977).
12. Weisz, P. B., W. O. Haag and P. G. Rodewald. "Catalytic Production of High-Grade Fuel (Gasoline) from Biomass Compounds by Shape-Selective Catalysis," *Science* 206:57–58 (1979).
13. Haag, W. O., P. G. Rodewald and P. B. Weisz. "Catalytic Production of Aromatics and Olefins from Plant Materials," paper presented at the Second Chemical Congress of the American Chemical Society, Las Vegas, NV, August 26, 1980.
14. Kohan, S. M., and D. J. Wilhelm. "Recovery of Hydrocarbon-like Compounds and Sugars from *Euphorbia lathyris,*" SRI International Report (1980).
15. Weisz, P. B., and J. F. Marshall. "High-Grade Fuels from Biomass Farming: Potentials and Constraints," *Science* 206:24–29 (1979).
16. Sachs, R., Department of Environmental Horticulture, University of California, Davis. Personal communication (1980).
17. Lipinsky, E. S. et al. "Systems Study of Fuels from Sugarcane, Sweet Sorghum, Sugar Beets and Corn," ERDA Report BM1-1957A/5 (1977).
18. Lipinsky, E. S. et al. "Sugar Crops as a Source of Fuels," DOE Report TID-29400/2 (1979).

CHAPTER 23

FLASH PYROLYSIS OF BIOMASS

D. S. Scott and J. Piskorz
Department of Chemical Engineering
University of Waterloo
Waterloo, Ontario
Canada

Maximum yields of product oil can be obtained from biomass by very short residence time heating in nonoxidizing atmospheres at moderate temperatures (400–700°C). This has been demonstrated in several recent articles [1–3]. The very high heating rates needed require an appropriate reactor design and the use of fairly fine biomass particles. In general, heating rates must be from 10^4 to 10^6 °C/sec, and particle size should be of the order of 500 μm or less. There are only a limited number of reactor designs which can achieve the required heating rates while yielding a vapor residence time of less than about five seconds. Entrained flow in hot wall reactors, or the use of a heat carrier solid, as in a fluidized bed, are the two easiest techniques. Both have been used for short-time flash pyrolysis of biomass to study gasification yields.

Relatively few studies have been done on the flash pyrolysis of biomass with the objective of maximizing liquid yields. Because of the variety of potential feedstocks, and the number of variables to be investigated, a small-scale reactor achieving rapid steady-state and meeting the requirement of rapid heating would allow screening of many materials with respect to pyrolysis yields in shorter times and with much less expense. The construction, operation and demonstration of reliability of such an apparatus was the principal objective of this work.

SMALL SCALE FLASH PYROLYSIS APPARATUS

Design Features

The basic design concept in our work was that employed by Tyler [4] in his work at the Mineral Processing Laboratories, CSIRO, Australia, for studies of the flash pyrolysis of coal. This apparatus was a small, heated, fluidized bed of sand, operating continuously with a coal feed rate of 1–3 g/hr. This minifluid-bed reactor has successfully predicted results obtained in a 20 kg/hr pilot plant fluid bed pyrolysis unit of similar design [5]. In adapting this concept to biomass pyrolysis, however, two problems immediately become apparent. First, the method of feeding solid particles used by Tyler was based on the fluidized bed feeders of Hamor and Smith [6]. As very few, if any, biomass small particles can successfully be fluidized, there was no known proven method of feeding pulverized biomass particles at uniform continuous rates at this low level. Second, the methods of analysis used by Tyler are only partially useful for the products of biomass pyrolysis. Chars from biomass have a lower density than coal char and tend to blow out of the bed more readily. Very few light liquids (boiling under 100°C) are produced in coal pyrolysis, whereas under most conditions in biomass pyrolysis, a significant fraction of the liquid product is of this type.

The design of a new feeder required some time, but a very successful new type was developed for this work. Its performance has been described in detail elsewhere [7]. Rates as low as 1–3 g/hr were not achieved, but steady feeding over the range of 5–100 g/hr was demonstrated. A rate of 10–15 g/hr was very conveniently obtained with ± 5% constancy over one to two hours, and the fluid bed apparatus was designed for this feed rate.

Low-Rate Entrainment-Type Biomass Feeder

A drawing of the feeder as finally used is shown in Figure 1. The sawdust or other biomass in the cylindrical hopper is stirred at speeds from 100 to 400 rpm by a Servodyne constant-torque motor. Constant torque is required because feed rate depends to some extent on stirring speed, so a constant speed is necessary. Metered entrainment gas entering the feeder is split into two streams, one of which goes to the freeboard above the bed. This flow is regulated by a needle valve in the inlet line and measured separately in a capillary before entering the feeder. The balance of the flow goes through a tube which traverses the bed of stirred

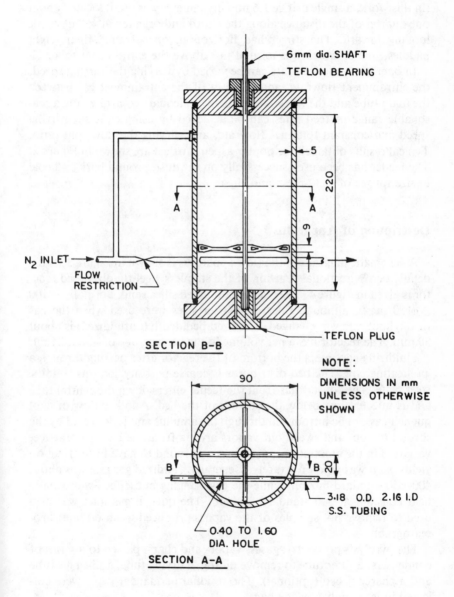

Figure 1.   Low rate entrained flow biomass feeder.

biomass particles near the bottom of the hopper. This tube is located along a chord at the half radial position. It contains one or more orifices (in this work, a single orifice 1.6 mm in diameter was used) located about one quarter of the distance along the chord and oriented at 45° upwards looking inward. The stirrer had horizontal, open frame, figure-eight paddles, one of which was located just above the entrainment tube.

In operation, the feed rate can be varied by varying the stirring speed, the entrainment flowrate, and the split of the entrainment gas between the main tube and the freeboard. Usually, adequate control giving a reasonable range of feed rates can be achieved by using constant stirring speed and constant total gas flowrate, and varying the flow split ratio. Typical results obtained for poplar-aspen sawdust are shown in Figure 2. The feeder has been used successfully on sawdust, ground bark and coal in size ranges of − 30 mesh and finer.

Description of Apparatus

A schematic diagram of the flow apparatus is shown in Figure 3, and a detailed cross-sectional drawing of the stainless steel fluidized bed reactor is given in Figure 4. The reactor contained silica sand, normally − 100 + 150 mesh, although finer or coarser sizes were used when the gas throughput rate was changed. The sand bed depth (unfluidized) is about 30 mm. The reactor has a net volume in the heated zone of about 24 ml.

Fluidizing gas enters the bottom of the reactor after passing through a preheating coil. The bed distributor is coarse-porosity, porous stainless steel plate. The "dusty" gas from the feeder enters down the central tube and is injected just below the surface of the bed. Another flow of cool quench gas can be introduced through the annulus and is deflected by the shroud to cool and sweep out vapors arising from the bed. In practice, variation in the amount of quench gas was found to have little effect on yields, so it was used only rarely. Generally, fluidized gas rate was about 400–650 standard ml/min while the gas bringing in the feed was usually held constant at 500 standard ml/min. The quench gas inlet was also used to remove hot samples of the vapors produced to an on-line chromatograph.

The pyrolysis products — gases, vapors and char — passed to a series of condensers, a filter tube to remove oil fog, a drying tube, a silica gel tube and a charcoal bed (optional). The residual permanent gases were collected in large polyethylene bags.

A typical run at 500°C lasted about 30 min. Because of the small volume, the system comes to steady state in 1–2 min. A 30-min period was

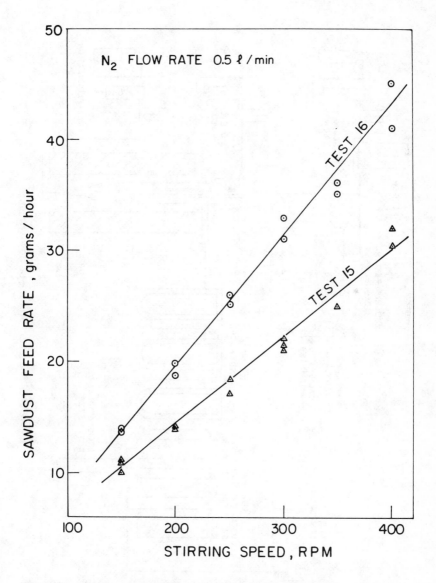

Figure 2. Feeder performance, aspen-poplar sawdust, 5% moisture, $-60 + 140$ mesh, N_2 flowrate 500 ml/min, tests 15 and 16 show effect of two different stirrer blade designs.

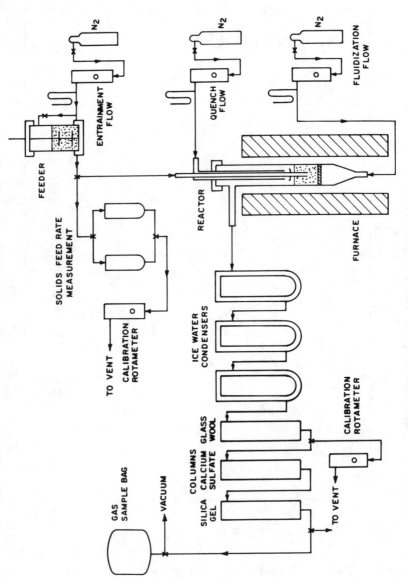

Figure 3. Schematic drawing of flow system.

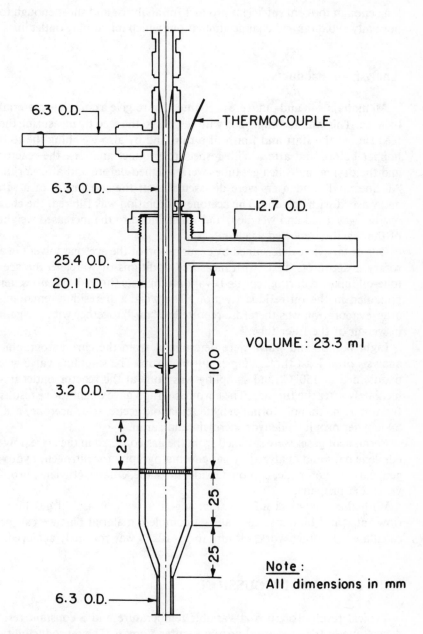

Figure 4. Details of fluidized bed reactor. Sand bed depth = 30 mm. All stainless steel type 304.

long enough to form sufficient product for analysis, and short enough to normally avoid tar and char accumulations which might plug outlet lines.

Analysis of Products

All materials in and out were accounted for to give a complete material balance. This was accomplished with respect to the feed by measuring the feed rate at the start and finish of a run, and by also weighing the feed hopper before and after. With respect to product analysis, the reactor and the drying and silica gel tubes were weighed before and after a run. All lines and condensers were disassembled at the conclusion of a run and washed with acetone. The acetone-oil solution was filtered, the char residue was dried and weighed, and when added to the increased weight of the reactor, reported as char.

Oils or tars were determined by evaporation of the acetone solvent in a rotary evaporator. Tests with measured additions of water to the acetone-oil mixture during repeated evaporation tests showed that no water remained in the oil residue fraction. The weight of residue remaining after evaporation was therefore reported as "tar," together with the gain in weight of the filter tube.

Light liquids and water were determined from the chromatographic analysis of at least three samples during a run. The sampling valve was maintained at 150°C, and sampling was done at the reactor outlet immediately after the furnace. The chromatograph gave quantitative results for water, methanol, formaldehyde, ethanol, acetic acid, acetone, acetaldehyde, propionaldehyde, acrolein and furan.

Permanent gases were sampled from the gas collected in the large polyethylene bags and analyzed by gas chromatography for nitrogen, hydrogen, carbon monoxide, carbon dioxide, methane, ethane, ethylene, propylene and propane.

With this analytical information, together with measured gas input flow rates and biomass feed rates, a complete material balance can be calculated. In other work, closure to 95–102% was routinely achieved.

RESULTS AND DISCUSSION

Typical results for runs at variable temperature and a constant residence time of 0.44 sec are shown in Figures 5 and 6. The reproducibility of results is shown by the data for tar and char yields taken five months apart. The high liquid yields obtainable from short residence time pyro-

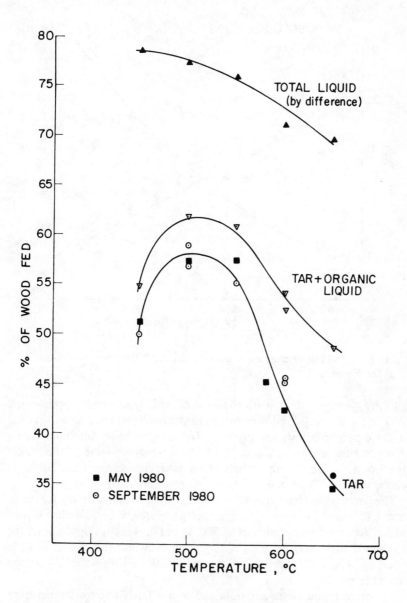

Figure 5.   Yields of liquids as a function of temperature. N_2 atmosphere, 0.44-sec residence time, $-60 + 140$ mesh poplar-aspen, 5% moisture, atmospheric pressure.

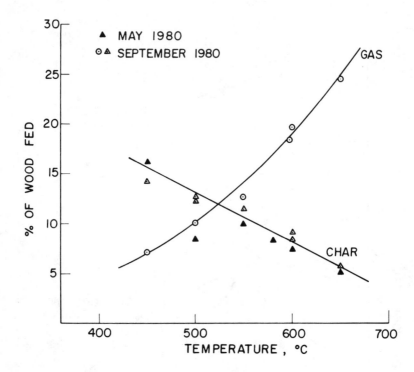

Figure 6.   Yield of char and gas with temperature. Reaction conditions as for Figure 5.

lysis are clearly evident, with the total organic liquid yield approaching 65% of the dry wood fed. The properties of the hybrid poplar-aspen used in these pyrolysis tests are given in Table I. Probably the outstanding characteristic of this species is its low lignin content—one of the lowest of all woods. The difference between organic liquids and total liquids is water, about 64% of which is water of reaction.

The decrease in char and increase in gas yield with increasing temperature is shown in Figure 6. At higher temperatures, very little char is produced. At optimum conditions of 500°C and 0.44 sec residence time, the product distribution by weight based on moisture-free wood is 61% wood oil, 4% light organics boiling under 100°C, 10% water, 14% char and 11% gas.

Optimum liquid yields are obtained over a fairly narrow temperature range of about 50°C. Yields of organic liquids decrease rapidly at both higher and lower temperatures, with char yields being the gainer at decreasing temperatures, and gas yield increasing sharply as temperature increases.

Table I.   Composition of Hybrid Poplar-Aspen Sawdust

	wt% dry wood
Cold Water Extractives	11.1
Hot Water Extractives	12.2
Acid-Insoluble Lignin	16.2
1% NaOH Solubility	14.2
Moisture Content	5.0
Ash Content	0.39

The gas compositions as functions of temperature are shown in Figures 7 and 8. From Figure 7, the rapid increase in decarboxylation reactions at higher temperatures is clearly evident. The corresponding increase in yields of hydrocarbon gases is shown in Figure 8. It is interesting to note the very small yield of hydrogen even at 650°C, despite a significant partial pressure of water in the gas.

Effects of residence time variation were measured over the relatively narrow range of variation conveniently possible in our fluidized bed (0.38–1.0 sec). As expected, liquid and char yield decreases and gas yield increases with longer residence time. The residence times given are those for the gases before leaving the reactor. Residence times of pyrolysis vapors in contact with char and hot fluidized solids would be about one-third of these values. The reactor was operated batch-wise with respect to solids; however, only 5–10% of the char formed remained in the bed and the remainder blew over with the pyrolysis vapors. Observations of char behaviour give a rough estimate of 2–3 sec as the char residence time in the bed before the density was such that the char particle was carried out of the bed. Additional tests showed that the particle size of $-250 \mu m$ to $+105 \mu m$ ($-60 + 140$ mesh) was close to the optimum for our test conditions. Both smaller and larger sizes gave results very similar to those shown in Figures 5 and 6 but with somewhat lower yields of liquids.

CONCLUSIONS

A small-scale continuous flash pyrolysis unit has been designed and routinely operated. Much of the reliability of the test unit is due to a new low-rate continuous feeder for fine biomass particles developed in this work. At feed rates of about 15 g/hr, a run requires 30 min, and all basic analyses can be completed in one day by one person. All products are accounted for quantitatively, and material balances generally close to better than ± 5%. The apparatus allows a wide range of raw materials and

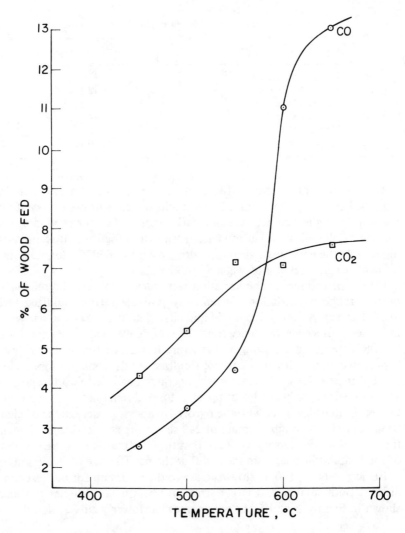

Figure 7.   Yield of carbon monoxide and carbon dioxide with temperature. Reaction conditions as for Figure 5.

operating conditions to be investigated rapidly and inexpensively. Reproducibility of results is very good.

Tests with hybrid poplar-aspen sawdust have demonstrated that at optimum conditions, high yields of organic liquid products are possible. A wider range of test conditions is being investigated, as are the chemical characteristics of the wood oils produced.

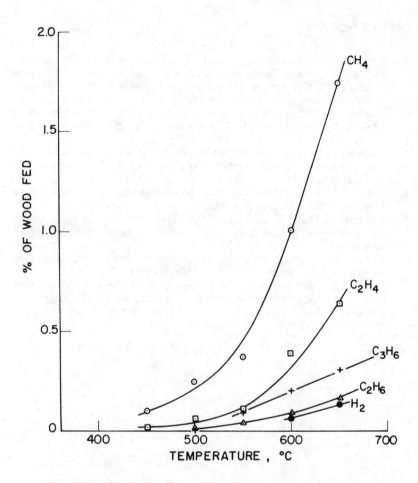

Figure 8. Yield of minor gas components with temperature. Reaction conditions as for Figure 5.

ACKNOWLEDGMENTS

This work was a part of ENFOR Project C-28 funded by the Canadian Forestry Service of the Department of the Environment, and the authors wish to express their thanks to the ENFOR program and to FORTIN-TEK Canada Corp., for financial and technical assistance. Thanks are also extended to cooperative engineering students R. Wagler, G. Slater and W. Selmici who assisted with the work.

REFERENCES

1. Rensfelt, E., G. Blomkvist, E. Ekstrom, S. Engstrom, B. G. Espensos and L. Liinsnki. "Basic Gasification Studies for Development of Bio-mass Medium BTU Gasification," in Symposium Papers, Energy from Biomass and Wastes, Inst. Gas Technology, (1978).
2. Kuester, J. L. "Conversion of Cellulosic and Waste Polymeric Material to Gasoline," Interim Report, U.S. DOE Contract No. EY-76-S-02-2982-000, (1979).
3. Shafizadeh, F., and A. G. W. Bradbury. "Thermal Degradation of Cellulose in Air and Nitrogen at Low Temperatures," *J. Appl. Polym. Sci.* 23:1431–1436 (1979).
4. Tyler, R. J. "Flash Pyrolysis of Coals: Devolatization of a Victorian Brown Coal in a Small Fluidized Bed Reactor," *Fuel* 58:680–686 (1979).
5. Edwards, J. H., I. W. Smith and R. J. Tyler. "Flash Pyrolysis of Coals: Comparison of Results from 1 g/hr. land 20 kg/hr. Reactors," *Fuel* (submitted).
6. Hamor, R. J., and I. W. Smith. "Fluidizing Feeders for Providing Fine Particles at Low, Stable Flows," *Fuel* 50:394–399 (1971).
7. Scott, D. S., and J. Piskorz. "A Low Rate Entrainment Feeder for Fine Solids," in *Proceedings of the Second Annual Bioenergy R&D Conference,* (Ottawa: National Research Council of Canada, 1980).

CHAPTER 24

PROCESS DEVELOPMENT FOR DIRECT LIQUEFACTION OF BIOMASS

D. C. Elliott

Battelle Pacific Northwest Laboratory
Richland, Washington

The U.S. Department of Energy (DOE) Biomass Liquefaction Experimental Facility at Albany, OR, was constructed for the purpose of developing processes to convert biomass to liquid fuels [1-3]. Facility equipment was sized to process 1-3 ton/day of wood chip equivalent [4]. The facility was dedicated in December 1976, and process development work has been under way since the summer of 1977. Reactant and product handling difficulties have required facility and process modifications. Bechtel National Inc., the original operating contractor for the facility, provides many of the details of these modifications in the reports describing the work performed during their contracting period [5,6]. Presently, there are two main versions of the CO-Steam process being tested at Albany under the direction of Wheelabrator Cleanfuel Inc., the facility operator. The original process, called the Bureau of Mines (BOM) process because it was developed by researchers at the former BOM station near Pittsburgh PA, [7-14], involves dried and ground woodchips slurried in a heavy product oil medium. The newer process is called the LBL process, since it was developed by staff members of the Lawrence Berkeley Laboratory [15,16]. In the LBL process, woodchips are broken down into a pumpable water slurry by acid hydrolysis. In either process, the slurry is then pumped into a high-temperature, high-pressure reactor wherein the product oil is formed through the action of carbon monoxide and steam under the influence of a sodium carbonate catalyst. The role of the Battelle Pacific Northwest Laboratory (PNL) in the process development effort has been twofold: to provide bench scale process development experimental support, and to provide analytical support as needed. The bench-scale work has involved for the most part batch autoclave tests and this work has been reported elsewhere [17–19]. This chapter provides the details of the analytical work completed on the biomass-derived product oil.

PROCESS DESCRIPTIONS

BOM Process

The Albany facility was originally constructed to develop the process of biomass conversion to a product liquid useful as a fuel oil, as shown diagrammatically in Figure 1. Woodchips are dried, ground and mixed with recycle product oil at 20–30% solids to provide a pumpable slurry. Anthracene oil, a coal tar distillate, is used as the startup oil. This slurry, along with carbon monoxide and aqueous sodium carbonate, are pumped at high pressure (13.8–27.6 MPa) through a scraped-surface preheater and into a stirred tank reactor. The average residence time in the reactor can be varied from 20 to 90 min at temperatures ranging from 300 to 370°C. After leaving the reactor, the product is cooled and the pressure is let down into a flash tank where fixed gases and most of the water is removed. A major change in the original process flow is the replacement of the centrifuge in the product cleanup stage with a vacuum still. After pressure letdown, the product is reheated and flashed in the still, where a light product oil is drawn off; a middle fraction is recovered and a portion is recycled for slurry makeup; heavy product, solids and catalyst residue are removed from the still bottom. This is the extent of the unit operations at Albany. However, the total process plan would have the still bottoms pumped to a gasifier for production of CO/H_2 gas feed for the liquefaction process. Sodium could be leached from the gasifier ash and could be recycled to the process probably after reaction with carbon dioxide from the offgas and gasifier product gas.

LBL Process

There are several basic differences between this process and that for which the Albany facility was designed. However, through modifications, the plant was made to operate in this mode, and in fact, the first large-scale product oil production was by this process. Woodchips (Fig-

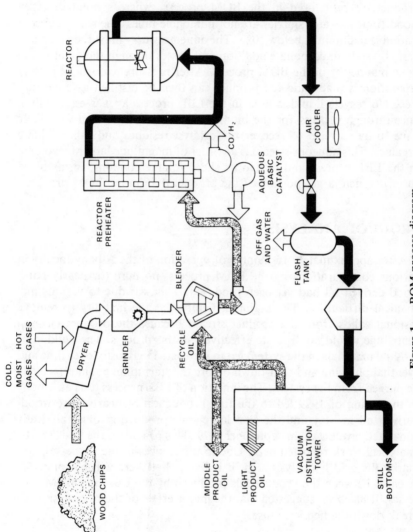

Figure 1. BOM process diagram.

ure 2) are reduced directly to a pumpable aqueous slurry through an acid hydrolysis step without preliminary drying and grinding. The wood-to-water ratio can be maintained at a level equivalent to the wood-to-product oil ratio used in the BOM process, whereas nonhydrolyzed wood flour in water slurries are not pumpable unless the solids concentration is maintained below 10%. The aqueous slurry must then be made basic by sodium carbonate addition and passes through the plant in the same manner as in the BOM process. After pressure letdown, a gravity separation is made and the product can then be distilled as a cleanup step. No recycle liquid is used in the LBL process as it is essentially a once-through process for the biomass. The aqueous stream will likely have to be recycled to recover the catalyst residues and other soluble organics. This aqueous stream is an order of magnitude larger by volume in the LBL process and due to its sheer volume contains, in dissolved form, a much larger portion of the product than in the BOM process.

PRODUCT ANALYSIS

After approximately 15 months of operation of the Albany facility in various configurations by the BOM process, no pure (or nearly pure) wood-derived oil had yet been produced. This was due to various mechanical difficulties. The major difficulty was the inability to remove residual solids from the product stream because the product cleanup centrifuge would not operate effectively in this process. The buildup of residual materials in the system led to increases in viscosity over time and eventual plugging and shut-down before the startup oil could effectively be purged from the system. The initiation of LBL process tests at Albany in the spring of 1979 led to the first production of nearly pure wood-derived oil in May and the first large-scale production of catalytically converted product oil in September 1979. This product is the basis for the analytical work reported here. Due to differences in the processes, primarily the acid hydrolysis step, it is likely that there will be some differences between the product oils produced by the LBL and BOM processes. It has been suggested that the major effect of the hydrolysis is to break down the hemicellulose.

Vacuum Distillation Procedure

A vacuum fractional distillation of wood-derived oil was performed by the use of an ASTM-D1160 distillation apparatus with a modified receiver which allows fraction collection while continuing the distillation

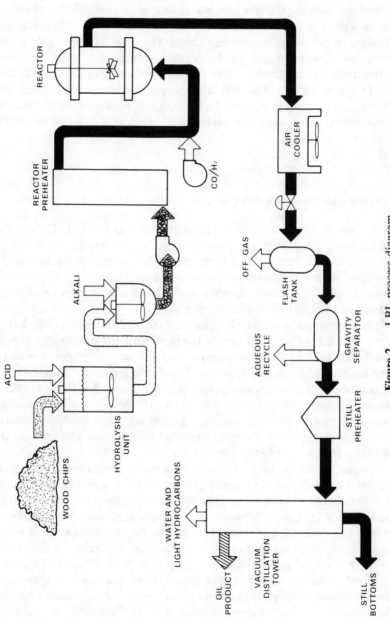

Figure 2. LBL process diagram.

under vacuum. The fractions collected are described in Table I. Fraction 1 includes both the water which was dissolved or emulsified in the wood oil as well as a light oil fraction which was immiscible with water and distilled in the same temperature range. The codistillation could be the result of similar boiling points or may also be the result of steam distillation. The atmospheric true boiling points were calculated based on the D1160 procedure. The distillation was discontinued at the point that decomposition of the product in the still pot became evident. The decomposition point is approximately 40°C below that experienced for petroleum crude oils.

Analysis of Distillate Fractions

A summary of the analytical data is presented in Table II. The elemental analyses, performed on a Perkin-Elmer 240 instrument, show a trend of increasing carbon content from the lighter to heavier fraction and a stronger reverse trend in hydrogen content. The hydrogen-to-carbon atomic ratio as a result shows a trend from nearly 2 in the lightest fraction to less than 1 in the still bottoms.

The oxygen content, which is directly determined on the P-E 240, is less patterned in that it is lowest in the light distillate, maintains a higher, nearly constant level through most of the distillate range and then drops to a lower level in the still bottoms. These data are mirrored in the heats of combustion results for the various oils. It is interesting to note that the nitrogen appears for the most part in two of the heavier distillate fractions, but not in the still bottoms. Elemental sulfur analysis puts the content at 0.006% for the total wood oil [20]; similar analyses for the distillate fractions were not performed.

The use of proton and ^{13}C nuclear magnetic resonance (NMR) spectrometry and infrared (IR) spectrophotometry has provided some insight into the chemical structure of the wood oil components. The ^{13}C NMR data show a fairly even balance between saturated and unsaturated carbon in the distillate oils. However, proton NMR shows a much larger amount of aliphatic hydrogen in proportion to aromatic hydrogen. There is essentially no olefinic hydrogen. Aromatic compounds, as a result of molecular bonding and structure, have a lower hydrogen-to-carbon ratio than aliphatics, (one or less for aromatic, greater than two for aliphatic). The disproportionately large amount of aliphatic hydrogen is an indication of the large amount of aliphatic substitution on the aromatic ring structures. This data is an average of dozens of

Table I. Vacuum Fractional Distillation of Wood Oil (ASTM-D1160) for Sample TR7-136

Fraction	Actual Amount	Relative Amount (%)	Color	Boiling Point at 1 atm (°C)	Boiling Point at 10 mm Hg (°C)
1	8 ml light oil 23 ml water	3 8	Clear	To 138	To 10
2	45 ml	18	Clear to yellow	138–266	10–132
3	35 ml	14	Green to orange	266–316	132–166
4	40 ml	16	Orange	316–382	166–243
5	20 g	8	Orange to brown	382–432	243–266
Residue	86.6 g	32	Dark Brown	Above 432	Above 266[a]

[a] Pot at 332°C, decomposition.

Table II. Analytical Data for Distillation Fractions

Fraction	C (wt %)	H (wt %)	N (wt %)	O (wt %)	Atomic H/C	HHV (MJ/kg)	^{13}C NMR Ali/Aro C	^{1}H NMR Ali/Aro H
1 (Oil Layer)	78.8	12.0	0.0	9.7	1.81	37.2	12	30
2	77.2	9.9	0.0	13.3	1.52	35.4	1.1	10.0
3	77.1	8.9	0.0	13.4	1.37	35.1	1.0	7.3
4	79.2	8.9	0.5	12.1	1.33	36.8	1.2	6.6
5	79.4	7.9	0.2	12.3	1.19	35.1	1.0	5.3
Residue	82.3	6.5	0.0	10.4	0.94	34.7		
TR7-136 (Including 8% Water)	72.3	8.6	0.2	17.6	1.41	33.7	0.53	

chemical compounds and, as such, shows a trend of decreasing amounts of aliphatic compounds and of aliphatic substitution on the aromatic rings through the distillation range. The proton NMR data also show the presence of other functional groups such as furans in Fraction 2 and naphthalenic and aromatic acid and ester compounds in Fractions, 4 and 5. The methoxy aromatic structure is very prominent in Fraction 2 but is also evident in the heavier fractions. Long-chain, oxygen-containing alkyl groups disappear from prominence after Fraction 2, however, the ethyl ether functional group remains prominent throughout. The IR spectra of these fractions do not provide nearly as definitive results as the NMR spectra, but they generally confirm the above-stated conclusions.

We have thus far been able to identify a significant number of the actual components of the distillate fractions of the wood oil through the use of gas chromatography/mass spectrometry (GC/MS). The components in Table III were identified by analysis of computer matched data from analyses on a J&W 60-m wall-coated open tubular (WCOT) SE54 capillary column operated in a temperature-programmed oven. Those compounds listed with a question mark could not be matched due to the limitations of the computer search library, but were determined by analysis of the mass spectra. In addition, the acid functional groups shown in fractions 4 and 5 were identified in derivatized (trimethylsily-lation) samples of the wood oil fractions. Work continues in this area as those compounds identified are not nearly all the compounds present.

Additional analytical results from petroleum crude oil test methods have also been produced for the wood oil [20]. These tests, performed at Southern Petroleum Laboratories, Inc., are indicative of the difference between LBL process wood-derived oil and crude petroleum. The figures in Table IV show that the wood oil is a heavy, nonaliphatic oil. The high solids and salt content will likely be reduced to nearly zero by the vacuum distillation step of product cleanup. Neutralization numbers for the distillable fractions of the oil ranged from 17.7 to 5.3 mg KOH/g. The existent gum ranged from 621 to 827 mg/100 ml of the same distillable fractions.

Ames assays have been completed on the crude product and five wood oil distillate fractions. These are all process streams derived from an LBL mode of operation at Albany and contain no coal tar distillate. The Ames assay measures the mutagenic tendency caused in a microorganism by addition of some foreign material. This test is a preliminary measure of the carcinogenic characteristic of the material tested. All the biomass liquefaction streams examined were determined to be inactive by the Ames assay. This is similar to the activity of crude petroleum and its fractions and is a much lower level of activity than coal tar, coal-derived

Table III.   Chemical Components of
Wood Oil Fractions by GC/MS

Fraction 1
  C_6-Diene
  Methyl Cyclopentene (two isomers)
  Methyl Hexadiene
  2-Pentanone
  Dimethyl Hexadiene
  2-Methyl Cyclopentanone
  Methyl Cyclopentadiene
  Ethyl Benzene
  Cyclooctane
  Dimethyl Heptene
  C_3-Benzene
  Indan
  Guaiacol
  Furfural
  Methyl Indan (three isomers)
  Dimethyl Indan (five isomers)
  Ethyl Styrene

Fraction 2
  Methyl Pentenal
  Formyl Dihydropyran
  Dimethyl Furan (two isomers)
  Trimethyl Furan
  Guaiacol
  Furfural
  Ethyl Styrene
  para-Cresol
  4-Methoxyphenol
  Methyl Indan
  Dimethyl Phenol
  Ethyl Phenol
  Dimethyl Indan
  Methyl Ethyl Phenol (two isomers)
  Trimethyl Phenol
  Dimethyl Ethyl Phenol
  Dihydroxy Acetophenone
  sec-Butyl Phenol
  Propyl Guaiacol

Fraction 3
  Propyl Guaiacol
  Dimethyl Methoxyphenol?
  Trimethyl Methoxyphenol?
  C_4-Methoxyphenol
  C_7-Phenol?
  C_8-Phenol?
  Dimethyl Naphthol
  Trimethyl Naphthol

Table III, continued

Fraction 4
Methyl Naphthol (two isomers)
Dimethyl Naphthol (seven isomers)
Trimethyl Naphthol
Alkylated Hydroxyphenyl Acids? (mol wt 138-206)

Fraction 5
Alkylated Hydroxyphenyl Acids? (mol wt 182-224)

Table IV. Analysis of LBL Process Wood-Derived Oil [20]

API Gravity at 16°C (60°F)	-4.93
Specific Gravity at 16°C (60°F)	1.12
Pentane-Soluble, vol %	3.25
Salt, kg/m³ (lb/1000 bbl)	0.227 (79.4)
Total Solids, BS&W	8.0

liquids or shale oil. This is not inconsistent with the results of the analysis of shale and coal oils since it was concluded that the detrimental activity in these products results from polynuclear and nitrogen-containing aromatics [21], neither of which is likely to be present in wood oil to any significant extent. However, even trace contaminants of the coal tar startup oil in the BOM product may be significant relative to activity as determined by the Ames assay.

Because of process differences, there are some minor characteristic differences between the BOM product and the LBL product. Prior to distillation, the LBL oil has a significantly larger amount of dissolved or emulsified water. There is also a slightly greater amount of oxygen in the LBL oil even when determined on a dry basis. As a result, the heating value of the LBL product is slightly lower than that for BOM product. Physically and chemically, the products appear to be similar. However, there is an apparently lower concentration of phenolics in the BOM oil. This difference may actually be a result of less dilution of the phenolic oils in the LBL product because the cellulosic components are more severely broken down to water soluble materials due to the hydrolysis pretreatment step.

ECONOMIC ANALYSIS

Numerous analyses have been performed on the economics of direct liquefaction of biomass in the past [22-24]. Presently, the Biomass

Energy Systems Division, DOE has assigned to SRI International the task of formulating a cost databank for biomass processes. Part of their updated results were recently included in a review of the biomass liquefaction program [25]. The evaluation of the LBL and BOM processes was undertaken with very few continuous process data available and as such, provides only a rough estimate of the projected economics. It does indicate that with the present technology, the product oil is expensive. Table V is a summary of the relevant data.

These calculations are for a commercial-sized plant including many unit operations which have not yet been demonstrated at the Albany scale of operations and are based to a significant degree on engineering judgment. The conclusion from the economic analysis is that the processes appear to be viable technically. However, the product presently produced is not equivalent to any commercial product and the extreme hydrotreatment required to produce an aliphatic hydrocarbon oil similar to petroleum would not be economically feasible. Therefore, the product must be considered on its own merits, and market substitution must be made based on consideration of the differences between wood-derived oil and petroleum hydrocarbons. There are many remaining questions relative to the Albany process. Process development work at DOE's experimental facility should provide answers to these questions. These answers will likely have a significant effect on process costs, however, it is not entirely clear whether the cost will increase or decrease.

An additional area that will require analysis will be the use of the wood oil as a petroleum substitute in chemical production. The distinctive set of functional groups (primarily phenolics) suggests that chemical utilization of the product could be feasible. Use of the product in phenol/formaldehyde resins and for surfactant purposes is being studied. The pay back of such chemical usage of the product or a portion thereof could well make biomass liquefaction economically viable, but this is speculation at this point in time.

Table V.   Cost Data for Wood-Derived Oil[a]

	LBL Process	BOM Process
Capital Cost for 1814 green metric ton/day $ million	39.5	56.1
Product Cost (100% equity)		
$/GJ	9.1	10.1
$/MBtu	9.6	10.7
l/m³	345	336
$/bbl	55.0	53.4

[a]Mid-1979 constant dollars; 15% DCF ROI on equity; debt interest rate, 9% long-term, 10% short-term; wood cost at $1.19/GJ ($1.25/MBtu).

Table VI.   Comparison of Some Fuel Oils

	C (wt %)	H (wt %)	N (wt %)	O (wt %)	S (wt %)	Ash (wt %)	Moisture (wt %)	HHV (MJ/kg)	Density (g/ml)
LBL Wood Oil	72.3	8.6	0.2	17.6	0.006	0.78	8.5	33.7	1.19
BOM Wood Oil	82.0	8.8	0.6	9.2	0-3.5[a]	0.66	3.1	36.8	
No. 6 Fuel Oil [26]	85.7	10.5	2.0	0-3.5	0-3.5[a]	<0.05	0.20	42.3	0.98
Pyrolytic Oil [26]	57.5	7.6	0.9	33.4	0.2	0.5-1.0	14	24.4	1.30

[a]Legal sulfur limit determined by use site, e.g., 0.35% maximum in Los Angeles County.

CONCLUSIONS

When considered for use as a substitute fuel oil, wood oil as produced at Albany appears to fall qualitatively somewhere between petroleum-derived No. 6 fuel oil and the synthetic oil derived from the Occidental flash pyrolysis process as shown in Table VI. Wood oil falls nearly half-way between the other two oils in nearly all categories, except that wood oil is very low in sulfur.

This comparison is valid on a chemical basis, but as stated earlier, the use of wood oil purely as a substitute fuel is not currently economically attractive. This process is at the development stage and new technology could have a significant impact on the process economics. The alternative use of wood oil as a chemical feedstock is also being studied.

ACKNOWLEDGMENTS

I wish to acknowledge the support provided by other staff members at PNL, but particularly J. A. Franz, who operated the NMR; R. E. Schirmer, who operated the GC/MS; and R. A. Pelroy, who supervised the Ames assays. Also, I wish to thank the Department of Energy Biomass Energy Systems Division, who provided the financial support for this work.

REFERENCES

1. "Study to Determine Optimal Use of Albany, Oregon, Pilot Plant for Advancement of Processes for Conversion of Cellulosic Materials to Liquids," prepared by Energy and Waste Technology Staff, Chemical Technology Department, Battelle Pacific Northwest Laboratory (1976).
2. "The Technical and Economic Desirability of Waste-to-Oil Liquefaction Process," prepared for NSF by Bechtel Corp. (1975).
3. Houle, E. H., S. F. Ciriello, S. Ergun and D. J. Basuino. "Technical Evaluation of the Waste-to-Oil Pilot Plant at Albany, Oregon," by Bechtel Corp, for U.S. ERDA Contract #E(04-3)-1194 (1976).
4. Del Bel, E., S. Friedman, P. M. Yavorsky and H. H. Ginsberg. "Design of a Wood Waste-to-Oil Pilot Plant," American Chemical Society Division of Fuel Chemistry Preprints 20(2):17–21 (1975).
5. Lindemuth, T. E. "Investigations of the PERC Process for Biomass Liquefaction at the Department of Energy, Albany, Oregon, Experimental Facility," ACS Symp Ser #76 (1978), pp. 371–91
6. "Final Technical Progress Report – Liquefaction Project, Albany, Oregon," by Bechtel National Corp for U.S. DOE Contract #EG-77-C03-1338 (1978).

7. Appell H. R., I. Wender and R. D. Miller. "Conversion of Urban Refuse to Oil," U.S. BOM, TP25 (1970).
8. Appell, H. R., Y. C. Fu, S. Friedman, P. M. Yavorsky and I. Wender. "Converting Organic Wastes to Oil, A Replenishable Energy Source," U.S. BOM, RI7560 (1971).
9. Friedman, S., H. H. Ginsberg, I. Wender and P. M. Yavorsky. "Continuous Processing of Urban Refuse to Oil Using Carbon Monoxide," paper presented at the 3rd Mineral Waste Utilization Symposium, Chicago, IL, March 14–16, 1972.
10. Fu, Y. C., S. J. Metlen, E. G. Illig and I. Wender. "Conversion of Bovine Manure to Oil," paper presented at Division of Fuel Chemistry, ACS National Meeting, New York, August 1972.
11. Fu, Y. C., E. G. Illig and S. J. Metlin. "Conversion of Manure to Oil by Catalytic Hydrotreating," *Environ. Sci. Technol.* 8(8):737–740 (1974).
12. Appell, H. R., Y. C. Fu, E. G. Illig, F. W. Steffgen and R. D. Miller. "Conversion of Cellulosic Wastes to Oil," U.S. BOM, RI8013 (1975).
13. Wender, I., F. W. Steffgen and P. M. Yavorsky. "Clean Liquid and Gaseous Fuels from Organic Solid Wastes," in *Recycling and Disposal of Solid Waste,* M. E. Henstock, Ed. (Elmsford, NY: Pergamon Press, 1975), pp. 43–99.
14. Appell, H. R. "The Production of Oil from Wood Waste," in *Fuels from Waste,* L. L. Anderson and D. A. Tillman, Eds. (New York: Academic Press, Inc., 1977), pp. 121–140.
15. Ergun, S. "Biomass Liquefaction Efforts in the United States," U.S. DOE report LBL-10456 (1980).
16. Schaleger, L. L., N. Yaghoubzadeh and S. Ergun. "Pretreatment of Biomass Prior to Liquefaction," in *3rd Annual Biomass Energy Systems Conference Proceedings,* SERI/TP-33-285 (Golden, CO: Solar Energy Research Institute, 1979), pp. 119–122.
17. Elliott, D. C. and P. C. Walkup. "Bench Scale Research in Thermochemical Conversion of Biomass to Liquids in Support of the Albany, Oregon Experimental Facility," U.S. DOE report TID-28415 (1977).
18. Elliott, D. C., and G. M. Giacoletto. "Bench Scale Research in Biomass Liquefaction in Support of the Albany, Oregon, Experimental Facility," in *3rd Annual Biomass Energy Systems Conference Proceedings,* SERI/TP-33-285 (Golden, CO: Solar Energy Research Institute, 1979), pp. 123–130.
19. Elliott, D. C. "Bench Scale Research in Biomass Liquefaction by the CO-Steam Process," in *Proceedings of the 29th Canadian Chemical Engineering Conference, Sessions on Synthetic Fuels from Coal and Biomass,* (Ottawa, Ontario: Canadian Society for Chemical Engineering, 1979), pp. 326–344.
20. Winfrey, J. C. "A Laboratory Evaluation of Biomass Oil (TR7-136)," prepared for Rust Engineering by Southern Petroleum Laboratories, Inc. Lab Report #77782 (1979).
21. "Biomedical Studies on Solvent-Refined Coal (SRC II) Liquefaction Materials," U.S. DOE report PNL-3189 (1979).
22. Dravo Corp. "Economic Feasibility Study for Conversion of Wood Wastes to Oil," for U.S. BOM, BKC-2515 (1973).
23. Ergun, S. "Economic Feasibility Assessment of Biomass Liquefaction," in *3rd Annual Biomass Energy Systems Conference Proceedings,* SERI/TP-33-285 (Golden, CO: Solar Energy Research Institute, 1979), pp. 147–148.

24.   Kam, A. Y. "Hydrocarbon Liquids and Heavy Oil from Biomass: Technology and Economics," IGT Symposium Energy from Biomass and Wastes IV, Lake Buena Vista, FL, (1980).
25.   Wilhelm, D. J., and J. W. Stallings. "Assessment of the Biomass Liquefaction Facility in Albany, Oregon, and Related Programs," by SRI International for U.S. DOE Contract DE-AC03-80C583008 (1980).
26.   Paber, K. W., and H. F. Bauer. "The Nature of Pyrolytic Oil from Municipal Solid Waste," in *Fuels from Waste,* L. L. Anderson and D. A. Tillman, Eds. (New York: Academic Press, Inc., 1977), pp. 73–86.

CHAPTER 25

MECHANISMS OF DIOL, KETONE AND FURAN FORMATION IN AQUEOUS ALKALINE CELLULOSE LIQUEFACTION

Rachel K. Miller, Peter M. Molton and Janet A. Russell

Battelle Pacific Northwest Laboratory
Richland, Washington

Within the large area of biomass conversion to synthetic fuels and chemical feedstocks, we examined aqueous alkaline liquefaction by heat and pressure. Liquids derived from biomass are particularly interesting now, since consumption of natural petroleum has created a demand for alternative energy resources. Studies have examined the conversion of cellulosic material to liquids since the 1920s [1-4]. However, several early studies on the degradation of cellulose did not focus on the formation of product liquids from cellulosic material but rather on the formation of coal [5-8]. Three publications review the literature on the following aspects of cellulose degradation: thermal degradation [9], alkaline degradation [10] and thermal aqueous degradation [11].

Although biomass conversion has been studied for more than 50 years, the mechanisms involved are not yet clearly understood. A better understanding of liquefaction reactions and compounds will help us choose reaction conditions and equipment designs that will minimize problems currently encountered in liquefaction attempts, such as high viscosity and corrosiveness [12]. Second, this understanding will help us to select an efficient means of separating problem compounds, such as potentially corrosive phenols, from the product. Third, we can better determine optimal uses for the conversion products.

There is no standard source or composition of biomass. For example, sewage sludge is a mixture of many compounds. This makes it hard to trace the complex chemical interactions from a given initial compound to a specific chemical product. To reduce the complexity of the reaction system, we chose to liquefy separately the major components of biomass. We first turned our attention to cellulose, since it is the major component.

We report formation pathways for three major groups of compounds among the identified products of cellulose liquefaction. The intermediates that we suggest are simple, initial, cellulose-degradation products such as acetone [13] and acrolein. Subsequently, these products undergo aldol condensation and Michael reaction to yield furan derivatives and cyclic ketones. Our current work with model compounds supports these postulated mechanisms.

CONVERSION PROCEDURE

For our work with cellulose, we used 8 and 38-liter autoclaves containing aqueous slurries of pure cellulose (23% Solka-Floc®*) and sodium carbonate (0.3–2.4 N). However, to avoid the long heating and cooling times needed for the large autoclaves, we used 7-ml reaction vessels in our most recent work. The small-scale reactions required only 5 min to reach the final temperature, while the large-scale liquefactions required 3 hr. Once at final temperature, which for 27 experiments ranged from 268 to 407°C, the temperature was maintained for 20 or 60 min. During this temperature hold, the resultant pressures ranged from 20.7 to 34.5 MPa (3000 to 5000 psi).

The conversion products of this process are product oil, char, an aqueous layer, and gas. The viscosity of the product oils produced by the 27 large-scale runs varied, depending on the conversion conditions. For example, using one set of conditions, the product oil had a viscosity of 0.550 Pa-sec (5.50 P) at 134°C, while other conversion conditions produced liquid product with a viscosity of 2.800 Pa-sec (28.00 P) at 132°C. The average liquid yield was 25%, expressed as weight percent of cellulose converted to acetone-soluble oil. In general, the char was less than 10% by weight, and the aqueous phase contained up to 5% of the organics. The gas was a mixture of up to 70% carbon dioxide and 7% hydrogen; the balance was nitrogen with trace amounts of carbon monoxide and methane.

PRODUCT ANALYSIS

Samples of the product oils were analyzed to identify compounds. The techniques used in preliminary sample cleanup were: filtration, prepara-

---

*Trademark of Brown Co., New York.

tive thin-layer chromatography, vacuum distillation and column chromatography. Figure 1 shows how these methods generated 15 of the samples.

The sample components were separated by gas chromatography (GC), with gas chromatography/mass spectrometry (GC/MS) providing the data for identification of compounds. Of the 15 samples in Figure 1, we analyzed 8 on a 30-m SP2100 glass capillary column, 6 on a 1.8-m Carbowax 20M packed column, and 1 on a 1.8-m OV-17 packed column. We used temperature programming for each analysis, varying the specific conditions as needed to give the best resolution. Comparison of the sample mass spectra with standards [14] enabled identification of 20% of the compounds evident in the gas chromatograms.

Of the 85 compounds that we have identified, 78 are summarized in Table I. A complete list of these 78 compounds has been published previously [15]. The additional seven compounds which we have since identified are:

- two aliphatic acids and esters (methyl formate and methoxy acetic acid);
- two cyclic ketones (formylcyclopentene and 2-methyl-5-isopropenyl-2-cyclohexenone);
- two furans (furfural and 5-methyl-2-furaldehyde); and
- one alkyl benzene (methyl isopropyl benzene).

Compounds whose mass spectra showed excellent correlation with standards were highly probable identifications. Probable identifications had good correlation with the standards, although two or three peaks failed to match closely. Finally, tentative identifications are likely to be isomers or other closely related compounds since only two or three of the peaks matched well. Half of the compounds in Table I are probable identifications, while the remainder are tentative. The large number of compounds in each sample resulted in incomplete separation by GC. Therefore, more than half of the mass spectra contain peaks from more then one compound, contributing to the high percentage of tentative identifications. The list includes 23 compounds that were identified in previously analyzed samples. Details of the analysis are described in an earlier publication [16].

PROPOSED MECHANISMS

Diols and Aliphatic Ketones

Two of the diols, ethylene glycol and diethylene glycol, are known products of the hydrogenolytic cleavage of carbohydrates [17,18]. Two

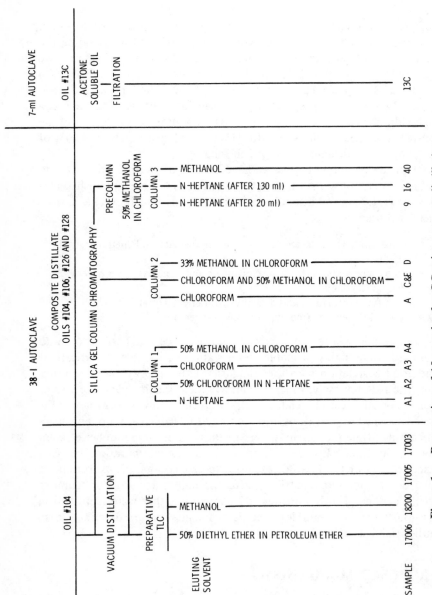

Figure 1. Preparation of 15 samples for GC using vacuum distillation, filtration, thin-layer chromatography and column chromatography.

Table I.   Compounds Identified as Products of Aqueous
Alkaline Cellulose Liquefaction

Class of Compounds	Number of Compounds Identified	Mol Wt
Open-Chain Aliphatics		
Aliphatic Hydrocarbons	8	84–296
Aliphatic Alcohols	6	62–162
Aliphatic Ketones	7	56–126
Cyclic Compounds		
Cyclic Hydrocarbons	7	110–160
Cyclic Alcohols	2	86 and 100
Cyclic Ketones	12	84–152
Furans	13	86–166
Aromatics		
Alkyl Benzenes	11	106–148
Phenol Derivatives	10	94–164
Polyfunctionals	2	112 and 162

of the aliphatic ketones, 4-hydroxy-4-methyl-2-pentanone and 4-methyl-3-penten-2-one, are commonly known acetone condensation products.

Cyclic Ketones

We identified 12 cyclic ketones as products of cellulose liquefaction, including 2,5-dimethyl-2-cyclopentenone. This compound was also produced in a model compound experiment that reacted propanal with acetoin (3-hydroxy-2-butanone) under the same reaction conditions used in cellulose liquefaction. Therefore, the experiment supported our predicted formation of 2,5-dimethyl-2-cyclopentenone by Michael reaction between propanal and dehydrated acetoin followed by cyclization of the intermediate, as shown in Figure 2. Our model compound data also supported a similar mechanism for the formation of 2,4-dimethylcyclopentanone by the aldol condensation of propanal and methylacrolein (2-methylpropenal).

Reaction of acetone and acrolein (propanal) supported one of two similar mechanisms for the formation of cyclohexanone. We had postulated that addition of the enol form of acetone to acrolein followed by cyclization would yield cyclohexanone and 2-methylcyclopentanone. Both of these compounds are produced by cellulose liquefaction; how-

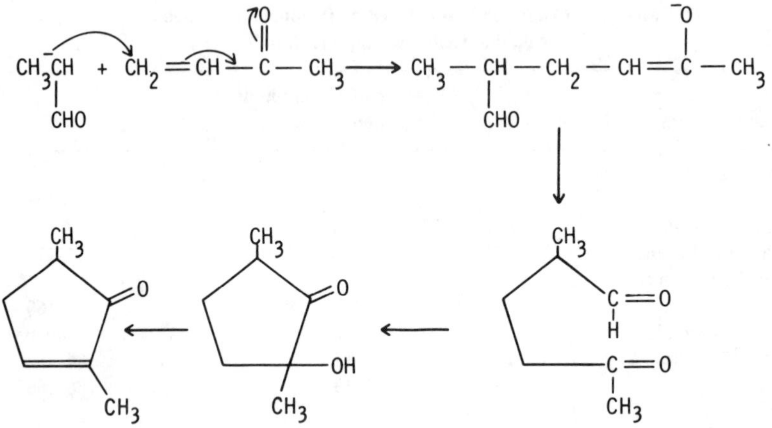

Figure 2.   Condensation of propanal and dehydrated acetoin (3-buten-2-one) followed by cyclization may lead to 2,5-dimethyl-2-cyclopentenone.

ever, 2-methylcyclopentanone was not produced when acetone and acrolein were used in model compound experiments. Rather, 2-cyclohexenone, cyclohexanone and phenol were found in the model compound reaction products which supported the mechanism outlined in Figure 3. Both phenol and cyclohexanone are products of cellulose liquefaction. In fact, we had considered the possibility that phenol was produced directly from cyclohexanone, but reaction of cyclohexanone did not yield phenol.

Furans

Aldol condensation may lead to furans in addition to ketones. Trimethylfuran, a product of cellulose liquefaction, was also produced when model compounds acetone and acetoin were reacted. Figure 4 shows a possible mechanism that begins with the condensation of acetone and acetoin; subsequently, the resultant seven-carbon molecule cyclizes, giving trimethylfuran. Condensation of similar small carbonyl compounds may yield other furan derivatives.

SUMMARY

Although liquefaction of pure cellulose is a simpler reaction than liquefaction of natural biomass, hundreds of compounds are still pro-

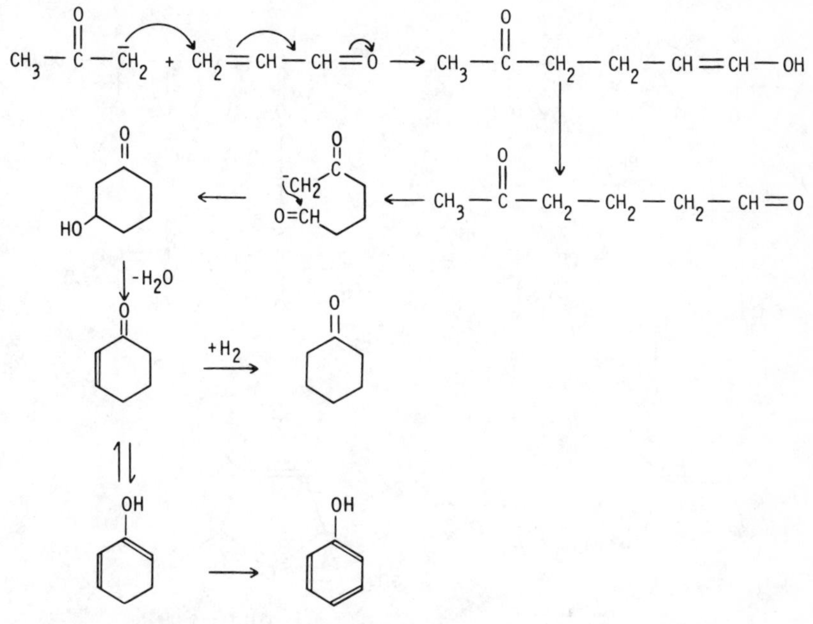

Figure 3. Acetone condensing with acrolein may eventually lead to cyclo-hexanone and phenol.

duced when cellulose is reacted in an alkaline aqueous slurry under high pressure and temperature. By using GC/MS data, we have identified 85 of these products, of which 39 are diols, ketones or furans. Hydrogenolytic cleavage of glucose leads to formation of diols. Our model compound experiments supported formation mechanisms involving the aldol condensation and Michael reaction of small carbonyl intermediates to yield ketones and furans. Though many compounds result from cellulose liquefaction, these basic mechanisms explain the formation of the three major groups of oxygen-containing compounds.

ACKNOWLEDGMENTS

The authors thank the following people for their help with the laboratory work: J. H. Campbell, J. M. Donovan, K. A. Grohs, S. D. Landsman, M. W. Stavig and D. M. Wall. We gratefully acknowledge support from the U.S. Department of Energy, Division of Chemical Sciences, Office of Basic Energy Sciences, under contract DE-AC06-76RLO 1830.

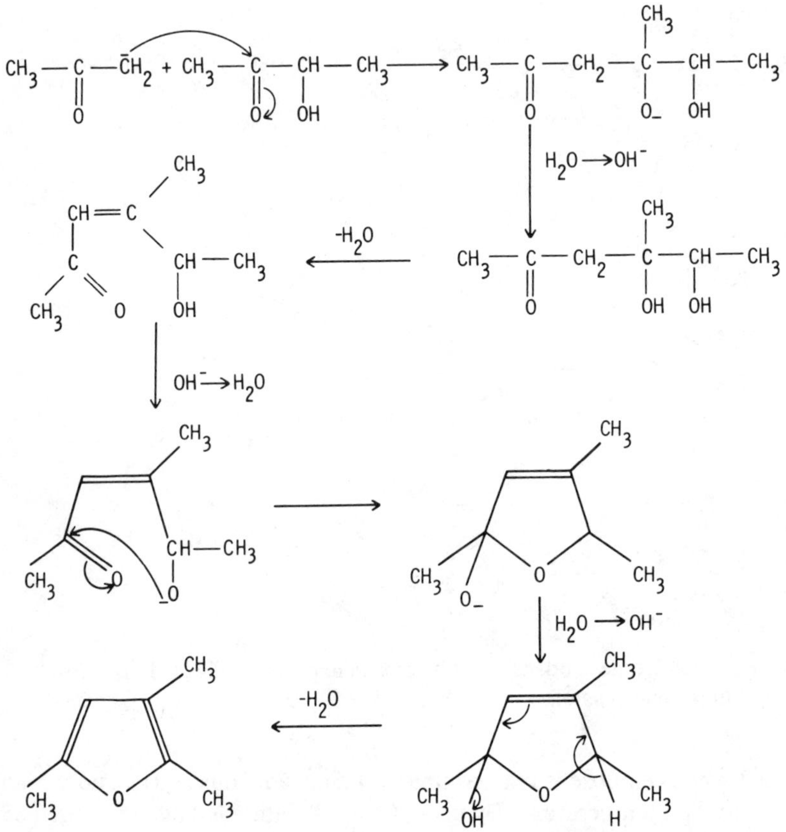

Figure 4.   Tremethylfuran may result from the condensation of acetone and acetoin.

REFERENCES

1.   Fierz-David, H. E. *Chem. Ind.* 44:942–944 (1925).
2.   Heinemann, H. *Petroleum Refiner* 33:161–163 (1954).
3.   Gillet, A., and P. Colson. *Ind. Chim. Belge, Suppl.* 1:402–412 (1959); also *Chem. Abstr.* 54:3941b (1960).
4.   Appell, H. R., Y. C. Fu, E. G. Illig, F. W. Steffgen and R. D. Miller. U.S. Bureau of Mines Rep. Invest. 8013 (1975).
5.   Tropsch, H., and A. von Philippovich. *Abhandl. Kennt. Kohle* 7:84–102 (1925); also *Chem. Abstr.* 21:2779 (1927).
6.   Bergius, F. *Mining J.* 163:1067–1068 (1928).
7.   Berl, E., A. Schmidt and H. Koch. *Z. Angew. Chem.* 43:1018–1019 (1930); *Chem. Abstr.* 25:899 (1931).

8. Berl, E., and A. Schmidt. *Annalen* 493:97–123 (1932); also *Chem. Abstr.* 26:2145 (1932).
9. Kilzer, F. J. In: *High Polymers, Vol. 5, Part 5,* N. M. Bikales and L. Segal, Eds., (New York: John Wiley and Sons, Inc., 1971), pp. 1015–1046.
10. Richards, G. N. In: *High Polymers, Vol. 5, Part 5* N. M. Bikales and L. Segal, Eds., (New York: John Wiley and Sons, Inc., 1971), pp. 1007–1014.
11. Molton, P. M., and T. F. Demmitt. *Polym.-Plast. Technol. Eng.* 11:127–157 (1978).
12. Bechtel Corp. "Final Technical Progress Report: Liquefaction Project at Albany, Oregon," prepared for U.S. DOE, Contract No. EG-77-C03-1338 (1978).
13. Urison, J., P. de Calignon and G. Pingard. *Chem. Abstr.* 45:3597g (1951).
14. Mass Spectrometry Data Center. *Eight Peak Index of Mass Spectra,* 2nd ed. (Palo Alto, CA: Pendragon House, 1974).
15. Molton, P. M., R. K. Miller, J. A. Russell and J. M. Donovan. In: *Biomass as a Nonfossil Fuel Source,* D. L. Klass, Ed., ACS Symposium Series No. 144 (Washington, DC: American Chemical Society, 1981), pp. 137–162.
16. Molton, P. M., R. K. Miller, J. M. Donovan and T. F. Demmitt. *Carbohydr. Res.* 75:199–206 (1979).
17. Herrick, F. W., and H. L. Hergert In: *Recent Advances in Phytochemistry, Vol. 11,* F. A. Loewus and V. C. Runeckles, Eds. (New York: Plenum Publishing Corporation, 1977), 485.
18. Maksimenko, O. A., L. A. Zyukova, E. V. Ignat'eva and R. M. Fedorovich. *Chem. Abstr.* 84:493 (1976).

SECTION 6

ENVIRONMENTAL EFFECTS

CHAPTER 26

ENVIRONMENTAL AND HEALTH ASPECTS OF BIOMASS ENERGY SYSTEMS

H. M. Braunstein, F. C. Kornegay, R. D. Roop and F. E. Sharples
Oak Ridge National Laboratory
Oak Ridge, Tennessee

The scope of biomass-to-energy systems is very broad, involving a wide variety of feedstock materials, a large number of production schemes and many different conversion technologies [1]. Thus, the environmental issues are also broadly based and touch on a multitude of concerns. These include competition for scarce land and water resources, alteration of the social and economic character of rural communities, disturbance of terrestrial and aquatic ecosystems, degradation of air and water quality, and disregard for the ethical and esthetic values of forest and wilderness areas. To ensure early incorporation of these concerns in decision-making, the Oak Ridge National Laboratory (ORNL), in 1978, carried out an environmental analysis of the biomass-to-energy options envisioned as promising at that time [1].

The greatest potential for negative impact arises primarily from two sources: the need for large-scale commitment of resources for biomass production, and uncontrolled widespread small-scale utilization. Biomass-derived energy is often regarded as benign; the technology is thought of as small, dispersed, simple, sustainable and appropriate. This is especially true of small-scale combustion. Because the technology has been employed since man's early discovery of fire, little thought has been given to controlling the air emissions and even less thought to the need for, or the magnitude and consequences of, growing and harvesting the biomass for fuel. Thus, this discussion, which is drawn from the overall study of biomass energy systems [1], specifically addresses the environmental concerns associated with biomass production in agricultural or forest systems and conversion of biomass to energy by residential combustion.

Biomass is fundamentally different from traditional energy sources. Sunlight is diffuse, and photosynthetic energy storage is not very efficient. Green plants convert only about 1% of incident solar energy into carbohydrates [2]. Thus, using green plants to gather even a moderate amount of energy requires a very large commitment of production resources. This would pose no problem if the optimum combinations of climate, soil, topography, etc., for biomass production were unlimited. However, limitations do exist, and most of the optimal resource base is already heavily exploited by agriculture and forestry, with discernible environmental costs. Two of the major questions about the future of biomass energy are, thus: "can the production resource base be shared so as to meet the needs of both food and fiber and energy production?" and "can large-scale biomass-for-energy production avoid intensifying the environmental effects characteristic of existing agriculture and forestry?"

Biomass materials are also unique as an energy source in that they are widely dispersed and bulky. Considerable monetary and energy costs will accrue to collect and transport them to centralized facilities. This clearly implies that numerous small-scale conversion facilities would be preferable to minimize transportation costs. Inherent in this solution is, however, the problem that small-scale facilities are subject to minimal environmental control at present. Although no one facility is usually perceived as producing overwhelming air and water emissions or solid waste disposal problems, the cumulative output of many such facilities could be significant.

Table I summarizes the potential negative impacts associated with the wide variety of production and conversion schemes that can be combined to form a biomass energy system. Small-scale refers to on-farm, residential or small commercial facilities, and large-scale implies industrial size. It is assumed that implementation of completely effective environmental control for either biomass production and harvesting or small-scale conversion will be difficult to attain, whereas industrial installations will be subject to existing or future regulation on air, water and solid waste emissions. This accounts in some cases for a greater severity of impact projected for small-scale application compared to industrial-scale deployment of the same technology. A note of caution is essential in interpreting Table I. Because biomass systems are not yet well defined and many of the issues are complex and far-reaching, assessment of the severity of environmental impact at this time must be considered only as an indicator of potential for negative effect, not as a predictor of unavoidable impact.

In this work, we examine some of the concerns that emerged as most important in the environmental assessment as indicated in Table I [1]. The systems selected include silvi- and agricultural production and harvesting as well as small-scale wood combustion. For each of these technologies, existing systems are detailed because energy-producing biomass systems will of necessity interact (sometimes negatively) with existing systems, and because much of the expected environmental behavior of proposed systems is deducible from existing systems. Although the focus of the work is on the potential impacts of proposed systems, an attempt is made to point out the availability of mitigation measures. No attempt is made to compare the potential environmental impacts of biomass systems with those of other energy-producing systems. Rather, the view is taken that with awareness, planning and proper management (but not necessarily without difficulty), biomass-produced energy can be made environmentally acceptable.

AGRICULTURAL BIOMASS FOR ENERGY

Existing Systems

Producing biomass for commercial energy requires soil, water and sunlight, i.e., a resource base unfamiliar to energy planners but well known and important to farmers. Agriculture in the United States today is a multibillion-dollar industry that provides food and fiber for both domestic and foreign markets. It is also a primary consumer of resources, making major demands on our land, water and energy, the same resource base that will be demanded of energy-crop production.

Land Use

Of the nation's total land area of 936.4 million ha (2313.7 million ac), approximately 20% is suitable for the cultivation of crops. The Soil Conservation Service (SCS) [3] estimates the present crop rotation as a-mounting to 162 million ha (400 million ac). Of this total, about 101 million ha (250 million ac) are considered prime lands. "Prime farmland" is defined as land that has the best combination of physical and chemical characteristics for producing food, feed, forage, fiber and oilseed crops and is also available for these uses (i.e., is not urban land or under water) [4]. The importance of prime farmland is that it is the most efficient, energy-conserving, environmentally stable land available for meeting food needs [5]. Therefore, withdrawal of prime land from food production can have serious consequences. It is unlikely that production of bio-

Table I.   Summary of Potential Levels of Negative Impacts Associated with Biomass Utilization [1]

			Impact Categories[a]				
	Land Use	Solid Waste Disposal	Soils	Air Quality	Water Quality	Institutional and Social	Wildlife and Ecosystems
Production and Harvesting Sources of Feedstock							
Wood from Residues, Marginal Forests	VL		M	M	M	H	M
Silvicultural Plantations	VH		H	M	M	VH	H
Agricultural Residues	VL		VH	H	H	L	L
Agricultural Plantations	VH		H	H	H	VH	H
Manure	VL		VL	VL	VL	L	VL
Freshwater Biomass	H		NA	VL	M	M	M
Marine Biomass[b]	H[c]		NA	U	H[d]	VH	H
Conversion Technologies							
Wood Combustion, Small-Scale		VL		VH	VL	M	NA
Wood Combustion, Large-Scale		L		M	VL	M	NA
Agricultural Residue Combustion		VL		H[e]	VL	VL	NA
Gasification, Small-Scale		L		L	M	L	NA
Gasification, Large-Scale[f]		M		L	M	L	NA
Liquefaction[f]		M		L	M	L	NA
Pyrolysis[f]		L		L	L	L	NA
Anaerobic Digestion, Small-Scale		VL		VL	VL	M	NA
Anaerobic Digestion, Large-Scale[f]		M		L	L	L	NA
Fermentation, Herbaceous, Small-Scale		VL		VL	VL	M	NA
Fermentation, Herbaceous, Large-Scale[f]		M		L	L	VH	NA
Fermentation, Lignocellulosic, Large-Scale[f]		M		M	M	L	NA

[a] Impact severity index: VH = very high; H = high; M = moderate; L = low; VL = very low; U = uncertain; NA = not applicable.

[b] Impacts are based on moderate development of marine biomass systems. However, estimates of potential for marine systems are widely divergent. Impacts in every category could vary from low to very high, depending on the eventual extent of deployment. See text for further discussion.

[c] Refers to use of ocean surface rather than land.

[d] Rather than specific severe negative impacts, this represents major changes in ocean surface water characteristics which may or may not have negative implications; however, because of the extent of deployment, any negative effects are apt to be very intense.

[e] Assumes that NO_x scrubbers are not yet commercially available in the U.S.

[f] Assumes large-scale facilities are subject to environmental control technology.

mass for energy on prime land can be justified, despite the fact that this land would be most appropriate in terms of efficient productivity.

Water Use

Although only 10% of the nation's cropland is irrigated (17 million ha, 42 million ac in 1974) [6], agriculture is nevertheless the largest consumer of water (Figure 1). About 85% of annual water consumption occurs in the 17 western states, and 90% of what is consumed in these areas is used by agriculture, largely for irrigation [7]. A severe water shortage currently exists in 21 Water Resources Council subregions, largely in the West (Figure 2), and potentially severe shortages are projected for an even larger area by the year 2000 if consumption grows as projected. The western United States has vast areas of public land, use of which for biomass production would engender little conflict. However, this land is also of low productivity unless water for expanded irrigation becomes available. Development of the West's substantial fossil fuel resources in the form of coal and oil shale [8] will also require large commitments of water. Vigorous competition for available water already exists in the West and it is likely to intensify in the future, precluding use for biomass production even if land is available.

Energy Use

Roughly 3% of the U.S. energy budget is accounted for by agriculture [9,10]. As an industry, production of raw agricultural commodities ranks third in energy consumption, after steel and petroleum refining. About half of this energy is consumed directly in the form of petroleum and electricity used in the production of crops and livestock. The remainder consists of purchased inputs: fertilizer, feeds and additives, farm machinery, and pesticides. Overall, the marked increase in per-acre yields of agricultural products that has occurred since World War II is largely attributable to inputs of fossil fuels [11]. For example, from 1945 to 1970, mean corn yields increased from about 34 to 81 bu/ac while the associated mean energy inputs increased from 3.8×10^9 to 1.2×10^{10} J/ac. Thus, although yields have increased, energy efficiency has simultaneously decreased, and agriculture has become heavily dependent on regular inputs of fossil fuel to maintain yields. There is also evidence that increased production as a function of increased energy input has leveled off and reached the point of diminishing return [12,13]. In any case, further increases in production may be expected to constitute a great energy expense unless alternative means of accomplishing this goal can be found.

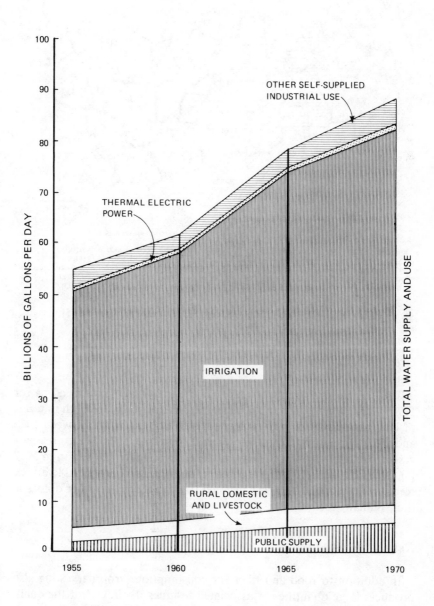

Figure 1.   Historic consumption of water for major uses [7].

CURRENT PROBLEMS

ARIZONA, NEW MEXICO, TEXAS, OKLAHOMA, KANSAS, NEBRASKA, NEVADA, SOUTHEASTERN WYOMING, CENTRAL AND SOUTHERN CALIFORNIA, EASTERN AND CENTRAL COLORADO, WESTERN UTAH

FUTURE PROBLEMS

MONTANA, SOUTH DAKOTA, WYOMING, SOUTHERN IDAHO, SOUTHEASTERN OREGON, SOUTHERN FLORIDA, NORTHERN CALIFORNIA, EASTERN UTAH, WESTERN COLORADO, WESTERN NORTH DAKOTA

Figure 2.   Water supply deficiencies by watershed region [5].

Crop Production

Table II summarizes production of the major agricultural crops for 1976. Corn, wheat, hay and soybeans are the major crops both in terms of land area devoted to them and the value of the crop. Between 1972 and 1977 agricultural exports accounted for about 30% of the area harvested annually; crop exports were valued at roughly $24 billion in 1977 [6]. An additional 60% of the acreage for field crops produced livestock feeds, whereas only 10% was devoted to production for other domestic (i.e., human) use.

Residue Production

In addition to food and fiber for consumption, crop harvesting also produces large quantities of associated residues. Estimates of the quantity of residue produced vary between 251 and 354 million dry metric tons (277 and 390 million dry tons) [14–16]. Corn, wheat and soybeans together account for 75% of the total residues produced [17].

Table II.   Land Use and Production of Major
Agricultural Crops in 1976 [6]

	Area Harvested (10³ ac)	Production (10³ Units)	Value of Production (10³ $)
Corn Grain	71,300	6,266,359 bu	13,471,796
Corn Grain Silage and Forage	12,130	117,813 ton	
Wheat	70,771	2,142,362 bu	5,851,443
Oats	11,946	546,315 bu	845,188
Barley	8,297	372,461 bu	829,716
Rice	2,480	115,648 cwt	811,358
Sorghum Grain	14,723	719,817 bu	1,450,085
Sorghum Grain Forage and Silage	2,655	7,168 ton	
Hay	60,311	120,006 ton	6,810,799
Soybeans	49,358	1,287,560 bu	8,768,979
Peanuts	1,522	3,750,890 lb	750,260
Sugarcane	747	28,120 ton	257,242
Sugar Beets	1,478	29,386 ton	616,813
Vegetables	3,296	Variable	3,120,261

Impacts and Problems of Existing Agriculture

Any discussion designed to provide background for understanding the implications of large-scale production of biomass for energy calls for consideraton of the problems and environmental effects of agriculture. Agricultural activities affect the environment as a whole and, most importantly, they affect the quality and productivity of the agricultural resource base. This is the resource base that will be expected to provide food and fuel efficiently, renewably, and indefinitely into the future.

Land and Soil Degradation. Continual cultivation leads to moderately serious changes in agricultural soils, such as decreased organic matter content, deterioration of physical structure, decreased aeration and water infiltration, and increased compaction [18]. These changes not only reduce yields, but also increase the amount of fuel which must be expended in farming. Much worse, however, is the fact that abusive exploitation of agricultural soils can and does lead to serious and sometimes permanent damage to productive land.

Erosion of soil by water presently results in delivery of about 3–4 billion tons of sediment to the nation's waterways each year. About 75% of this comes from agricultural land [5]. Wind erosion accounts for the loss of another billion tons of agricultural soil annually. Dust storms similar to those of the 1930s recurred on the Great Plains in 1976 and 1977, with erosion damage affecting an acreage comparable to that of the worst year (1938-1939) of the "Dust Bowl" [19]. Unfortunately, serious erosion losses are not now confined to the Dust Bowl region. More than one-third of all U.S. cropland is estimated to be suffering soil loss too great to be sustained without an ultimately disastrous decline in productivity [20] and two-thirds of all crop and pasture land is in need of additional soil protection measures [5]. The threat of loss of productivity is real. Since 1935 about 40 million ha (100 million ac) of land have been degraded beyond the point of sustaining further cultivation. Another 40 million ha have lost more than 50% of their original topsoil [5]. Soil conservation measures adequate to sustain long-term agricultural productivity have not yet been achieved, despite the expenditure of nearly $20 billion on soil conservation since the 1930s [21].

The causes of this situation are both political and economic. Contradictory federal policies have encouraged marginal land set-aside and other conservation measures on the one hand, while calling for all-out crop production on the other. In response to large increases in commodity prices, particularly for exported food, some farmers have maximized short-term crop yields at the expense of longer-term soil conservation efforts. Practices such as abandonment of crop rotation for continuous planting of wheat, removal of shelter belts, and plowing of highly erodible marginal land have been identified as destructive to soil conservation efforts.

In addition to cropland loss caused by agricultural abuse, about 1.2 million ha/yr (3 million ac/yr) of agricultural land is converted to urban, highway and reservoir use. Prime farmland in particular is being withdrawn from food and fiber production at a disproportionate rate amounting to some 0.4 million ha/yr (1 million ac/yr), or somewhat over 4 mi^2/day [5]. Costle [22] cites "soil loss caused both by the conversion of prime agricultural land to urban use and by shortsighted management practices" as one of the two biggest environmental problems to deal with in the 1980s.

To make up for loss of cropland, land now in pasture and range will have to be converted. The SCS had to revise its estimate of land available for conversion from the 1967 figure of 107 million ha (265 million ac) to 45 million ha (111 million ac) [3]. Only 14 million ha (34.9 million ac) of this land has high potential for conversion to cropland without requiring installation of costly conservation measures. Thus, the United States has a declining active agricultural land base, its cropland reserves are questionable, and the economic feasibility of future conversions is uncertain [20,23]. Disagreement has arisen over how soon and how severely our cropland resource base will be tested. Some analysts believe this could occur as early as 1985 [24]. A recent U.S. Department of Agriculture (USDA) study [25] suggested that a combination of increased export demand, constraints on use of water, energy and agricultural chemicals, and unfavorable weather conditions could result in a tremendous need for additional cropland in the near future. An economic projection model incorporating only one of these factors—increased export demand—projected a 33-million-ha (81-million-ac) increase in harvested cropland by 1985.

Nonpoint-Source Pollution. Runoff of soils, fertilizers and pesticides from farmland and irrigation return flows is among the most significant sources of water pollution in the nation [26]. Agriculture is the most extensive source of nonpoint-source pollution, affecting more than half of the water basins in all geographic regions of the United States [5]. The 3 billion tons of waterborne sediment from agricultural land that is settling annually in reservoirs, rivers and lakes shortens the usable lifespan of rural watersheds and municipal water supplies and smothers aquatic life. Sediment carries with it fertilizers and pesticides that load watercourses with toxins and high-biochemical oxygen demand (BOD) materials, which deplete oxygen levels and make water uninhabitable for aquatic organisms. If not curbed, nonpoint pollution will prevent attainment of the national water quality goals mandated by the Clean Water Act. To date, progress in control has been minimal [5]. Funds for implementing a program of best management practices (BMP) on rural lands were authorized under the Clean Water Act Amendments of 1977. These funds were never appropriated by Congress, and the program remains inactive [21].

Proposed Energy Systems and Potential Impacts

The biomass sources from which energy could be harvested comprise three distinct categories of material—wastes, residues and new production. Wastes are biomass materials that must currently be disposed at an economic cost. Residues are products that have a current use, and therefore, some value, although use may not entail active collection, transportation or sale. New production denotes energy crops, i.e., biomass specifically grown for its energy content. Crop agricultural systems can supply biomass for energy as either residues or new production. The residues of food and fiber crops could be collected for fuel recovery, or a wide variety of conventional and unconventional plant species could be grown on "energy plantations" to provide both cellulosic materials and special chemicals (oils, latices and alcohols).

Residue Collection

Crop residues include stems, leaves, roots, chaff and any other plant parts that remain after agricultural crops are harvested. Although energy density in joules per cubic meter of residue is only 10–20% that of wood [14], large quantities of these materials are produced, resulting in an overall gross energy content that is large—estimated at about 5 quads (approx. 5×10^{18} J).

The great attractiveness of crop residues for fuel is that production costs have already been paid in producing the primary crop. However, only a minor fraction (2%) of residue is considered to be "waste"; traditionally, these materials have functioned in maintenance and protection of soil. When returned to the soil, crop residue decomposes to form humus. Residue and the resulting humus protect soils by: (1) retaining plant nutrients; (2) keeping soils porous and in condition for easy tillage and good plant growth; (3) increasing infiltration of water into the soil and hence increasing water storage; (4) reducing soil detachment caused by wind and raindrop impact; and (5) reducing velocity of runoff and the associated soil transport potential [27]. The fraction of crop residue which can be removed on a sustained basis without significant soil deterioration is variable and poorly known for most sites. Thus, the uncontrolled use of crop residue for fuel recovery will in many cases compete directly with important soil conditioning uses.

Erosion Impacts. Perhaps the most extensive constraints on removal of crop residues arise from the need to protect soil from erosion. Effects of erosion are difficult to quantify because they are influenced by factors such as crop variety, soil nutrients, soil structure, topsoil depth, drainage, temperature, moisture and pests. Evidence suggests, however, that each 1 cm loss of topsoil (or a loss of about 150 metric ton/ha) from a base of 30 cm (12 in.) represents a decrease in productivity that is conservatively estimated at 250 kg/ha of corn, 86 kg/ha of oats, 110 kg/ha of wheat and 175 kg/ha of soybeans [28]. With soil loss levels commonly 20–40 metric ton/yr (or 0.13 to 0.27 cm/ha-yr), yield reductions may be considerable, given cumulative effects over many years. Soil is reformed at much lower rates than typical loss rates. Under ideal soil management conditions, 1 cm of soil could reform in about 12 years, but under normal agricultural conditions, 1 cm reforms in about 40 years.

Protection against erosion is considered adequate when soil removed by wind or water is less than or equal to soil-loss tolerance limits [27]. These limits are defined by the SCS as the maximum allowable soil loss that is safe for maintaining long-term soil productivity. Maximum soil-loss tolerance ranges from 2 to 10 metric ton/ha-yr (1 to 5 ton/ac-yr), depending on soil properties, soil depth, topography and prior erosion [29]. Use of the universal soil loss equation or the wind erosion equation has allowed estimates of permissible residue removal levels for some regions. For example, the need for water erosion protection would limit residue removal to 58% of the total residue produced in the Corn Belt [30]; 40% in Georgia, the Carolinas and Virginia; and 10% in Mississippi and Alabama [31]. Wind erosion protection would leave only 21% of produced residues available for removal from the Great Plains. If, however, water erosion is considered with wind erosion, available residues from this region are even lower [32].

Soil Nutrient Removal. Removal of residues may also conflict sharply with maintenance requirements for soil nutrients. Annual residues from nine leading crops (barley, corn, cotton, oats, rice, rye, sorghum, soybeans and wheat) are estimated to contain about 4×10^6 metric tons of nitrogen, 0.5×10^6 metric tons of phosphorus and 4×10^6 metric tons of potassium [17]. This is the portion of nutrients extracted from the soil by crop vegetation that can be recycled after harvest [33]. These amounts are roughly equivalent to 40% of the nitrogen, 10% of the phosphorus and 30% of the potassium currently applied in inorganic fertilizers [17]. Hence, residues contain an appreciable portion of the nutrients applied to cropland. Removal of these nutrients would necessitate their replacement via energy-expensive inorganic fertilizer. The annual gross fertilizer value of nitrogen, phosphorus, and potassium in total available crop residues is estimated at $3.3 billion [34]. The energy required to manufacture additional fertilizer could substantially decrease the net energy gain obtainable from residues as fuel. Furthermore, if removal of residue resulted in accelerated erosion, even larger amounts of nutrients would be lost indirectly because eroded surface soil contains more nutrients by weight and is thus more fertile than the underlying material from which it is derived [34]. Hence, nutrient depletion could be the direct and indirect result of removing residue for fuel recovery. Depending on the conversion processes employed, however, use of residues for fuel recovery need not render them valueless as fertilizer; for example, anaerobic digestion would preserve substantial portions of nutrients for reapplication to the land. Pyrolysis or direct combustion would, however, result in nutrient loss. Figure 3 summarizes the potential effects on soils and nutrients of residue removal. Site-specific conditions will determine the severity of

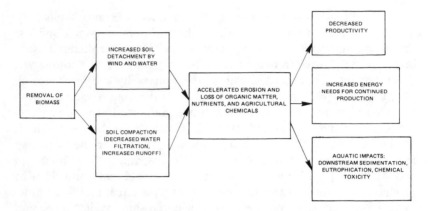

Figure 3.   Potential impacts on soils and nutrients of agricultural residue
removal or whole-plant harvesting.

the impacts depicted. Current research is directed toward understanding
the critical interactions between residue removal rates and soil fertility so
that safe levels of biomass harvest may be predicted on a site-specific
basis.

Energy Farming

The ultimate scheme for new energy crop production is the "energy
plantation"—a systematic approach to the large-scale production of
economic fuel from plant matter. Such a proposed plantation is envisioned, in one case, to be one or more units of minimum area of 11,500
ha (28,500 ac) [35]. For comparison, the average size of all farms in the
United States was 214.5 ha (529.7 ac) in the 1970 census, while the average size of farms that grew primarily corn was only 159.7 ha (394.4 ac)
[36]. The energy plantation would be located close to the fuel-converting
facility so that the average fuel transport distance is 17.6 km (11 mi). The
farm itself would consist of dense monocultural stands of one or more
high-yield perennial plant species that are harvested almost year-round.
Other schemes for energy plantations vary in characteristics such as
size, for example, from as small as 1600–3200 ha (4000–8000 ac) [37] to
as large as 58,000 ha (144,000 ac) [38]. Most energy-plantation schemes,
however, share the following features:

1. They will be large relative to conventional agriculture.
2. They will grow perennial crops when possible, utilizing intensive man-
    agement practices and advanced harvesting systems.

3. They will rely on resprouting after harvest for reestablishment when possible.

4. They will be established on sites where precipitation is at least 50–64 cm/yr (20–25 in./yr), slope is less than 30%, and land is currently used as forest or pasture range and is producing forage as opposed to field crops [39].

According to the Stanford Research Institute [38], the general objective of such a plantation is "to produce the greatest amount of biomass possible per unit of time and space, at the lowest possible cost, and with a minimum of energy expenditure."

Land Use Conflicts. In general, the land for siting such plantations would not presently be used as cropland. Most proponents of energy plantations envision the use of marginal land now in pasture and range. However, the economics of producing biomass for energy make it unlikely that the use of marginal land, of lower productivity than cropland, would be cost- or energy-effective. For example, cultivation of corn requires at least 50% more energy input per bushel on marginal land than when grown on average (good quality) cropland [40]. Biomass would therefore be most efficiently produced on land in the higher quality categories (Classes I and II) not now used by agriculture. This land is, however, the reserve cropland to which food and fiber production may need to expand in the future. The potential for competition for the higher-quality cropland reserve between the energy producers and the food system is evident.

If, however, biomass production for energy is relegated to lower-quality pasture and rangeland, other potential problems could surface. Conversion of pasture or rangeland to biomass production could compete with the raising of livestock. Range-fed meat production is more than ten times as energy-efficient as confinement (feedlot) production because of the large amount of rangeland currently available [10]. Should this availability substantially decrease, requiring increased confinement of beef cattle with the concomitant required increase in the production of cattle feed, conflict over the agricultural land resource could be generated indirectly.

Erosion. When cultivated, marginal lands were generally more susceptible to erosion than are present croplands. For example, use of marginal lands may entail dust dispersion through wind erosion, a problem that is especially difficult to control. Wind erosion in the Great Plains has been a major source of air pollution in the form of recurrent severe dust storms in the 1930s, 1950s and 1970s [41]. Average dust concentrations were found to be 10–15 mg/m^3 during dust storms in the driest areas of the southern Great Plains, while average dust concentrations ranged from about 5 to 10 mg/m^3 even in the least erosive areas. These concentrations are very large compared with the 1.0-mg/m^3 concentrations considered severe in urban air pollution episodes [42]. Atmospheric particulate mass from wind erosion in the United States probably already exceeds the combined output from all other primary sources which emit 35.2 million ton/yr [43]. The magnitude of potential increase in suspended particulate pollutants with increased use of marginal land cannot accurately be estimated at the present. Data from the Great Plains during the period of "all-out production" of wheat in the mid-1970s suggests, however, that there is a direct relationship between marginal acreage planted and acreage damaged by wind erosion. If this relationship is true for other regions, a proportional increase in general atmospheric particulate pollutant load and acreage degraded by soil loss may be expected.

Runoff. Similarly, runoff of sediments, fertilizers and pesticides may be expected to increase at least in proportion to the amount of new land brought under cultivation for biomass. Annual erosion losses from United States soils now average about 20 metric ton/ha [27]. New land brought into use for energy farming would be no less erodible than cropland, and soil loss is likely to occur on at least the same average level if conventional agricultural practices are employed. Use of marginal land could result in considerable intensification of erosion and nonpoint-source pollution, particularly if whole-plant harvest is employed, leaving soils bare and unprotected.

Loss of Wildlife Habitat. Conversion of large new areas of land to cultivation of biomass for fuel could significantly change terrestrial habitats and biota. Agriculture itself has had perhaps the most profound impact on wildlife of all human activities in the United States. Agriculture has affected huge areas over a brief time span, a combination of factors that has allowed wildlife little or no time to adapt and no room for retreat [44]. Habitat availability and diversity have recently decreased to their lowest level as huge monocultures replace the varied crops and landscapes of small family farms to maximize efficient operation of large specialized equipment. The establishment of huge monocultural biomass plantations could have particularly serious effects. Monocultural stands of vegetation represent drastically simplified biological environments, which are likely to be unattractive to most animal species because few are adapted to a single type of cover and a single type of food. This is especially true if the vegetation in question is not native to a location. As a result, most wildlife would avoid such areas. Those animals that could survive in such a simplified environment would be apt to flourish unchecked and become pests. A monoculture would present an enormous supply of resources to the pest species, would likely harbor few or no predators, and would pose no obstacles to continuous successful pest dispersal. Extensive areas of high-quality habitat have been well identified as the highest risk areas for initiation of pest outbreaks [45]. Such outbreaks may, in turn, require massive application of pesticides that further simplify, and even sterilize, both target and nontarget environments. A monocultural plantation can, in the long run, become a system that can only be maintained by continual inputs of energy. This situation could be severely detrimental to biomass production economics as well as wildlife.

Mitigation Measures

How can the detrimental effects of using agricultural biomass for energy be avoided? To some degree, adoption of new agricultural techniques will help reduce impacts of both energy use and crop production itself.

Conservation Tillage

Conservation-tillage or no-tillage techniques have only recently begun to be recognized as a widely applicable means of counteracting soil erosion. No-tillage agriculture, in which a crop is planted with little or no tillage into soil covered with the residues of a previous crop, has been found to be extremely effective at controlling soil loss, even on highly erodible soils [46]. Combining minimum-tillage with the practice of allowing some residue to remain can increase the protection given to cropland while still allowing some residue harvest for energy recovery. For example, the estimated percentage of cropland adequately protected against water erosion in the Corn Belt increases from 36% under conventional (plow-disk-harrow) tillage with all residue removed to 78% for no-tillage with 3920 kg of residue left per hectare [30].

Ecological Cropping

Energy plantation designs could incorporate production strategies which enhance, rather than defy, the resiliency of nature. Choice of plantation species can be guided by ecological as well as economic factors, tailoring the tolerances and resource-using abilities of the chosen plants to the ecological factors of a site. Plants with high demand for water would not, for example, be appropriate for arid or semiarid regions where water is scarce. Sod-forming grasses, rather than row crops, would minimize exposure of erodible soil in the Great Plains. A more balanced approach for ensuring ecological stability and meeting wildlife needs is to replace the monoculture idea with designs for networks of diverse vegetation types. Numerous plant types, preferably native species, could be integrated into a system of patches. This would provide a diversity of resources for more kinds of beneficial animals (e.g., predators of pest species), thereby helping to control the spread of pests and to minimize intraspecific competition among the crop plants themselves. Integrated pest management systems can be used to reduce the use of chemical pesticides and thereby the energy expenditures for a plantation. Similarly, tilling and harvest techniques which minimize soil disturbance and maximize vegetative cover at times when seasonal wind and water erosion potentials are highest can be used to great benefit both in resource and energy conservation terms.

Resource Management

Although the basic methods of maintaining agricultural productivity are known, the protection of resource quality has too often been sacrificed to short-term economic gain. Attention to the maintenance of long-term productivity is, however, essential if biomass is to be regarded as a source of truly renewable fuels. In addition, careful delineation of national priorities is needed to ensure that the shared resources available for both energy and food and fiber production are partitioned in accord with the social and economic needs of all interest groups.

FOREST BIOMASS FOR ENERGY

Almost one third of the United States is forested — out of a total area of 936 million ha (2300 million ac), forests cover 300 million ha (740 million ac). While these forested lands do have tremendous potential for energy production, there are many constraints. The demand for wood by the pulp, paper and timber industries is the principal limitaton on the use of wood for energy. In addition, large-scale reliance on our forests for energy could cause significant environmental impacts.

There are five categories of wood which could be used for energy: (1) mill wastes, (2) forest residues, (3) wood from forests that would not ordinarily be cut, (4) biomass derived from more intensive forest management, and (5) wood from silvicultural energy farms (SEF). These sources of wood are listed in order of increasing intensity of management and thus in order of increasing potential for environmental impact.

Existing Systems

In 1976 wood fuels were estimated to be supplying 1.67×10^{18} J (1.58 quads) of energy [47]. The pulp and paper industry used 62% of this total, while residential use accounted for 25%.

Mill Wastes

Mill wastes include spent pulping liquor, bark, chips, sawdust, shavings, slabs, trim pieces and similar materials. Coarse residue, large debris, defective products and otherwise unusable wood are "hogged" (chipped or ground up) to provide chips for wood pulping or "hogged fuel." In the production of lumber plywood, 55-66% of the total weight of logs brought to the mill appears as refuse [48]. The Forest Service [49] reported that in 1970, 75% of the wastes at primary wood processing plants was used, mostly for pulping but partly for fuel. Wood and bark wastes from processing provided roughly 1 quad of energy in 1970 [48]. Steinbeck [50] reported that the use of wood for energy is accelerating in the southeastern United States at such a pace that in some areas "old sawdust piles have disappeared and competition for sawmill residues is becoming stiffer." In all of the timber-producing areas of North America, the use of mill waste for energy has recently increased. Examples include use of bark in pulp mill boilers (Calhoun, TN, Rumford, ME), steam production for a textile mill using woodchips purchased from 30 small sawmills (Alexander City, AL), cogenerating electricity and space heat from hogged fuel from nearby mills (Eugene, OR) and plywood veneer drying by use of trim scraps (Omak, WA) [51]. The Weyerhauser Corporation is developing improved methods of picking up, cleaning and burning debris in the log yards. Boilers are being modified to accept a wider range of wood-particle sizes and qualities; for example, cyclone burners are being added to burn sander dust. However, Weyerhauser perceives the wood-energy resource base as limited in the long term; therefore, energy systems that can use alternative fuels (such as coal) are being planned [52].

Forest Residues

Forest residues are the unused materials from thinning and harvesting activity, and include understory biomass, trees from stand thinning, cull trees, branches and leaves. Residues left after logging are the most readily available materials for fuel supply, because most forest product manufacturing requires only the bole or stem of the tree. Residue mass varies from about 25 to 45% of wood cut [53]. Logging residues, which are most easily obtained in connection with clear-cutting and site preparation for replanting, increasingly are hogged and used as fuel or pulp feedstock. The most economically attractive residues are generated by logging old growth stands such as those in the Pacific Northwest. This produces not only the highest density of residues, but also large-size materials which are easier to handle. For handling smaller, lower-density residues, mobile hogs and other types of equipment are becoming available or are under development. The development and capital costs of such equipment are barriers to increased use of residues.

Low-Grade Standing Forest

A potentially large source of biomass for energy is wood from standing forest that would not ordinarily be cut. The United States has large areas of low-grade forests that receive little or no management because of poor site quality and fragmented ownership. Although currently it is not economically feasible to log these forests for timber or pulpwood, cutting could occur to supply fuel wood and/or to remove low-quality forests and replace them with higher-quality forests. Much of the low grade forest occurs in the northern and eastern U.S.

Proposed Systems

Increased Intensity of Natural-Forest Management

Most existing forests could be managed more intensively. Techniques for improving forest growth include: (1) site clearing and immediate reforestation; (2) conversion of stands to faster growing or genetically improved species; (3) site drainage; (4) fertilization; (5) regular thinning, and (6) fire, insect and disease control [47]. Such management could substantially increase forest productivity and the availability of fuelwood, pulpwood and timber. At the same time, this type of management can lead to environmental impact.

Silviculture Farms

Because wastes and residues constitute somewhat limited resources for energy production, the concept of silvicultural energy farms has been advanced. At a SEF, trees would be intensively managed to produce an energy crop. A SEF would use tree species that were (1) selected for high yield; (2) well suited to the local site conditions; and (3) responsive to management [53]. Species that would coppice, or sprout from stumps, would be desirable to eliminate the need to reestablish the stand after harvest. Short-rotation management (4–8 years) would probably be used to get maximum yield, rapid return on investment, and increased efficiency through mechanization. Establishing a SEF would involve intensive site preparation and planting of high-density stands every few rotations. Management would be intensive and would include fertilization, weed control, pest control and, perhaps, irrigation.

Potential Impacts

The development of new sources of wood requires increasing intensity of production management and the concomitant possibility of substantial environmental impacts (Table III). Some of the most significant management activities are whole-tree removal, the removal of cull trees, building of new logging roads, reopening of old logging roads, intensive site preparation, and intensive silvicultural practices such as fertilization and pest control. Such practices could cause significant soil degradation, changes in land use, effects on wildlife and alteration of water quality. Some (but not all) impacts can be decreased or eliminated by planning and administration to ensure application of proper forestry practices.

Impacts to Soils

Harvest, or production, of woody biomass for energy on a large scale may affect soils in three ways: (1) soil disturbance, (2) loss of organic matter, and (3) nutrient losses. These impacts are synergistic and cumulative; if sufficient losses of organic matter and nutrients occur, the productivity of the site may decline (Figure 4).

Erosion and Compaction. Soil disturbance can lead to erosion, which is a major contributor to soil depletion and may adversely affect downstream environments. Equipment for handling forest biomass runs on rubber tires or tracks. Both types of equipment can compact soils and

Table III.   Summary of Environmental Impacts from
Producing and Harvesting Wood for Energy

Wood Source	Activities	Potential Impacts
Mill Wastes	Collection	Negligible or beneficial impacts; will alleviate problems with waste disposal
Forest Residues	Residue removal; small increase in tree cutting	Depletion of soil nutrients due to residue removal. Destruction of habitat for cavity-nesting birds and other wildlife due to removal of cull trees
Forests Not Ordinarily Cut	Possible road-building; large-scale cutting and wood removal	Depletion of soil nutrients and organic matter due to whole-tree removal; accelerated erosion due to new logging roads; impacts to wildlife resulting from habitat modifications
Increased Intensity of Natural-Forest Management	Site clearing and reforestation; cultivation; fertilization; thinning; pest control	Accelerated erosion and soil depletion; alteration of water quality due to runoff containing pesticides and fertilizer; harvesting impacts; alteration of wildlife habitat
Silvicultural Energy Farms (SEF)	Changes in land use; removal of existing vegetation; intensive site preparation; planting; cultivating; fertilizing; possible irrigation; frequent harvesting	Competition with grazing, forestry and agriculture for use of land; accelerated erosion and soil depletion due to site preparation and cultivation; alteration of water quality caused by use of pesticides and fertilizer; harvesting impacts; reduction of wildlife populations because of simplified habitat

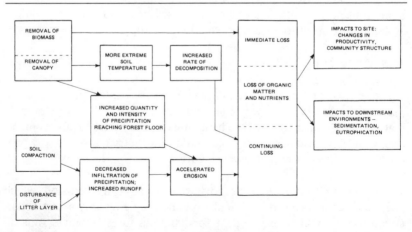

Figure 4.   Potential impacts of biomass harvesting on soils and nutrients.

seriously disturb soil litter. Disturbance of soil litter exposes the mineral soil and increases the ability of raindrops to detach particles. Soil compaction decreases the airspace within soils, thus decreasing the moisture-holding capacity, the ability of roots to penetrate soil, and the success of seed germination and seedling survival. Both soil compaction and disturbance of the soil litter contribute to decreased infiltration of precipitation, increased runoff and the risk of accelerated erosion [54].

Loss of Organic Matter. Loss of organic matter can occur directly through erosion and removal of wood, twigs and leaves which form soil organic matter. Indirect loss can occur after tree harvesting if higher soil temperatures increase microbial respiration of organic matter [55].

Nutrient Loss. Nutrients are lost both through direct biomass removal and indirectly through leaching or erosion of soils. Schemes for residue collection and harvest of low-grade forests or SEF are likely to involve utilization of the whole tree or at least a larger fraction of the total tree biomass. The decision to harvest branches and foliage is an important factor in determining direct nutrient removal. Nutrients are most concentrated in the extremities of trees and least concentrated in the stemwood [56,57]. The estimates in Table IV indicate that whole-tree harvesting can increase removal of various nutrients from 2 to 16 times the rates for harvesting stemwood only [58]. More conservative estimates indicate two- to fourfold increases in nutrient removal [59].

Table IV.   A Comparison of Estimated Nutrient Losses from Conventional Whole-Tree Logging (Tons of Nutrient in the Vermont Forest Standing Crop) [58]

Element	Whole-tree Logging	Conventional Logging	Ratio, Whole-tree/Conventional
Ca	700,790	88,909	7.9
Mg	82,607	11,480	7.2
K	37,045	24,908	1.5
P	57,200	3,483	16.4
S	63,746	6,651	9.6
N	550,620	48,736	11.3
Na	4,295	672	6.4
Fe	6,792	716	9.5
Zn	8,310	807	10.3
Cu	623	64	9.8
Mn	87,590	11,731	7.5
Ash Weight	2,743,084	316,390	8.7

In general, impacts to soils can vary greatly depending on forest type, soils, slope, the equipment used and the forestry practices used. In SEF, for instance, nutrient depletion might be severe, except that management will probably include fertilization to maintain high growth rates. Much additional research, both general and site-specific, is needed to monitor the long-term effects of intensive harvest.

Land Use Conflicts

The large-scale production of woody biomass for energy could affect land use and forestry practice on millions of hectares in the United States. Assuming that SEF produce 10–35 dry metric ton/ha-yr (4 to 14 dry ton equivalents/ac-yr), the land requirements to produce 10^{18} J of wood would range from 1.6 to 5.7 million ha. To establish SEF would require conversion of land use from grazing, forestry or agriculture. An analysis of land use by SEF [60] suggested that impacts would be greatest in the Great Lakes states, Southern Plains, Delta states, and Southeastern regions. Land in the Corn Belt would probably be used for intensive agriculture, whereas climate, terrain, ownership patterns and urbanization would limit use of land for SEF in the Northeastern and Appalachian regions. The conversion of forest, pasture and rangeland to SEF could create land-use conflicts with agriculture, both directly and because such noncroplands are the areas most likely to be converted to cropland if necessary. In addition to conflicts with forestry, agriculture and, perhaps, agricultural biomass-energy production, establishment of large SEF would require acquisition or control of large land areas. Thus, changes in the patterns of land ownership or leasing could be important. The land-use impacts of SEF clearly require further investigation, including assessments for various regions and specific projects.

Loss of Wildlife Habitat

The large-scale use of forest biomass for energy would bring about changes in forest habitat and wildlife. Habitat alteration can produce long-lasting impacts to wildlife, and, if large areas of habitat are affected, changes in the types, distribution and density of wildlife could be profound. Impacts would be most noticeable with SEF. Management with short rotation periods would maintain vegetation predominantly in xeric, brushy habitat. This would cause the loss of wildlife species preferring forest interiors but would favor brush-dwelling species. An open forest canopy promotes the growth of ground cover and increases the availability of some types of food for species such as deer, rabbits and quail. However, large clear areas may reduce suitable habitats for deer, because of this species' tendency to remain near the forest edge [61,62] and because of the loss of annual mast crops. Turkeys would be adversely affected, because they prefer extensive mature open forests [63] and open grassy fields such as pastures [64]. The absence of treetops eliminates habitat for species such as squirrels and cavity-nesting birds.

Silvicultural energy plantations may be planted with only one or two tree species, perhaps even a highly selected variety or clone within a species. The habitat created and maintained in this way could be structurally and perhaps genetically highly simplified. In contrast with the mature forest, the simplified habitat of a silvicultural plantation would not support a wide variety of wildlife [65]. If large areas are brought under moderate to intense management for the single purpose of maximizing biomass harvest, it is likely that the resulting systems will be less stable than natural forests or forest ecosystems managed for multiple use. Silvicultural energy farms that do not provide sufficient genetic heterogeneity may be susceptible to outbreaks of pests or disease organisms.

Deterioration of Water Quality

In the harvest or production of forest biomass for energy, water quality could be affected by sedimentation, logging debris in stream channels, changes in stream temperatures, and release of nutrients, organic matter or silvicultural chemicals [66]. The most severe impacts are likely to occur with opening low-grade unharvested forests to fuel wood cutting and the establishment of SEF. The water quality impacts of opening unharvested forest would depend heavily on whether logging roads are built. Roads create the largest portion of the erosion problem, perhaps as great as 90% in many cases [67]. The construction of new roads and the possible use of old roads in poor condition create the possibility of significant increases in erosion and stream sedimentation.

Much of the unharvested forest land in the United States is privately owned and receives little or no management. Because of this, one cannot assume that the best management practices would be uniformly applied in harvesting this land. The opening of inaccessible forests for energy harvest could cause 10- to 100-fold increases in erosion from the affected area, with subsequent impact on streams. Good planning and adequate control measures could restrict this impact greatly. Where planning and good practice are lacking, however, the rate of erosion could cause temporary sediment yields approaching those from agricultural cultivation.

Other Considerations

Silvicultural energy farms would incur the impacts of site preparation and stand establishment in addition to the impacts of harvesting. The removal or reduction of residues and the scarification of soils might allow short-term runoff, erosion and sedimentation. If revegetation were slow, the impacts of site preparation might extend over several years. The uses of mechanical cultivation or herbicides to control vegetation that competes with the biomass species might retard the tendency for erosion rates to return to normal levels. The rates of erosion and sedimentation for SEF would depend on the erodibility of the soils and the management practices employed. Quantification of the long-term impacts of SEF will require monitoring and research on a regional and site-specific basis.

The intensive management of SEF might also affect water quality through the use of fertilizers, herbicides and insecticides [68]. Water uses such as public supply and irrigation could be affected, and impacts to aquatic organisms may occur. The impacts of using silvicultural chemicals depend greatly on site-specific circumstances, such as the chemical applied, local topography, the season of application and weather conditions at the time of application. Impacts will also depend on management practices, such as whether strips of vegetation are left along streams or whether operators avoid direct aerial application of chemicals to stream channels. Predicting and minimizing impacts of using silvicultural chemicals will require extensive continuing analysis for individual sites and projects.

Mitigation Measures

Wood represents an available energy resource of modest size. The environmental impacts of using this resource, while potentially significant, are considerably less than the impacts of producing agricultural biomass for energy. In forestry, as in agriculture, there are techniques, both established and newly emerging, which can minimize and control some adverse impacts. The 1977 Amendments to the Clean Water Act and the Soil and Water Resources Conservation Act of 1977 call for the development of BMP which would define mitigation measures specifically tailored to various regions and their characteristics. Employing such techniques, however, may reduce productivity from the highest potential level. Thus, the challenge is to combine economically feasible strategies with careful management which ensures renewability of the resources.

RESIDENTIAL COMBUSTION FOR ENERGY

To effect the release of the energy in biomass materials, they must be converted either directly, by combustion, or indirectly, by thermochemical or biochemical conversion. This involves a wide variety of technologies, many different processes and various-size operations. Of the resulting array of biomass-to-energy options, one of the most familiar and readily available is residential wood combustion. Unfortunately, little definitive data exist on the environmental impacts of this use. Because wood burning is considered relatively free of some of the most serious environmental problems associated with coal combustion such as solid waste disposal and sulfur dioxide emission, and because environmental control is difficult to implement at the homeowner level, little attention has been focused on environmental management of this biomass application. However, home wood burning, which is becoming increasingly popular and widespread, produces air emissions which, if uncontrolled, can pose a threat not only to the environment but also to human health.

Existing Systems

Although wood-burning stoves used in the home are becoming more common, knowledge about their properties, efficiency of operation and pollution potential is still scarce. Wood stoves come in a wide variety of sizes, shapes and designs, but most can be classified within the four types described below, depending on the amount of control applied to the combustion air [69].

Fireplaces

Open fireplaces allow no control of combustion air. As a result they have a very low efficiency when operating ($\sim 0\%$). If the damper is left open when the fireplace is not in use, as it often is overnight, the fireplace will have negative efficiency. In homes with a thermostatically controlled furnace, one study showed that the net overall efficiency, as determined by the amount of furnace fuel burned during fireplace use, was zero at best, when the fireplace was covered at night, and negative otherwise [70]. The increased ventilation of cold air provoked by use of a fireplace is a major contributor to nullifying its heating effect.

In addition to low efficiency, a fireplace has a high emission rate for particulates, carbon monoxide and hydrocarbons. Despite the zero or negative efficiency and high pollution potential, fireplaces seem to be desirable; they undoubtedly have a high, overriding esthetic appeal.

Nonairtight Stoves

These are freestanding stoves with loose-fitting doors through which air can leak readily into the firebox. They include the well-known "pot belly" stove and the ordinary Franklin stove. Unlike the fireplace, which heats primarily by direct radiant energy from the flames and coals, stoves heat by radiant energy emitted from the stove walls, chimney and firebox, as well as by convection of heated room air in contact with the hot stove walls. For most stoves, more energy is transferred by radiation than by convection (as in a fireplace), but the esthetic pleasure of seeing the dancing flames and glowing coals is absent.

Like a fireplace, a stove, which draws air from the house to supply the necessary oxygen for combustion, pulls outside air into the house and discharges heated house air into the chimney. In a nonairtight stove, the air flow is not controllable, and thus the rate of combustion will not be controllable. The fuel will burn rapidly, inefficiently and with a hot flame; and the fast-flowing, very hot combustion gases will be vented to the outside. Also, this occasions the production and venting of particulates, carbon monoxide, and hydrocarbons.

Nearly Airtight Stoves

An example of a stove that is nearly airtight is the Franklin-type stove with welded seams. This kind of stove has an essentially airtight body with tightly fitted, but nongasketed, doors. Nearly airtight stoves can be damped to kill the flames but not the burn. Thus, they will burn for a longer time at a lower rate. Additionally, they draw so little air, compared with the normal air exchange rate for a house, that they do not appreciably add to the heating needs of the house the way a fireplace or nonairtight stove can. The emission rates for this type of stove have not been reported.

Airtight Stoves

This type of stove has an airtight body and close-fitting gasketed doors. Known as circulating stoves and space heaters, airtight stoves use convection rather than radiation as the dominant mechanism for transferring heat to the room. Either natural convection or small blowers are employed to move warm air into the room. The range of efficiencies covered by all kinds of airtight stoves is reported as probably between 40 and 65%. This efficiency is compared with advertised new-furnace efficiencies of 80–85%, which are thought to probably be closer to 75% [70].

If wood burning is to become an effective home heating method, studies will clearly be needed to determine designs that combine combustion efficiency with optimal heat transfer and emissions control.

Emissions

The major emissions from almost any means of combustion of biomass materials are air pollutants—notably particulates and carbon monoxide, with lesser amounts of nitrogen oxides, sulfur oxides and hydrocarbons. The amount of these depends on a number of variables, such as the composition and moisture content of the fuel, the furnace or stove design, and the combustion conditions (e.g., feed rate and amount of excess air).

Emissions depend strongly on operating conditions. For example, wood should be dried prior to combustion; otherwise, moisture in the fuel will cause the combustion temperature to be lowered, resulting in incomplete combustion with high emissions of particulates and carbon monoxide. Additionally, unless sufficient secondary air is supplied above the fuel bed to completely burn the volatiles, a high hydrocarbon emission will also result. Wood is estimated to contain 80% volatiles by dry weight. For this reason, home fireplaces that have no provision for burning volatiles not only are inefficient heating devices (less than 15% efficiency) but can be serious polluters. Similarly, open burning, in which low temperatures and poor air circulation are usual, can lead to high particulate and carbon monoxide emissions (Table V).

Smoke

The important thermal decomposition products of wood are smoke, volatile hydrocarbons and carbon monoxide. Significantly, even when wood burning produces low concentrations of smoke, large quantities of carbon monoxide may be produced [71]. Additionally, the conditions that promote abundant emission of both smoke and carbon monoxide are exactly those prevalent in the conventional residential wood stove or fireplace.

Wood smoke is a complex mixture of solid particulates, condensed liquid particulates and volatiles, and is produced most abundantly by combustion of moist wood in a restricted flow of air [72]. Figure 5 shows

Table V. Air Emission Factors from Combustion of Biomass (lb/ton) [78]

Source	Particulates	Carbon Monoxide	Hydrocarbons	Sulfur Oxides	Nitrogen Oxides
Wood/Bark in Boilers, 50% Moisture	30[a]	2–60	2–70	1.5	10
Wood in Residential Fireplaces	20[b]	120	5	0	1
Small Wood Stoves, Oak and Pine	3.4–20.0				
Waste Open Burning, Agricultural and Wood Refuse					
Unspecified	21	117	23		
Corn	14	108	16		
Rice (dry)	9	83	10		
Rice (wet)	29	161	21		
Sorghum	18	77	9		
Sugarcane	7	71	10		
Oats	22–44	136	18–33		
Wheat	13–22	108–128	11–17		
Forest Residues, Unspecified	17	140	24		
Bagasse Burning	22				
Basasse Boilers	16				

[a] Bark only, 50 lb/ton.
[b] Contains as much as 70% condensate.

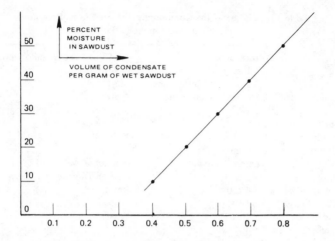

Figure 5.   Relationship of sawdust moisture to volume of smoke condensates [72].

the amount of wood smoke condensates produced at different wood
moisture contents for hickory sawdust. Clearly, the volume of airborne
condensate increases rapidly and linearly with increasing levels of moisture. Some of the compounds reported to be present in hardwood smoke
are given in Table VI.

The relationship between wood smoke and the emission of carbon
monoxide is important. A common myth is that, when the concentration
of smoke particles is low, so is the concentration of combustible gases
[71]. However, in a controlled study that compared the formation of
smoke and carbon monoxide from burning red oak with their formation
from burning other flammable organic materials, red oak liberated large
quantities of carbon monoxide under nonflaming burning conditions.
The most rapid rate of carbon monoxide generation (80 ppm/min for a
4.3-g sample heated at 2.5 W/cm^2) occurred with the onset of glowing
combustion. Unlike the other flammables, which produced far more
smoke than carbon monoxide under the test conditions, red oak had the
highest carbon monoxide emission rate and the lowest smoke production
[71]. The smoke was found to be a mixture of solid particles and ellipti-
cal-shaped liquid residue, with particle size ranging from 0.2 to 0.7 μm.

A noteworthy aspect of home woodburning is the potential hazard
from creosote formation. The black, tar-like substances that condense in
flue pipes are the products of pyrolysis and combustion. Stoves with high
energy-efficiencies are more likely to have chimney creosote problems
because they are so effective at extracting heat from the flue gases [70]. In

Table VI.   Some of the Compounds Reported to be Present in Hardwood Smoke [72]

Acids	Hydrocarbons	1,3-Dimethyl pyrogallol derivatives:
Formic	3,4-Benzpyrene	5-Methyl
Acetic	1,2,5,6-Dibenzanthracene	5-Ethyl
Propionic	20-Methylcholanthrene	5-Propyl
Butyric	Phenols	1-0-Methyl-5-methyl pyrogallol
Aconitic?	Cresols	Veratrole
Tricarballylic?	Creosol	Xylenols
Ketoglutaric?	Guaiacol	Others
Alcohols	Guaiacol derivatives	Ammonia
Methanol	4-Ethyl	Carbon Dioxide
Ethanol	4-Propyl	Resins
Carbonyls	6-Methyl	Water
Formaldehyde	6-Ethyl	Waxes
Acetaldehyde	6-Propyl	
Acetone	Pyrogallol ethers	
Diacetyl	1,0-Methyl	
Furfural	1,0–3,0-Dimethyl	
Methyl furfural		

addition to creating hazards from chimney fires (if flue pipes are not cleaned regularly), creosote formation allows exposure to potentially hazardous compounds produced in pyrolysis. These substances may pose a hazard because of possible toxicity and carcinogenicity.

Carbon Monoxide

Carbon monoxide is readily formed in the combustion of almost any carbonaceous fuel in a limited supply of air or oxygen according to the reaction:

$$C + \frac{1}{2} O_2 = CO \tag{1}$$

If excess air or oxygen is available at the combustion site, the reaction will be driven to complete combustion, producing carbon dioxide:

$$C + O_2 = CO_2 \tag{2}$$

However, contrary to common belief, feeding excess air through a solid-fuel fixed bed will not prevent the copious production of carbon monoxide. In fact burning wood, wood waste or packed biomass in a wood-burning stove or fireplace with an excess of air through the fuel bed will increase carbon monoxide production. This occurs because the CO_2 formed as indicated in Equation 2 is quickly reduced by contact with the glowing coals in the bed [73] according to the reaction:

$$C + CO_2 = 2CO \tag{3}$$

Equation 3, which represents an equilibrium between the two gas-phase species, CO and CO_2, and the solid phase carbon (C), is slightly endothermic as written, absorbing about 170 kJ/mol (160 Btu/mol). The rate of the reaction is strongly temperature-dependent, proceeding rapidly and efficiently to completion (to form carbon monoxide) above 1000°C, but proceeding more slowly below 800°C. The temperature range 800–900°C corresponds to a bed of embers glowing bright cherry red to orange. The reaction is reversible, so that as the products of combustion cool and come into contact with a solid surface, some of the carbon monoxide can disproportionate, depositing carbon on the internal surfaces of stovepipes and chimneys. Thus, if the carbon monoxide is not ignited and burned above the bed, it will be emitted either as the toxic gas or, if converted, as particulate carbon (soot).

Air Quality Impacts

Attempts to estimate the air quality impacts of small-scale residential combustion are hindered by the lack of standard techniques such as exist for assessing the emissions associated with large, centralized sources. The latter commonly employ sophisticated emission abatement devices, uniform fuels and carefully designed, operated and maintained combustion devices. None of these conditions apply to the residential wood stove. Nonetheless, the rapid pace with which wood stoves are replacing more conventional heating sources and the consequent potential for environmental impact, demands an evaluation.

Available predictive techniques were utilized to predict ground-level concentrations of pollutants from wood combustion devices. These techniques [74] assume Gaussian distributions of pollutants, and are most applicable in flat to gently rolling terrain. Concentrations can be calculated for a variety of pollutant emission rates, wind speeds, and wind directions. Using this typical Gaussian dispersion approach, the values in Tables VII and VIII were obtained. The time and distance dependence was determined for ground-level concentrations of emissions from one

Table VII.   Maximum One-Hour Ground-Level
Concentrations of Emissions from One Wood-Burning
Device (μg/m^3)

| Downwind Distance (m) | Stove | | | Pine Particulates[d] | Fireplace Particulates[e] |
| | Oak | | | | |
	Particulates[a]	Hydrocarbons[b]	CO[c]		
25	1.5	1–32	114	8.8	13.2
50	4.6	3–98	354	27.2	40.7
75	4.2	3–87	320	24.6	36.8
100	3.3	2–70	254	19.5	29.2
150	2.1	1–45	161	12.4	18.6
200	1.5	1–31	111	8.5	12.8
250	1.1	0.7–23	82	6.3	9.4
300	0.8	0.5–17	62	4.8	7.2

[a] From Butcher and Buckley [76]. Emissions: 1.7 g/kg wood.
[b] From "Compilations of Air Pollutant Emissions Factors" [77]. Emissions: 1–35 g/kg wood.
[c] From Milliken [78]. Emissions: 130 g/kg wood.
[d] From Butcher and Buckley [76]. Emissions: 10.0 g/kg wood.
[e] From Feldstein [79]. Emissions: 15.0 g/kg wood.

Table VIII.   Peak Concentrations from One Wood-Burning
Device (μg/m^3)

| Time[a] (hr) | Stove | | | Pine Particulates[e] | Fireplace Particulates[f] |
| | Oak | | | | |
	Particulates[b]	Hydrocarbons[c]	CO[d]		
1	4.7	3–99	356	27.4	41.1
3	3.2	2–67	242	18.6	18.0
24	1.5	1–32	116	8.9	13.3

[a] Time scaling factors from Montgomery and Coleman [80].
[b] From Butcher and Buckley [76]. Emissions: 1.7 g/kg wood.
[c] From "Compilations of Air Pollutant Emissions Factors" [77]. Emissions: 1–35 g/kg wood.
[d] From Milliken [78]. Emissions: 130 g/kg wood.
[e] From Butcher and Buckley [76]. Emissions: 10.0 g/kg wood.
[f] From Feldstein [79]. Emissions: 15.0 g/kg wood.

wood combustion device burning 3 kg of wood per hour under the following typical meteorological conditions:

- wind speed = 2 m/sec
- effective emission height = 10 m
- atmospheric stability = stable, class E

Prevention of Significant Deterioration Increases

In the Clean Air Act Amendments of 1977 (PL 95-95), the U.S. Environmental Protection Agency (EPA) defined the allowable amount of air quality deterioration by promulgating the prevention of significant deterioration (PSD) regulations. Presently, only large facilities such as industrial sources and power plants are reviewed for PSD compliance. However, the allowable deterioration represents levels of air quality degradation deemed reasonably acceptable. PSD increments for particulates (none exist for carbon monoxide or hydrocarbons) are given for Class I and II areas [75]. For presently pristine (Class I) areas, increments are $10\mu g/m^3$ maximum for 24-hr period with an overall mean maximum of 5 $\mu g/m^3$, for typical (Class II) areas, particulate concentrations are limited to a 24-hr maximum increment of 37 $\mu g/m^3$ and a mean of 19 $\mu g/m^3$. Clearly, values in Table VII show that a single fireplace, or several woodburning stoves located within 100 m of each other, would exceed allowable increases in pristine areas and would consume much of the allowable increases in typical areas.

Particulate Emissions

If a community of houses converted to wood heat, under adverse meteorological conditions (inversion with a 15-m stable layer), the air quality impact could be severe, as indicated in Table IX. The study area in Table IX represents a typical small community, and the dispersion conditions are representative of small Appalachian valleys and areas throughout New England. The assumed adverse conditions represent a realistic local worst-case situation that could be expected to occur nightly in portions of the southeastern United States and in New England, occasionally lasting for up to 24 hours. If all 247 houses in the community heat with wood for four months, the total particulate emissions would be 18 tons during those 122 days. The comparable total particulate emissions from a 1000-MW coal-fired power plant serving the needs of approximately 500,000 people using a state-of-the-art particulate removal

Table IX.   Short-Term Worst-Case Estimates of Emissions
from a Study Area[a] ($\mu g/m^3$)

| Downwind Distance (m) | Stove | | | Pine Particulates[e] | Fireplace Particulates[f] |
| | Oak | | CO^d | | |
	Particulates[b]	Hydrocarbons[c]			
25	53	31–1071	4017	309	463
50	52	30–1070	3991	307	459
100	51	30–1050	3939	303	453
150	51	30–1040	3900	300	455
200	50	29–1029	3809	293	439
250	49	29–1009	3718	285	428
300	48	28–988	3666	282	422

[a] Study area: size, 0.5 × 0.5 km; housing density, 4 unit/ac = 247 dwellings; wood use, 3 kg/hr-house.
[b] From Butcher and Buckley [76]. Emissions: 1.7 g/kg wood.
[c] From "Compilations of Air Pollutant Emissions Factors" [77]. Emissions: 1–35 g/kg wood.
[d] From Milliken [78]. Emissions: 130 g/kg wood.
[e] From Butcher and Buckley [76]. Emissions: 10.0 g/kg wood.
[f] From Feldstein [79]. Emissions: 15.0 g/kg wood.

system (99.5% efficient) would emit approximately 13 tons of particulate during the same time period [81]. Also, particles emitted from tall stacks at a large generating facility are dispersed throughout a large area, resulting in much smaller peak ground-level concentrations than those resulting from short low-velocity home chimneys. In addition, although little is known about the potential health effects of long-term exposure to wood combustion particulates (smoke), many of the identified hydrocarbons are known carcinogens [72]. Additionally, preliminary studies indicate the potential for absorption of polycyclic aromatic hydrocarbons onto the surface of respirable-size wood ash particles [82].

Carbon Monoxide Emissions

The level of carbon monoxide downwind of the small community (Table IX) is within the current ambient 8-hr maximum standard of 10,000 $\mu g/m^3$ [75]. However, the 4000-$\mu g/m^3$ should perhaps not be considered inconsequential. Animal studies indicate that exposure to low levels of carbon monoxide for periods as short as 4 hr converts the myocardium from aerobic to anaerobic metabolism leading to ultrastructural heart damage [83].

SUMMARY

Biomass-to-energy systems, by utilizing a renewable resource, can make an important contribution to our overall energy needs. However, assurance of environmental acceptability will require close attention to the possible impacts of rapid, impulsive and uncontrolled implementation.

To produce even a modest amount of energy from biomass specifically grown for energy applications would require a very large area of land. Production of 10^{18} J (0.958 quad) of wood energy would require between 1.6 and 5.7 million ha devoted to silvicultural energy farms. Producing 10^{18} J of ethanol from grain would require 16–20 million ha of cropland. These levels of production each represent roughly 1% of our current energy consumption. Clearly, allocating resources from agriculture and forestry to biomass energy production has vast economic and biological implications which have yet to be fully appreciated. Because of the large land areas involved, one long-term issue assumes critical importance: will biomass energy production be managed to ensure renewability?

Residential wood combustion represents a very different problem. While biomass production and harvest pose difficulties because of the very large resource commitments required, woodstoves present problems because of their very small scale. Our habit has been to accommodate dispersed, small-scale environmental alterations and to consider local effects as insignificant relative to the large centralized source. Implicit in this accommodation is the doctrine that environmental dilution is equivalent to environmental dissipation. But because a renewable technology is a long-term technology, an in-depth evaluation will require knowledge about low-level, long-term effects. Unfortunately, this is an area that we know little about. Thus, until this information is available, it may be difficult to assess long-term effects of the large number of relatively small, dispersed disturbances that can arise from this broad-based technology.

REFERENCES

1. Braunstein, H. M., R. D. Roop, F. E. Sharples, J. Tatum, P. Kanciruk and K. M. Oakes. *Biomass Energy Systems and the Environment* (New York: Pergamon Press, 1981).
2. Calvin, M. "The Sunny Side of the Future," *Chemtech* 7:352–363 (1977).
3. "USDA Potential Cropland Study," SCS Statistical Bulletin 578, U.S. Department of Agriculture, Washington, DC (1977).
4. "Prime and Unique Farmlands," *Federal Register* 43(31):4030 (1978).
5. "Environmental Quality: The Ninth Annual Report of the Council on Environmental Quality," U.S. Government Printing Office, Washington, DC (1978).

6.   "Agricultural Statistics," U.S. Department of Agriculture, Washington, DC (1978).
7.   "Water Requirements, Availabilities, Constraints, and Recommended Federal Actions," Federal Energy Administration Project Independence Blueprint, Final Task Force Report, Water Resources Council, Washington, DC (1974).
8.   Harte, J., and M. El-Gassier. "Energy and Water," *Science* 199:623–634 (1978).
9.   Heichel, G. H. "Agricultural Production and Energy Resources," *Am. Sci.* 64:64–72 (1976).
10. Green, M. B. *Eating Oil—Energy Use in Food Production* (Boulder, CO: Westview Press, 1978).
11. Pimentel, D., L. E. Hurd, A. C. Bellotti, M. J. Forster, I.N. Oka, O. D. Sholes and R. J. Whitman. "Food Production and the Energy Crisis," *Science* 182:443–449 (1973).
12. Steinhart, J. S., and C. E. Steinhart. "Energy Use in the U.S. Food System," *Science* 184:307–316 (1974).
13. Jensen, N. F. "Limits to Growth in World Food Production," *Science* 201:317–320 (1978).
14. Burwell, C. C. "Solar Biomass Energy: An Overview of U.S. Potential," *Science* 199:1041–1048 (1978).
15. Klass, D. L. "Wastes and Biomass as Energy Resources: An Overview," in *IGT Symposium Papers, Clean Fuels From Biomass, Sewage, Urban Refuse and Agricultural Wastes,* (Chicago, IL: Institute of Gas Technology, 1976), pp. 21–58.
16. *Crop, Forestry, and Manure Residue Inventory, Vols. 1–8* (Menlo Park, CA: Stanford Research Institute, 1976).
17. Larson, W. E., R. F. Holt and C. W. Carson. "Residues for Soil Conservation, Drop Residue Management Systems," American Society of Agronomy Spec. Publ. 31, Madison, WI (1978).
18. Cox, G. W., and M. D. Atkins. *Agricultural Ecology: An Analysis of World Food Production Systems* (San Francisco: W. H. Freeman & Company, 1979).
19. Lockeretz, W. "The Lessons of the Dust Bowl," *Am. Sci* 66:560–569 (1978).
20. Carter, L. J. "Soil Erosion: The Problem Persists Despite Billions Spent on It," *Science* 198:409–411 (1977).
21. "Environmental Quality: The Tenth Annual Report of the Council on Environmental Quality," U.S. Government Printing Office, Washington, DC (1979).
22. "Progress and Challenges—An Interview with Douglas M. Costle, EPA Administration," *EPA J.* (January, 1980).
23. Doering, O. C. "Future Competitive Demands on Land for Biomass Production," in *IGT Symposium, Clean Fuels from Biomass and Wastes,* (Chicago, IL: Institute of Gas Technology, 1977).
24. Schiff, S. D. "Land and Food: Dilemmas in Protecting the Resource Base," *J. Soil Water Conserv.* 34:54–59 (1979).
25. Lee, L. K. "A Perspective on Cropland Availability," *Agric. Econ.* Rep. 406, U.S. Department of Agriculture, Washington, DC (1978).
26. Train, R. E. "The Environment Today," *Science* 201:320–324 (1978).

27. Larson, W. E. "Crop Residues: Energy Production or Erosion Control?" *J. Soil Water Conserv.* 34:74–76 (1979).
28. Pimentel, D., E. C. Terhune, R. Kyson-Hudson, S. Rochereau, R. Samis, E. A. Smith, D. Denman, D. Reifschneider and M. Shepard. "Land Degradation: Effects on Food and Energy Resources," *Science* 194:149–155 (1976).
29. Gupta, S. C., W. E. Larson, L. D. Hanson and R. H. Rust. "Area Delineation of Possible Corn Residue Removal for Bioenergy in Four Minnesota Counties," Minnesota Agric. Expt. Sta. Paper 9631, Minneapolis, MN (1976).
30. Lindstrom, M. J., S. C. Gupta, C. A. Onstad, W. E. Larson and R. F. Holt. "Tillage and Crop Residue Effects on Soil Erosion in the Corn Belt," *J. Soil Water Conserv.* 34:80–82 (1979).
31. Campbell, R. B., T. A. Matheny, P. G. Hunt and S. C. Gupta. "Crop Residue Requirements for Water Erosion Control in Six Southern States," *J. Soil Water Conserv.* 34:83–85 (1979).
32. Skidmore, E. L., M. Kumar and W. E. Larson. "Crop Residue Management for Wind Erosion Control in the Great Plains," *J. Soil Water Conserv.* 4:90–93 (1979).
33. Holt, R. F. "Crop Residue, Soil Erosion, and Plant Nutrient Relationships," *J. Soil Water Conserv.* 34:96–98 (1979).
34. "Improving Soils with Organic Wastes," 0-623-484/770, U.S. Department of Agriculture, U.S. Government Printing Office, Washington, DC (1979).
35. Henry, J. F., M. D. Frazer and C. W. Vail. "The Energy Plantation: Design, Operation, and Economic Potential," in *Thermal Uses and Properties of Carbohydrates and Lignins*, F. Shafizadeh, K. V. Sarkanen, and D. A. Tillman, Eds. (New York: Academic Press, 1976).
36. *Systems Studies of Fuels from Sugarcane, Sweet Sorghum, Sugar Beets and Corn, Vol. 5, Comprehensive Evaluation of Corn*, BMI-1957 (Columbus, OH: Battelle Columbus Laboratories, 1977).
37. "Solar Program Assessment: Environmental Factors, Fuels from Biomass," ERDA 77-47/7, U.S. Energy Research and Development Administration, Washington, DC (1977).
38. *Effective Utilization of Solar Energy to Produce Clean Fuel* (Menlo Park, CA: Stanford Research Institute, 1974).
39. "Systems Descriptions and Engineering Costs for Solar and Related Technologies, Vol. 9: Biomass Fuels Production and Conversion Systems," Tech. Rep. 7485, Mitre Corporation, McLean, VA (1977).
40. Hertzmark, D. I. "A Preliminary Report on the Agricultural Sector Impacts of Obtaining Ethanol From Grain," Solar Energy Research Institute, Golden, CO (1979).
41. Lockeretz, W. "The Lessons of the Dust Bowl," *Am. Sci.* 66:560–569 (1978).
42. Hagen, L. J., and N. P. Woodruff. "Air Pollution from Duststorms in the Great Plains," *Atmos. Environ.* 7:323–332 (1973).
43. Hagen, L. J., and N. P. Woodruff. "Particulate Loads Caused by Wind Erosion," *J. Air Poll. Control Assoc.* 25:860–861 (1975).
44. Burger, G. V. "Agriculture and Wildlife," in *Wildlife and America,* H. P. Brokaw, Ed. (Washington, DC: Council on Environmental Quality, 1978).

45.   Clark, C., D. D. Jones and C. S. Holling. "Patches, Movements and Population Dynamics in Ecological Systems: A Terrestrial Perspective," in *Spatial Pattern in Plankton Communities,* J. S. Steele, Ed. (New York: Plenum Press, 1978).

46.   Phillips, R. E., R. L. Blevins, G. W. Thomas, W. W. Frye and S. H. Phillips. "No-Tillage Agriculture," *Science* 208:1108–1113 (1980).

47.   Tillman, D. A. *Wood as an Energy Source*, (New York: Academic Press, 1978).

48.   Howlett, K., and A. Gamache. "Silvicultural Biomass Farms, Vol. 6: Forest and Mill Residues as Potential Sources of Biomass." Mitre Corporation Tech. Rep. 7347, McLean, VA, (1977).

49.   "The Outlook for Timber in the U.S., 1970," Forest Service Res. Rep. 10, U.S. Department of Agriculture, Washington, DC (1973).

50.   Steinbeck, K. "Short-Rotation Hardwood Forestry in the Southeast," in *Proc. 2nd Annual Symposium on Fuels From Biomass,* U.S. Department of Energy, Washington, DC (1978), pp. 175–184.

51.   "Biomass Energy Success Stories, A Portfolio Illustrating Current Economic Uses of Renewable Biomass Energy," NTIS HCP/TO285-01, Biomass Energy Institute, Inc., Prepared for the U.S. Department of Energy, Washington, DC (1978).

52.   Jamison, R. L. "Trees as a Renewable Energy Resource," in *Clean Fuels from Biomass and Wastes* (Chicago, IL: Institute of Gas Technology, 1977), pp. 169–183.

53.   Howlett, K., and A. Gamache. "Silvicultural Biomass Farms, Vol. 2: Biomass Potential of Short Rotation Farms," Mitre Corp. Tech. Rep. 7347, McLean, VA (1977).

54.   DiNovo, S. T., W. E. Ballantyne, L. M. Curran, W. C. Baytos, K. M. Duke, B. W. Cornaby, M. C. Matthews, R. W. Weing and B. W. Bigon. "Preliminary Environmental Assessment of Biomass Conversion to Synthetic Fuels," EPA-600/7-78-204, Industrial Environmental Research Laboratory, Cincinnati, OH (1978).

55.   Armson, K. A. *Forest Soils: Properties and Processes* (Buffalo, NY: University of Toronto Press, 1977).

56.   Duvigneaud, P., and S. Denaeyer-de Smet. "Biological Cycling of Minerals in Temperate Deciduous Forests," in *Analysis of Temperature Forest Ecosystems,* D. E. Reichle, Ed. (New York: Springer-Verlag, 1970), pp. 199–225.

57.   Fortescue, J. A. C., and G. G. Marten. "Micronutrients: Forest Ecology and Systems Analysis," in *Analysis of Temperature Forest Ecosystems,* D. E. Reichle, Ed. (New York: Springer-Verlag, 1970), pp. 173–198.

58.   "The Feasibility of Generating Electricity in the State of Vermont Using Wood as a Fuel: A Study," JPR Associates, Inc., submitted to Vermont Agency of Environmental Conservation, Department of Forests and Parks, Montpelier, VT, (1975).

59.   Morrison, I. K., and N. W. Foster. "Biomass and Element Removal by Complete-Tree Harvesting of Medium Rotation Forest Stands," in *Impact of Intensive Harvesting on Forest Nutrient Cycling* (Syracuse, NY: State University of New York, College of Environmental Science and Forestry, 1979), pp. 111–129.

60. Salo, D. J., R. E. Inman, B. J. McGurk and J. Verhoett. "Silvicultural Biomass Farms, Vol. 3: Land Suitability and Availability," MTR-7347, The Mitre Corp., McLean, VA, (1977).

61. Segelquist, C., and M. Rogers. "Use of Wildlife Forage Clearings by White-Tailed Deer in the Arkansas Ozarks," *Proc. Ann. Conf. SE Assoc. Game Fish Comm.* 28:568–573 (1975).

62. Blymyer, M. J., and H. S. Mosby. "Deer Utilization of Clearcuts in Southwestern Virginia," *South J. Appl. Forestry* 1:10–13 (1977).

63. *Wildlife Habitat Management Handbook,* Southern Region, U.S. Forest Service, U.S. Department of Agriculture, Atlanta, GA (1971).

64. Hamrick, W. J., and J. R. Davis. "Summer Food Items of Juvenile Wild Turkeys," *Proc. 25th Ann. Conf. SE Assoc. Game Fish Comm.* (1972), pp. 85–89.

65. Leopold, A. S. "Wildlife and Forest Practice," in *Wildlife and America— Contributions to an Understanding of American Wildlife and Its Conservation,* H. P. Brokaw, Ed. (Washington, DC: Council on Environmental Quality, 1978), pp. 108–120.

66. Stone, E. "The Impact of Timber Harvest on Soils and Water," in *Report of the President's Advisory Panel on Timber and the Environment,* U.S. Government Printing Office, Washington, DC (1973), pp. 427–467.

67. Megahan, W. F. "Logging, Erosion, and Sedimentation—Are They Dirty Words?" *J. For.* 40(7):403–407 (1972).

68. Fredriksen, R. L., D. G. Moore and L. A. Norris. "Impact of Timber Harvest, Fertilization and Herbicide Treatment of Streamwater Quality in Western Oregon and Washington," in *Forest Soils and Land Management, Proc. 4th North American Forest Soils Conference,* B. Bernier and C. H. Winget, Eds., (Quebec, Quebec, Canada: Presses de l'Universite Laval, 1975), pp. 283–313.

69. Wortman, D. "Heat Output of Wood-Burning Stoves," in *Alternative Sources of Energy No. 35,* (1978).

70. Shelton, J., and A. B. Shapiro. *The Woodburners Encyclopedia* (Waitsfield, VT: Vermont Crossroads Press, 1978).

71. King, T. Y. "Smoke and Carbon Monoxide Formation from Materials Tested in the Smoke Density Chamber," National Bureau of Standards, Washington, DC (1975).

72. Jahnsen, V. J. "The Chemical Composition of Hardwood Smoke," PhD Thesis, Purdue Univeristy, Lafayette, IN (1961).

73. Kreisinger, H. "Combustion of Wood-Waste Fuels," *Mech. Eng.* 61:115–119 (1939).

74. *Recommended Guide for the Prediction of the Dispersion of Airborne Effluents* (New York: The American Society of Mechanical Engineers, 1973).

75. "National Primary and Secondary Ambient Air Quality Standards, EPA," *Federal Register* 36(84):8186–8187 (1971).

76. Butcher, S. S., and D. I. Buckley. "A Preliminary Study of Particulate Emissions from Small Wood Stoves," *J. Am. Poll. Control Assoc.* 27(4):346–348 (1977).

77. "Compilations of Air Pollutant Emission Factors," 3rd ed., U.S. EPA Rep. AP-42, Research Triangle Park, NC (1977).

78. Milliken, J. E. "Airborne Emissions from Wood Combustion," paper presented at the Wood Energy Institute Wood Heating Seminar IV, March 22–24, Portland, OR (1979).

79. Feldstein, M. In: *Combustion-Generated Air Pollution,* E. S. Starkman, Ed. (New York: Plenum Press, 1971), pp. 291–318.

80. Montgomery, T. L., and J. H. Coleman. "Empirical Relationships Between Time-Averaged SO_2 Concentrations," *Environ. Sci. Technol.* 9(11):953–957 (1975).

81. Dvorak, A. S. et al. "The Environmental Effects of Using Coal for Generating Electricity," NUREG-0252, Argonne National Laboratory, Chicago, IL (1977).

82. Golembiewski, M. "Environmental Assessment of a Waste-to-Energy Process Wood and Oil-Fired Power Boiler," Midwest Research Institute, Kansas City, MO (1979).

83. *Carbon Monoxide* (Washington, DC: National Academy of Sciences, 1977).

CHAPTER 27

ENVIRONMENTAL ASSESSMENT OF WASTE-TO-ENERGY CONVERSION SYSTEMS

K. P. Ananth and M. A. Golembiewski

Midwest Research Institute
Kansas City, Missouri

H. M. Freeman

U. S. Environmental Protection Agency
Cincinnati, Ohio

Increased emphasis on energy and material recovery, and the need for alternatives to solid waste disposal in landfills have generated growing interest in waste-as-fuel processes. The processes include, on a generic basis, waterwall incinerators, pyrolysis systems, combined fuel-fired systems (coal plus refuse-derived fuel (RDF), RDF plus municipal sewage sludge, coal plus wood waste) and biochemical conversion of waste to methane.

The Fuels Technology Branch of the U.S. Environmental Protection Agency (EPA) Industrial Environmental Research Laboratory in Cincinnati has sponsored a program at Midwest Research Institute (MRI) to conduct environmental assessments of some of the above waste-to-energy conversion processes. The overall objective of this program is to evaluate the potential multimedia environmental impacts resulting from use of combustible wastes as an energy source, and thereby identify control technology needs. As part of this program, MRI has undertaken fairly extensive sampling and analysis efforts at the following waste conversion facilities.

- 181 metric ton/day (200 short ton/day) partial oxidation system for refuse;

- 109 metric ton/day (120 short ton/day) municipal incinerator fired with municipal solid waste (MSW);
- 10-MW power plant boiler fired with wood waste and No. 2 oil;
- 31,752-kg/hr (70,000-lb/hr) steam boiler fired with coal and densified RDF (DRDF); and
- 20-MW power plant boiler fired with RDF.

A description of the facility, the sampling and analysis methods used, and the results obtained are individually presented below for each of these facilities.

REFUSE PYROLYSIS SYSTEM

The Union Carbide process (PUROX®) at South Charleston, WV, was designed to convert 181 metric ton/day (200 short ton/day) of RDF. The RDF was produced by shredding MSW to a 7.6-cm (3-in.) size and removing magnetic materials from the shredded waste. The PUROX system is a partial oxidation process that uses oxygen to convert solid wastes into a gas having a higher heating value (HHV) of about 13.8 MJ/standard m³ (370 Btu/scf).

Figure 1 is a schematic illustration of PUROX. Raw refuse is received by truck in the plant's storage building. It is moved and stacked in the storage area by a front-end loader. The same loader picks up the stored waste, weighs it on a platform, and dumps it on a conveyor leading to the shredder, where it is shredded to a 7.6-cm (3-in.) size. Ferrous material is removed by a magnetic recovery system. The refuse fuel is fed into the top of the reactor, the principal unit on the process, by two hydraulic rams. There are three general zones of reaction within the reactor (drying, pyrolysis and combustion). The reactor is maintained essentially full of refuse, which slowly descends by gravity from the drying zone through the pyrolysis zone into the combustion zone. A counterflow of hot gases, rising from the combustion zone at the bottom, dries the incoming moist refuse. As the material progresses downward, it is pyrolyzed to form fuel gas, char and organic liquids. Oxygen is injected into the bottom hearth section at a level of about 20 wt % of incoming refuse. The oxygen reacts with char formed from the refuse to generate temperatures of 1370–1650°C (2495–3002°F) in the lower zone, which converts the noncombustibles into a molten residue. The residue is discharged into a water quench tank where it forms a slag.

The hot gases from the hearth section are cooled as they rise through the zones of the reactor. After leaving the reactor, the gases are passed through a recirculating water scrubber. Entrained solids are separated

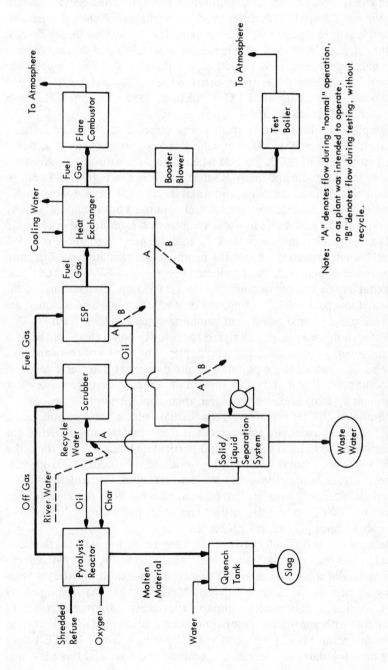

Figure 1. Flow diagram for PUROX process.

Note: "A" denotes flow during "normal" operation, or as plant was intended to operate.
"B" denotes flow during testing, without recycle.

from the scrubber water in a solid-liquid separator, and recycled to the reactor for disposal. The water product discharged from the separator system is sent to a plant treatment system. The gas leaving the scrubber is further cleaned in an electrostatic precipitator (ESP) and then cooled in a heat exchanger prior to combustion in a flare combustor. The gas was burned in a package boiler transported to the site for these tests. The fuel gas consisted of about 40% CO by volume, 23% CO_2, 5% CH_4, 26% H_2, and small amounts of N_2, C_2H_2, etc.

Sampling at the PUROX facility was directed to the three effluent streams; slag, scrubber effluent and gaseous emissions from a boiler when fired with PUROX gas and when fired with natural gas. An overview of the sampling and analysis scheme is shown in Figure 2. As can be seen in this figure, sampling and analysis of each stream were rather complex, being directed to conventional pollutants but including, among others, priority pollutants in water samples and sampling of both liquid and gaseous emissions for most of the analyses prescribed under the EPA Level 1 environmental assessment protocol. Particulate emission sampling in the boiler stack was conducted according to EPA Method 5, except that a high-volume sampling system (HVSS) was used because of the expected low particulate loading. Boiler stack sampling also included use of the Level 1 source assessment sampling system (SASS) train.

Water samples were also analyzed for priority pollutants, but the data are too lengthy for inclusion in this chapter. The results of these analyses showed that few of these pollutants were present at detectable levels in the scrubber effluent. UNOX® the Union Carbide aerobic wastewater treatment system, effectively reduced their concentrations.

Results of the testing effort showed that, of the criteria pollutants, only NO_x and particulate emissions increased when burning PUROX gas as compared to natural gas. NO_x and particulate levels were of the order of 350–400 ppm and 0.0046–0.011 g/standard m^3 (0.002–0.005 gr/scf), respectively. SO_2 emissions averaged 70 to 100 ppm. Particulate and SO_2 emissions were below present standards, whereas NO_x required further reduction. Also, analysis for metals and other pollutants indicated that these should not present any problems.

Because of the difficulty involved in interpreting much of the data collected in this test, especially the Level 1 analysis results, the environmental assessment work was extended to include application of the methodology known as the Source Analysis Model (SAM/1A) developed by EPA. Basically, this model compares the measured concentrations of pollutants with approximate emission concentration guidelines known as minimum acute toxicity effluents (MATE) values. These MATE values have been tabulated for several compounds or classes and there is a spe-

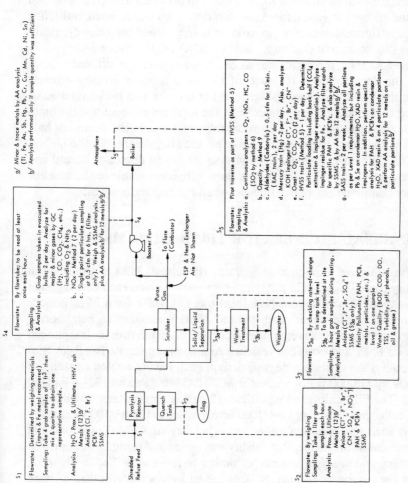

Figure 2. Sampling and analysis scheme for PUROX process.

cific MATE concentration for each compound and for each type of effluent stream (solid, liquid or gaseous). The MATE values are used to compute the ratio of the measured concentration to the MATE concentration, and this ratio is termed the "degree of hazard." The degree of hazard for each pollutant is then summed to provide the degree of hazard for the effluent stream under consideration. This value, when multiplied by the effluent flowrate, in specific units (e.g., liters per second), establishes the "toxic unit discharge rate" (TUDR) for the stream.

The SAM/1A methodology, as described above, was utilized to analyze the data obtained for each of the three primary effluent streams from the PUROX process (slag, scrubber effluent and boiler stack gas). Based on the SAM/1A methodology, the scrubber effluent had the highest degree of hazard, being considerably greater than the degree of hazard for the input river water. However, the slag stream had the highest TUDR. The boiler flue gas effluent had the lowest degree of hazard and the lowest TUDR. Both of these values were comparable to the baseline values computed for boiler flue gas when burning natural gas.

MUNICIPAL INCINERATOR FIRED WITH MSW

The Braintree municipal incinerator (Braintree, MA) is a mass-burn facility consisting of twin waterwall combustion units, each with a design capacity of 109 metric tons (120 short tons) of MSW for a 24-hr period. A portion of the steam produced (20–35%) is supplied to neighboring manufacturers and the remainder is condensed. Each furnace is equipped with an ESP, and both ESP exhaust to a common stack. The Riley Stoker boilers are of the single pass design, each having a rated capacity of 13,608 kg (30,000 lb) of steam/hr at 204°C (400°F) and 1724 kPa (250 psig). The ESP units are single-field, 12 passage precipitators with a specific collection area of 413 m²/1000 actual m³-min (125 ft²/1000 actual ft³-min); each has a design collection efficiency of 93%. Environmental assessment of the incinerator facility was conducted using EPA approved sampling and analysis procedures similar to those identified in Figure 2.

Of the criteria pollutants, SO_2, NO_x and hydrocarbon emissions were low. However, CO levels were high and could not be explained, considering the large quantities of excess air that were used. The average particulate concentration was 0.55 g/standard m³ (0.24 gr/scf), corrected to 12% CO_2. This level exceeded federal and state regulations. However, subsequent tests for compliance had an outlet particulate loading of 0.17 g/standard m³ (0.074 gr/scf), which shows compliance. Elemental analysis of the glass- and metal-free bottom ash revealed an overall increase in the elemental concentrations when compared to the refuse feed. The collected fly ash contained levels of chlorides, sulfates and some trace metals which may be of concern. Polychlorinated biphenyls (PCB) were not detected in the collected fly ash; four polycyclic aromatic hydrocarbons (PAH) were identified. Levels of biochemical oxygen demand (BOD), chemical oxygen demand (COD), oil, grease, total suspended solids (TSS) and total dissolved solids (TDS) in the bottom ash quench water do not appear to be of concern. The phenolic content was found to be <0.1 mg/l in all samples. Levels of gaseous chlorides and other halides were low. The presence of PCB was confirmed only in the SASS train XAD-2 resin at a concentration of 3.6 μg/m^3.

Results of the SAM/1A environmental assessment procedure showed the incinerator stack emissions to have the highest apparent degree of health hazard. Further analysis is needed to determine the exact composition of the organic components of the stack emissions to better ascertain the hazard potential. SAM/1A also showed that the bottom ash effluent had the largest toxic unit discharge rate, due primarily to the abundance of phosphorus and metals in this stream.

POWER PLANT BOILER FIRED WITH WOOD WASTE AND FUEL OIL

The No. 1 unit at the Burlington Electric Plant (Burlington, VT) was originally a coal-fired boiler which was modified to fire woodchips with supplementary No. 2 fuel oil. Because of the high moisture content of the chips, the boiler cannot provide the desired steam output on wood alone. Therefore, No. 2 fuel oil is used. Steam production is rated at 45,360 kg/hr (100,000 lb/hr), which powers a 10-MW turbine generator. Residual ash from the boiler is discharged at the end of the grate into a hopper and is then transported pneumatically to an emission control system consisting of two high-efficiency mechanical collectors in series. For a flue gas flowrate of 102,000 standard m^3/hr (60,000 ft^3/min) at 166°C (330°F), the collectors were designed for an overall pressure drop of 1.6 kPa (6.5 in. H$_2$O) and a collection efficiency of 97.75%. Sampling and analysis were based on the matrix shown in Figure 3.

On a heat input basis, wood accounted for 80% of the boiler fuel, and oil the remainder. The heat of combustion of wood was 13.65 MJ/kg (5870 Btu/lb) as received, and for oil, the heat of combustion was 45.36 MJ/kg (19,500 Btu/lb).

Bottom ash analysis indicated that most elements were more concentrated in the ash relative to the input fuels. No PCB were detected in

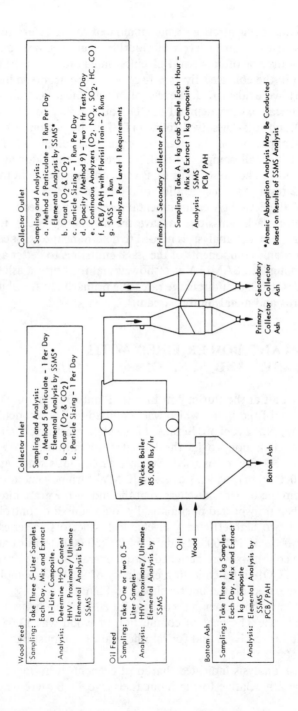

Figure 3. Test matrix for the Burlington Electric wood- and oil-fired power plant.

bottom ash, but one PAH compound, phenanthrene, was present at a concentration of 0.89 $\mu g/g$. Primary and secondary collector ash contained no PCB, but several PAH compounds were identified in the secondary ash, with one sample containing 10 $\mu g/g$ of phenanthrene.

Particle sizing at the collector inlet and outlet could not be established due to constant plugging of the optical counter's dilution system. Stack concentration of particulates averaged 0.18 g/standard m³ (0.08 gr/scf), and the collector had a particulate removal efficiency of 94.2%. NO_x and SO_2 concentrations averaged 66 and 138 ppm, respectively. CO averaged 213 ppm and hydrocarbons 9 ppm. Analysis of Method 5 particulate indicated concentrations approaching 100 μg/standard m³ for Pb, Ba, Sr, Fe and Ti in the stack gases. PCB and PAH tests of the stack gases were negative.

The EPA SAM/1A analysis indicated that the secondary collector ash contained the highest degree of hazard, although all three ash streams were similar in the magnitude of their hazard values. Stack emissions showed a low degree of hazard. The primary collector ash had the highest toxic unit discharge rate.

STEAM BOILER FIRED WITH COAL AND DRDF

Emission tests were conducted on the GSA/Pentagon facility No. 4 boiler in Arlington, VA during a test burn program coordinated by the General Services Administration (GSA) and the National Center for Resource Recovery (NCRR). The No. 4 unit is an underfeed-retort stoker boiler with a rated steam capacity of 31,752 kg/hr (70,000 lb/hr) at 862 kPa (125 psig) and 177°C (350°F). During the tests, the boiler was equipped with a multicyclone collector for removal of particulates from the exhaust gases.

The test burn program included three fuel firing modes: 100% coal (baseline conditions), 20% DRDF/80% coal, and 40% DRDF/60% coal. Samples of coal, DRDF and the coal/DRDF mixtures were collected hourly by NCRR and analyzed for moisture, ash, heating value and chemical composition. Several daily samples of bottom ash were also collected by NCRR and analyzed for loss on ignition and chemical composition. MRI conducted sampling and analysis of the stack effluent. Parameters measured included particulate concentration, gaseous criteria pollutants (SO_2, NO_x, CO and total hydrocarbons), and chlorides. The particulate samples were further analyzed for lead content.

Results of the emission tests showed that:

1. Particulate emissions were reduced by 22–38% when DRDF was blended with the original coal fuel. Filtrable particulate emissions were

lowest when using the 20% DRDF blend and rose again when the proportion of DRDF was raised to 60%. This finding may not be conclusive, however, since the boiler load was held steady during the 20% RDF firing but not during the 60% mode.

2. The amount of particulate lead emitted when buring DRDF with coal is substantially higher than that from combustion of coal alone (an average of 1000 $\mu g/m^3$ with 20% DRDF and 2260 $\mu g/m^3$ with 60% DRDF, versus 330 $\mu g/m^3$ with coal only).

3. Chloride emissions showed no definite trend that could be used to correlate chloride emissions with RDF modes, although slightly higher concentrations of HC1 were observed in two of the samples collected during combustion of the 60% DRDF blend.

4. Concentrations of sulfur dioxide, nitrogen oxides and carbon monoxide all appeared to decrease slightly when the RDF was used with coal. Because of the very low sulfur content of DRDF, SO_2 emissions were reduced progressively as the proportion of DRDF with coal was increased. However, the reduction in NO_x and CO levels may or may not have been the direct result of burning DRDF, since they are highly dependent on boiler combustion conditions.

POWER BOILER FIRED WITH RDF

The Hempstead Resource Recovery Plant (Long Island, NY) receives MSW, produces RDF and converts the fuel to electrical power. The facility consists of two distinct segments: a refuse processing operation, utilizing the Black Clawson Hydrasposal system; and a power house, which contains two steam boilers and two 20-MW electrical turbine generators, plus the associated control equipment.

Tests were conducted by MRI on the No. 2 unit of the power house, which is an air-swept spreader stoker, waterwall boiler with a nominal capacity of 90,720 kg/hr (200,000 lb/hr) of steam at 4309 kPa (625 psig) and 399°C (750°F). The boiler was fired with 100% RDF, although auxiliary oil burners were used for startup and during fuel feed interruptions. Air pollution controls for the boiler consisted of a bank of 12 mechanical cyclones followed by an ESP.

The purpose of the assessment was primarily to investigate organic constituents of the stack gases and to quantify odorous components. However, other tests were also included. Emission streams evaluated included the boiler bottom ash, cyclone ash, ESP ash and the stack effluent gases. Samples of the RDF were also collected and analyzed for moisture plus chemical and elemental composition. The three ash streams were analyzed for elemental composition. Stack emissions were continuously monitored for SO_2, NO_x, CO, O_2 and total hydrocarbon concentrations, and were also tested to determine levels of vaporous mercury and aldehydes. In addition, a sample was collected using the EPA SASS for analysis under the EPA Level 1 protocol.

Initial results of the test program did not indicate any pollutant emissions of major concern. Stack gases contained relatively low concentrations of SO_2, NO_x and hydrocarbons. Carbon monoxide levels were slightly greater than anticipated. Emissions of carbonyl compounds (aldehydes) were detected at a maximum level of 7 ppm (2.95 kg/hr). Mercury vapor concentrations in the stack effluent were very low (< 0.12 mg/m^3), and it appears that mercury levels are greatest in the fly ash collected by the electrostatic precipitator. The concentration of mercury in samples of the RDF was constant at about 3 μg/g.

Several trace metals were detected in the stack gases at relatively high concentrations. Of these, lead, antimony, chromium and arsenic were most notable. Their respective concentrations in the SASS sample were 580, 460, 640 and 560 μg/m^3. Elemental analysis of the bottom ash, cyclone ash and ESP ash streams also indicated that many of the more volatile elements were associated with the smaller-sized particles. Organic analysis of the SASS sample, using EPA Level 1 and additional gas chromatography/mass spectrometry (GC/MS) analytical techniques, showed a variety of organic constituents. No single compound group appeared to predominate, although several PAH were detected. All organic results were qualitative.

Compounds consistently observed in all SASS component extracts included naphthalene, fluoranthene, acenaphthylene, pyrene, phenanthrene/anthracene, *bis*(2-ethylhexyl) phthalate and diphenylamine. The majority of additional compounds were found in the XAD-2 resin extract and included two chlorobenzenes, hexachlorobenzene, fluorene and dibutylphthalate.

CONCLUSIONS

Based on the tests undertaken of the various waste-to-energy conversion systems described in this chapter, control of criteria pollutant emissions, such as particulates, SO_2 and NO_x, should not pose any special problems. Other pollutants, such as specific hazardous organics, halides (chlorides) and some trace metals, could be a problem. Further sampling and analysis should be undertaken to draw definitive conclusions for individual systems. Also, the lack of emission standards and health-risk information for some of these pollutants makes it difficult to interpret observed emission data with regard to degree of hazard.

SECTION 7

SYSTEMS AND CASE STUDIES

CHAPTER 28

THE POTENTIAL OF
BIOMASS ENERGY FOR INDIANA

Donald L. Klass

Institute of Gas Technology
Chicago, Illinois

Excluding most of the contribution made by biomass and wastes, the United States consumed about 78.2 quads of primary energy in 1979 [1]. When the energy consumed in the form of biomass (1.8 quads) is added to this figure, the contribution of biomass energy is about 2.3% of total primary energy consumption, or the equivalent of about 850,000 bbl of oil/day [2].

The purpose of this chapter is to examine biomass, in the form of plant material and organic wastes within Indiana, as an energy resource for the state. Raw materials are considered from the standpoint of availability and maximum potential impact on a statewide basis, particularly with respect to liquid fuel needs for transportation. Unless otherwise indicated, the year 1978 is used as the baseline for data comparisons, because Indiana's fuel consumption statistics are available for that year.

INDIANA FUEL CONSUMPTION

To evaluate the potential of biomass energy, it is necessary to know the energy consumption for Indiana. The figures are summarized in Tables I to III [3]. Table I shows that total primary Indiana energy consumption in 1978 was 2.514 quads, or about 3.2% of the total consumed by the United States. On a per-capita basis, this corresponds to 80.7 bbl oil-equiv, or 131% of the per-capita energy consumption for the United

Table I.   Primary Energy Consumption in Indiana in 1978

	Quads	10^6 Units	10^6 Units	% of U.S.
Coal	1.04	41.1 metric ton	45.3 ton	7.32
Petroleum	1.02	29.4 m^3	185 bbl	2.70
Natural Gas	0.45	11,800 normal m^3	441,000 scf	2.24
Hydroelectric	0.004	1,300 MJ	361 kWh	0.12
Nuclear Electric	0	0	0	0
Total	2.514			3.21

States as shown in Table IV. This table also shows other pertinent statistics which will be referred to later. Overall, coal was the largest source of energy for the state, and petroleum was a close second.

Table II shows fossil fuel consumption by end use. In order of decreasing end-use consumption, the sequence is: industrial, electric, transportation, residential and commercial energy uses. Coal is the largest contributor to the industrial and electric sectors; petroleum is the largest contributor to the transportation and commercial sectors; and natural gas is the largest contributor to the residential sector.

Petroleum contributed almost all of the energy needs of the transportation sector, 0.459 quads or 18.2% of Indiana's total fossil fuel consumption. All of this contribution was in the form of liquid fuels, the breakdown of which is shown in Table III. Motor gasoline is the largest liquid fuel, as expected.

The reserves of native fossil fuel deposits in Indiana are very low, with the exception of coal, as shown by the reserves-to-consumption ratios in Table V. The greatest impact of indigenous biomass is therefore as a substitute for petroleum and gaseous fuels.

POTENTIAL CONTRIBUTION OF WASTES

To examine the energy potential of organic wastes in Indiana, it is necessary to estimate the amounts of residues produced and their energy contents. The results of these projections are shown in Tables VI to XII. The energy inputs required to collect, dry (if necessary) and convert the wastes to other products were not included. Thus, the energy potential of each waste would actually be lower in real systems than the calculated values. It is also obvious that all of the wastes in a given category could not be collected and made available for energy applications. For example, cattle manure collection is only feasible in locations where animal populations are concentrated, such as in confined feedlots or barns.

Table II. Fossil Fuel Consumption in Indiana by End Use in 1978

	Industrial		Residential		Transportation		Commercial		Electric	
	10^{12} kJ	%	10^{12} kJ	%	10^{12} kJ	%	10^{12} kJ	%	10^{12} kJ	%
Coal	446	47.6	1.82	0.6	0.03	Nil	3.25	1.5	651	95.9
Petroleum	282	30.1	148	44.8	484	98.9	141	64.3	24.7	3.6
Natural Gas	209	22.3	180	54.7	5.22	1.1	75.1	34.2	3.43	0.5
Total	937		330		489		219		679	

Table III.   Petroleum Liquids Consumption in Indiana for Mobile Applications in 1978

	10^{12} kJ	10^{12} Btu	10^6 bbl	%
Motor Gasoline	391	371	70.6	80.9
Diesel Fuel	75	71	12.2	15.5
Jet Fuel	16	15	2.8	3.3
Aviation Gasoline	1.7	1.6	0.3	0.3
Total	484	459	85.9	

Table IV.   Selected Indiana Statistics for 1978[a]

		Percent of U.S.	Percent of Indiana
Population	5,374,000	2.46	
Land Area (10^6 ha)	9.3491	1.02	
Water Area (10^6 ha)	0.0502	0.25	
Income per Capita ($)	7,696	23[b]	
Income/km^2 ($)	442,471	15[b]	
Energy Consumption per Capita (bbl oil-equiv)	80.7	131[c]	
Land Usage (10^6 ha)			
Cropland			
For Crops	4.847		51.8
For Pasture	0.5759		6.16
Idle	0.2744		2.93
Grasslands	0.6018		6.44
Forestland	1.566		16.75
Special Use	0.8260		8.83
Other	0.6576		7.03
Farms			
Number (approx.)	95,000	3.5	
Total Area (10^6 ha)	6.9	1.6	73

[a] Adapted from the statistical information in Reference 4. The income per capita and square kilometer, and the energy consumption per capita were calculated.
[b] State rank.
[c] Percent of bbl oil-equiv/capita for U.S.

Table V.   Fossil Energy Reserves/Consumption Ratio for Indiana in 1978 [4-6]

	Consumption (quads)	Proved Recoverable Reserves (quads)	Reserves/ Consumption
Coal	1.04	249[a]	239
Petroleum	1.02	0.151	0.15
Natural Gas	0.45	0	0

[a]Proved recoverable in 1976.

Table VI.   Indiana Grain Crop as Percent of State Areas and U.S. Production in 1978 [7]

	Area Planted		Production	
Crop	10^6 ha	Indiana (%)	10^6Bu	% (U.S.)
Corn	2.47	26.3	637.2	9.00
Soybeans	1.68	17.9	140.4	7.62
Wheat	0.36	3.9	31.78	1.77
Oats	0.089	0.9	8.910	1.48
Rye	0.016	0.2	0.225	0.86
Sorghum	0.0101	0.1	0.975	0.13
Barley	0.0036[a]	0.04	0.320[a]	0.08
Total		49.3		

[a]For 1977.

Table VII.   Estimated Energy Content of Indiana Grain Crop Residues in 1978

		Energy Equivalent	
Crop	Residue (10^6 dry metric ton)	10^{12} kJ	10^{12} Btu
Corn	14.27	249	236
Soybeans	2.92	50.9	48.3
Wheat	1.27	22.1	21
Oats	0.33	5.7	5.4
Rye	0.03	0.47	0.45
Sorghum	0.04	0.63	0.60
Barley	0.01	0.16	0.15
Total	18.87	329	312

Table VIII.   Estimated Energy Content of Indiana Farm Manures in 1978

	Average Population	Residue (10^6 dry metric ton)	Energy Equivalent	
			10^{12} kJ	10^{12} Btu
Cattle	2,025,000	3.22	56.2	53.3
Hogs	4,100,000	0.73	12.7	12.0
Sheep	159,000	0.01	0.2	0.2
Chickens	20,900,000	0.29	5.1	4.8
Total			74.2	70.3

Table IX.   Estimated Energy Content of Indiana Forest Residues in 1976

Wood Type	Residue (10^6 dry metric ton)	Energy Equivalent	
		10^{12} kJ	10^{12} Btu
Hard	0.395	7.89	7.48
Soft	0.002	0.04	0.04
Total		7.93	7.52

Table X.   Estimated Energy Content of Indiana Sawmill Residues in 1978

Residue Type	Residue (10^6 dry metric ton)	Energy Equivalent		Used for Fuel (%)	Available for Fuel (%)
		10^{12} kJ	10^{12} Btu		
Slabs and Edgings	0.149	2.97	2.82	12.6	11.3
Sawdust	0.132	2.63	2.49	19.2	36.3
Bark	0.087	1.74	1.65	32.5	40.7
Total		7.34	6.96		

Table XI.   Estimated Energy Content of Indiana Municipal Wastes in 1978

Waste Type	Residue (10^6 dry metric ton)	Energy Equivalent	
		10^{12} kJ	10^{12} Btu
Refuse	2.40	27.9	26.5
Industrial	0.36	3.2	3.0
Sewage	0.18	3.0	2.8
Total		34.1	32.3

Table XII.   Estimated Energy Content of Indiana Wastes in 1978

| Waste Type | Energy Equivalent | | Percent of Total Energy Consumption |
	10^{12} kJ	10^{12} Btu	
Grain Crop	329	312	12.4
Manures	74.1	70.3	2.8
Municipal	34.1	32.3	1.3
Forest	7.93	7.52	0.3
Sawmill	7.34	6.96	0.3
Total	452	429	17.1

Manure collection from individual animals in open pasture is not economically or energetically practical. Thus, the estimates of biomass energy in the form of wastes in this chapter should be regarded as upper limits; the estimates were made simply to ascertain the maximum potential of waste-derived energy and to compare this potential with energy demands.

Grain Crop Wastes

It is clear from the data presented in Table IV that Indiana is a farm state; more than 60% of the state area was devoted to cropland in 1978, and about 52% was under active cultivation. The major grain crops, areas planted and production for 1978 are listed in Table VI. The seven crops shown in this table account for about 95% of the land under cultivation in the state; corn is by far the largest commercial crop. The residues were estimated by applying existing factors [8]. The energy contents were estimated by assuming a heating value of 17.43 GJ/dry metric ton (15 million Btu/dry ton). The results of the calculations are shown in Table VII.

Farm Manure Wastes

The estimates of farm manure residues are listed in Table VIII. They were calculated by use of population estimates [7], average values of dry manure produced per animal per day [9], and an assumed heating value of 17.43 GJ/dry metric ton (15 million Btu/dry ton).

Forest Residues

The estimates of the quantities and energy contents of Indiana forest residues are shown in Table IX. They were calculated from reported amounts of hardwood and softwood residues for 1976 [10], typical moisture contents and weights per cubic foot for oak (hardwood) and pine (softwood) [11], and typical fuel values for birch (hardwood) and pine [12].

Sawmill Residues

Reported estimates for the quantities of Indiana sawmill residues are presented in Table X [13]. They were converted to energy values by assuming the residues were all derived from hardwood and had a heating value of 20.0 GJ/dry metric ton (17.2 million Btu/dry ton) [12].

Municipal Wastes

The results of the calculations for municipal wastes are shown in Table XI. The quantity of urban refuse was estimated by assuming that 1.8 kg (4 lb) of refuse/person-day [14] was collected for the metropolitan population of Indiana [4]. The quantity of industrial waste was assumed to be equivalent to 15% of the urban refuse generated, as in the case of the wastes generated in the United States [14]. The sewage produced was estimated by assuming 0.14 kg (0.3 lb) of solids/person-day was generated by all of the metropolitan population served by sewer systems [14]. The heating values assumed for the municipal residues were 11.6 GJ/dry metric ton (10 million Btu/dry ton) for refuse [14], 8.7 GJ/dry metric ton (7.5 million Btu/dry ton) for industrial waste [14], and 16.3 GJ/dry metric ton (14 million Btu/dry ton) for sewage solids [15].

Energy Potential of Wastes

Based on these estimates, the energy potential of Indiana wastes is equivalent to about 17% of total primary energy consumption as shown in Table XII. Grain crop residues are the largest energy resource, but it is evident that even if all the organic wastes generated in the state were available for energy applications, only a relatively small fraction of total energy demand could be met, especially after energy deductions are made for operating integrated systems consisting of waste collection, transport, storage and utilization components. Waste utilization as an energy resource in Indiana will therefore be site-specific and localized, as suggested by the data on sawmill residues already used for fuel by the sawmills (Table X) [13]. Each case must be evaluated on its own merits and intrinsic characteristics. For example, urban refuse can probably best be used in the city where generated as a supplemental source of solid fuel or, after suitable conversion, as fossil fuel substitutes [14,16]. The overall costs are less because an operating collection system usually exists; the tranportation costs for the waste are minimized; and the larger energy markets are in the city which already had existing fuel and electric power delivery systems.

Another approach to the assessment of the energy potential of organic residues is to assume that they are converted to products which displace fuels that are projected to be in short supply over the long term. The substitution of biomass-derived ethanol for petroleum-based liquid fuels is an example of technology under development now where organic residues can play an important role [17]. Power ethanol production for use in 10/90 vol% ethanol/unleaded gasoline blends (gasohol) is currently estimated to be about 7100 bbl/day and has been targeted for significant production increases up to 1990 (Table XIII). The technology already exists for the conversion of starchy raw materials such as corn and high-

Table XIII.   U.S. Power Ethanol Production Estimates

Year	liter/day	gal/day	bbl/day	Remarks
1979	378,500	100,000	2,400	Gasohol first marketed in U.S. in 1979 [17]
1980	1,135,000	300,000	7,100	Estimated from literature
1981	5,185,450	1,370,000	32,600	Administration goal in January 1980 [17]
1982	9,462,500	2,500,000	60,000	Goal of 1980 Energy Security Act, June 30, 1980 [18]
1990	15,897,000	4,200,000	100,000	Administration goal in January 1980 [17]
	113,550,000[a]	30,000,000[a]	720,000[a]	Goal of 1980 Energy Security Act, June 30, 1980 [18]

[a] Title II of the Energy Security Act calls for a level of fuel alcohol production of 10% of gasoline production in 1990. These estimates assume gasoline production will be 7.2 million bbl/day in 1990.

sugar materials such as sugar beets to ethanol, and is under development for conversion of high-cellulose materials such as refuse and wood. Assuming that processes exist for conversion of all Indiana organic residues to ethanol, calculations were made to assess the potential of ethanol to displace the petroleum liquids consumed in Indiana. The results of these calculations are presented in Table XIV. Ethanol from conversion of all the corn grown in 1978 and corn that could have been grown on idle farmland, assuming it was suitable for corn growth, is also included for comparison purposes.

These estimates indicate that Indiana organic wastes, particularly the grain crop residues, could make a substantial contribution to displacement of petroleum liquids and afford captive supplies of liquid fuels for vehicular applications.

Similar results would have been obtained if the calculations had been performed based on the assumption that substitute natural gas was produced from organic wastes instead of ethanol, because the amount of natural gas energy consumed in 1978 (0.45 quads) was about the same as the energy content of the petroleum liquids used by the transportation sector (0.46 quads). All organic wastes can be converted to methane-containing gas by a variety of processes [16].

Table XIV.   Potential Contribution of Ethanol to Indiana
Energy Consumption in 1978 Using Existing Biomass

Route to Ethanol	Ethanol Production (10^6 bbl)	Percent of Total Energy Consumption[a]	Percent of Petroleum Liquids Consumption[a]
All Corn Production	39.4[b]	4.98	45.9
Corn Residues Only	37.1[c]	4.69	43.2
Idle Cropland in Corn and All Corn Converted	4.20[d]	0.53	4.9
All Grain Residues	49.1[c]	6.21	57.2
All Organic Residues	67.5[c]	8.54	78.6

[a]Total energy consumption is 2.514 quads, and total petroleum liquids consumption is 85.9 million bbl.
[b]Yield based on 9.8 liter/bu (2.6 gal/bu) of corn.
[c]Yield based on overall thermal efficiency of 50%.
[d]Yield based on 245 bu/ha (100 bu/ac) corn yield on idle cropland and ethanol yield of 9.8 liter/bu of corn.

Thus, Indiana's organic residues have the potential of serving as supplemental sources of energy, which, in suitable form, can displace significant quantities of fuels that are now mainly supplied by sources outside the state.

LAND BIOMASS

The other possibility for using renewable biomass as an energy resource within the state is to plant selected water- or land-based species specifically for their energy content, and then to harvest and process them to manufacture synfuels or to produce heat by direct combustion.

Examination of the statistical data on land and water distribution in Indiana (Table IV) indicates there is little water area in the state that might be dedicated to biomass energy. The potential of aquatic biomass as an energy resource in Indiana appears small. The potential of using idle farmland for growth of new corn and its conversion to ethanol at state-of-art corn and ethanol yields is also small as illustrated by the calculations in Table XIV.

Although use of idle farmland for energy applications is a source of supplemental fuels that should be considered in the context of any statewide or local programs for biomass energy, the grasslands and forestlands appear from present land usage patterns to offer a higher energy potential. These areas comprise almost 25% of the state.

To obtain approximate values for the energy potential of various size areas of the state when used for biomass energy, calculations were made for three total areas under cultivation (1, 5 and 10% of state), at three

Table XV.   Estimate of Biomass Contribution to Primary
Indiana Energy Consumption[a]: Biomass Energy Yield
as Percent Total Energy Consumption at Given
Yield Level of Biomass

State Area Planted (%)	dry metric ton/ha-yr (dry ton/ac-yr)		
	6.73 (3)	22.4 (10)	56.0 (25)
1	0.4%	1.4%	3.5%
5	2.1	6.9	17.3
10	4.2	13.9	34.7

[a] Assumptions: biomass has heating value of 17.43 GJ/dry metric ton (15×10^6 Btu/dry ton); energy consumption is 2.514 quads; no adjustment made for energy inputs on production or processing of biomass.

biomass yields (3, 10 and 25 dry ton/ac-yr). These yield levels are believed to span the state-of-the-art and also include future advances expected to be made as biomass yields are improved through research programs directed to specific plant and tree species as energy resources [2]. The results of the calculations are shown in Table XV. It is apparent that biomass grown specifically for energy applications could contribute to energy demand by serving as a major resource provided sufficient land could be devoted to biomass energy. This route should be evaluated, especially if an overall program is designed for the state to develop its own biomass energy supplies.

CONCLUSIONS

The organic residues generated in Indiana, particularly the grain crop residues, have the potential of serving as energy resources to supplement fuel needs which are currently supplied by fossil fuels. Energy from wastes represents a small fraction of Indiana's total energy consumption, but has the potential of replacing a significant portion of petroleum-based liquid fuels or natural gas. Land-based biomass grown specifically for energy applications has the potential of making a significant contribution to Indiana's total energy demand provided sufficient land can be used for this purpose.

REFERENCES

1. Institute of Gas Technology, *Energy Statistics* 3 (1) (1980).
2. Klass, D. L. "Energy From Biomass and Wastes: 1979 Update," Symposium Paper—Energy From Biomass and Wastes IV, Lake Buena Vista, FL, (1980), pp. 1–41.
3. U.S. Department of Energy, "State Energy Data Report," DOE/EIA-0214 (78), Dist. Cat. UC-13 (1980), pp. 135-141.
4. "Statistical Abstracts of the United States," U.S. Department of Commerce (1978-1979).
5. The American Gas Association. *Gas Facts 1978 Data,* p. 13.
6. U.S. Dept. of the Interior, "Demonstrated Coal Reserve Base of the United States on January 1, 1978," Mineral Industry Surveys (1977), p. 6.
7. U.S. Department of Agriculture, "Agricultural Statistics 1979" (1979).
8. Welch, L. F. "An Inventory of the Energy Potential From Biomass in Illinois," Petroleum and Natural Gas in Illinois, Proceedings of the 7th Annual Illinois Energy Conference (1979), pp. 224-236.
9. Ghosh, S., and D. L. Klass. "Conversion of Animal Feedlot Wastes to Fuel, Fertilizer, and Feed by the BIOGAS® Process," Research Proposal, Institute of Gas Technology (1974).

10. "Forest Statistics of the U.S., 1977," U.S. Department of Agriculture, pp. 124–125.
11. *Engineering Materials Handbook*, 1st ed., C. L. Mantell, Ed. (New York: McGraw-Hill 1958).
12. Klass, D. L. "A Perpetual Methane Economy—Is It Possible?" *Chemtech* 4 (3):161–168 (1974).
13. Cassens, D. L., and D. McGuire. "Wood Residue Survey of Primary Manufacturers in Indiana," Station Bulletin No. 238, Dept. of Forestry and Natural Resources, Agricultural Experiment Station, Purdue University, West Lafayette, IN (1979).
14. Klass, D. L. "Energy Production From Municipal Wastes," paper presented at 33rd Annual Meeting, Soil Conservation Society of America, Denver, CO, July 30–August 2, 1978.
15. Klass, D. L. "Anaerobic Digestion for Methane Production—A Status Report," Proceedings Bio-Energy 80 (1980).
16. Klass, D. L., and S. Ghosh. "Fuel Gas From Organic Wastes," *Chemtech* 3(11):689–698 (1973) November.
17. Klass, D. L. "Alcohol Fuels for Motor Vehicles: An Overview," *Energy Topics*, (1980).
18. "U.S. Energy Security Act of 1980," PL 96–294, Title II.

CHAPTER 29

BIOMASS ENERGY IN THE CARIBBEAN BASIN: A CASE STUDY OF THE DOMINICAN REPUBLIC

Charles Peterson

Gershman, Brickner & Bratton, Incorporated
Washington, DC

The 25 countries that comprise the Caribbean basin (Figure 1) are dependent on imports for conventional energy, such as coal, natural gas and petroleum. Trinidad and Tobago, which is a net exporter of petroleum products, is the lone exception. The ability of these countries to maintain their trade balances has been strained by the dependence on and cost of conventional energy imports. To compensate for their energy imports, these countries rely primarily on agricultural exports, especially sugar. However, as a commodity the value of agricultural products fluctuates on international markets, whereas the cost of conventional energy has increased steadily since 1973.

Countries in the Caribbean basin have begun to examine the potential for the use of unconventional domestic sources of energy to reduce the consumption of imported energy. Among the alternatives under consideration are biomass resources. Some of these resources have been an important part of the energy supply in the region for many years. Firewood and charcoal are commonly used for domestic cooking in most Caribbean countries. To a lesser extent, wood is used to power commercial and industrial operations. Bagasse, the fibrous residue of sugarcane processing, is used as a fuel by the cane mills.

Current interest in biomass includes efforts to improve the efficiency with which fuel wood and bagasse are used. Excess consumption of fuel wood can lead to deforestation. The energy content of bagasse exceeds the amount required to process sugar, if efficiently used. The excess bagasse

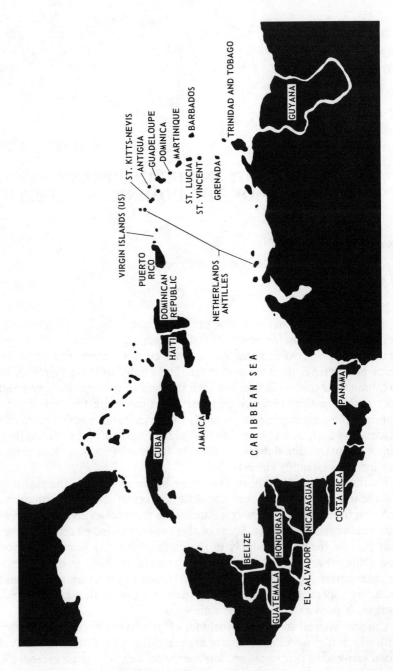

Figure 1. Map of the Caribbean basin.

could be used to produce energy for other uses. There also is interest in other types of biomass. Production of ethanol from sugarcane has generated the most interest throughout the region. This interest has arisen from the success of the ethanol program in Brazil, which is based primarily on sugarcane. Fuel wood plantations, the growing of trees for fuel, also has been a subject of interest. Other biomass resources which have received attention are animal manures, municipal solid waste (MSW), and aquatic plants.

In the background section of this chapter, information on the population, economic status and commercial energy use in the countries of the Caribbean basin is presented. The next section gives an overview of biomass resources in the region. Actions to develop biomass energy in the Dominican Republic are examined in the third section.

BACKGROUND

The Caribbean basin can be divided into four areas: (1) Central America, (2) the Greater Antilles, (3) the Lesser Antilles, and (4) Guyana. The Greater Antilles include the large islands on the northern boundary of the Caribbean Sea. Those islands which form an arc from the U.S. Virgin Islands to the southern group of islands in the Netherlands Antilles have been classified as the Lesser Antilles. Within this group are the windward and leeward islands. Those islands from Dominica south, except for the Netherlands Antilles, are the windward group. Plentiful rainfall characterizes these islands (234 cm/yr). Conversely, the leeward islands are much drier (124 cm/yr). Guyana, which is located in South America, was included because it is a member of the Caribbean Community (CARICOM), an organization of British Commonwealth nations.

Numerous differences exist between the Caribbean basin countries. However, in relation to the other countries in the Western Hemisphere, these differences are relatively small. Four characteristics the basin countries generally have in common are: (1) small size, (2) high population density, (3) low economic prosperity, and (4) lack of conventional energy resources. There are differences in these characteristics within the region, and these differences have an impact on the type of biomass energy suitable for the basin countries. In general, these differences tend to follow the four country groupings that were listed above.

On an average size basis, the rank of the country groups is Guyana, Central America, the Greater Antilles and the Lesser Antilles (Table I). For a majority of the countries in the region, size alone is an unimportant variable in terms of the potential for biomass energy development. Size is

Table I.   Land Area, Population and Gross National Products of the Countries in the Caribbean Basin, 1978 [1]

Country	Land Area (km²)	Population			Gross National Product		
		Total (10³)	Density (people/km²)	Growth Rate[a] (%)	Total (10⁶ $)	Per Capita ($)	Growth Rate[a] (%)
Central America							
Belize	22,960	132	6	0.9	110	840	4.7
Costa Rica	50,900	2,110	42	2.5	3,250	1,540	3.2
El Salvador	20,920	4,382	209	3.1	2,810	600	2.1
Guatemala	108,890	6,627	61	2.0	6,040	910	3.3
Honduras	112,090	3,441	31	3.3	1,650	480	0.0
Nicaragua	139,000	2,490	18	3.3	2,100	840	2.5
Panama	75,650	1,826	24	3.1	2,350	1,290	-0.1
Average	75,780	3,001	56	2.6	2,616	930	2.2
Greater Antilles							
Cuba	110,920	9,718	88	1.6	7,860	810	-1.2
Dominican Republic	48,442	5,128	106	3.0	4,680	910	4.6
Haiti	27,750	4,831	174	1.7	1,240	260	2.1
Jamaica	10,962	2,131	194	1.7	2,350	1,110	-2.0
Puerto Rico	8,897	3,365	378	2.8	9,150	2,720	0.1
Average	41,394	5,034	188	2.2	5,056	1,162	0.7
Guyana	214,970	836	4	2.0	460	550	0.4
Lesser Antilles							
Antigua	443	74	167	1.3	70	950	-3.7
Barbados	430	250	581	0.7	490	1,770	2.6
Dominica	749	77	103	1.2	30	440	-4.1
Grenada	344	106	308	1.8	60	530	-3.2
Guadeloupe	1,761	319	181	0.2	910	2,850	2.9

Martinique	1,101	321	292	0.1	1,270	3,950	5.7
Netherlands Antilles	995	246	247	1.3	780	3,150	0.5
St. Kitts-Nevis	262	50	191	0.9	30	660	1.6
St. Lucia	616	120	195	2.3	80	630	0.7
St. Vincent	338	105	311	2.3	40	380	-2.2
Trinidad and Tobago	5,128	1,137	222	1.2	3,310	2,910	1.5
Virgin Islands	352	101	287	3.6	540	5,350	0.6
Average	1,043	242	257	1.4	634	1,961	0.2

[a]Growth rate is from 1970 to 1977.

important for many of the islands in the Lesser Antilles. The limited land area of these countries restricts and even precludes the development of forestry and agricultural based biomass energy.

A high population density can have an influence on biomass energy. For example, in a country in which firewood and charcoal are used for cooking, which is common in a majority of the countries of the region, the potential exists for deforestation to become a serious problem, as it has in El Salvador and Haiti. In addition, as the population grows, the competition for land will be intensified. This can decrease the opportunities for such biomass developments as fuel wood plantations. Large quantities of MSW, however, are generated in areas of high population density.

Of the four country groupings, the Lesser Antilles have the highest average population density, but the lowest population growth rate (Table I). In terms of population density, the Greater Antilles, Central America and Guyana follow in order after the Lesser Antilles. Population growth is fastest, on average, in Central America, followed by the Greater Antilles and Guyana.

The economic status, as measured by gross national product (GNP), of the countries in the Caribbean basin is, on average, low. On a per-capita basis, no consistent pattern in GNP was found among the four island groups. On an average GNP per-capita basis, the four groups ranking are: the Lesser Antilles, the Greater Antilles, Central America and Guyana (Table I). Significant variations exist among the countries in these groups, however. The growth in GNP also was found to vary within the groups. Eight countries were reported to have no growth or a negative growth during the period 1973-1977. Only three countires — Belize, Dominican Republic and Martinique — showed a growth rate greater than 4.5% per year. In terms of growth, the Lesser Antilles fared the worst, on the average, from 1970 to 1977. Central America had the best average growth record (2.2%/yr), followed by the Greater Antilles and Guyana.

To the extent that the economic prosperity of a country and its inhabitants is dependent on energy consumption, an improvement in the economic status of the Caribbean basin will require the consumption of increased amounts of energy, or an improvement in the efficiency with which energy is used. Even to maintain the existing level of prosperity will require the use of more energy, as the population expands. As mentioned in the introduction, the countries of the region generally lack conventional energy resources.

In terms of conventional energy, all the countries in the Caribbean basin except Trinidad and Tobago were dependent on imports in 1978

(Table II). Only three countries — Barbados, Cuba and Guatemala — in addition to Trinidad and Tobago, were producers of fossil fuel during 1978. During the period 1973–1978, fossil fuel production rose in these countries, except in Cuba, which experienced a 50% decline in output. The remaining 21 countries had to import all of their conventional energy. Eleven of these countries had hydroelectric generating capacity. Ten countries in the region obtained no energy from domestic commercial resources (hydroelectric plus fossil fuels) between 1973 and 1978. Biomass energy consumption, which is substantial in most of the countries in the Caribbean basin, was excluded from Table II.

OVERVIEW: BIOMASS IN THE CARIBBEAN BASIN

Firewood, charcoal and bagasse are commonly used throughout the region for energy. In addition, as previously mentioned, there is interest in the use of other biomass resources for energy, including fuel wood plantations, agricultural products, MSW and aquatic plants. The last category, aquatic plants, is still too experimental to be suitable for commercial-scale operations and, thus, was excluded from this review.

Forest Products

Firewood and Charcoal

Domestic cooking is the primary use of firewood and charcoal. Some fuel wood is also used for commercial and industrial operations. Even so, firewood and charcoal are an important part of the energy supply in the region. In 1974, for example, forest products supplied the equivalent of 10% of the commercial energy consumed that year [3].

The energy value of the fuel wood consumed in selected countries of the Caribbean basin during 1977 is shown in Table III. The Central American countries had the highest per-capita use of firewood and charcoal. The islands in the Greater Antilles were next in order based on average per-capita consumption. However, Haiti had a consumption level that closely resembled the Central American countries. Conversely, fuel wood consumption in Jamaica was similar to that in Guyana and Trinidad and Tobago, which had very low per-capita use of wood for energy. Fuel wood consumption has been forecast to grow through 1990 in a majority of the countries listed in Table III. Conversely, per-capita use

Table II. Domestic and Total Commercial Energy Use in the Caribbean Basin Countries, 1978 [2]

Country	Domestic Energy Production Fossil Fuels (10^{15} J)	Domestic Energy Production Hydroelectric (10^{15} J)	Total Energy Consumption[a] (10^{15} J)	Per Capita Energy Consumption (10^9 J)	Percent Domestic to Total Energy Consumption
Central America					
Belize	0.0	0.0	2.7	17.5	0.0
Costa Rica	0.0	5.6	34.9	16.5	16.0
El Salvador	0.0	3.5	35.1	7.8	10.0
Guatemala	1.3	1.3	50.4	7.6	5.2
Honduras	0.0	1.8	27.9	8.3	6.5
Nicaragua	0.0	1.3	36.3	15.1	3.6
Panama	0.0	0.3	53.0	29.0	0.6
Average	0.2	2.0	34.3	14.5	6.0
Greater Antilles					
Cuba	3.1	0.3	345.6	34.2	1.0
Dominican Republic	0.0	0.3	69.7	13.6	0.4
Haiti	0.0	0.8	8.0	1.7	10.0
Jamaica	0.0	0.4	113.0	53.4	0.4
Puerto Rico	0.0	0.3	410.7	136.7	0.1
Average	0.6	0.4	189.4	47.9	2.4
Guyana	0.0	0.0	25.7	31.4	0.0
Lesser Antilles					
Antigua	0.0	0.0	3.0	50.7	0.0
Barbados	1.5	0.3	8.0	32.3	22.5
Dominica	0.0	0.1	0.5	6.0	20.0
Grenada	0.0	0.0	0.6	6.1	0.0
Guadeloupe	0.0	0.0	8.1	21.7	0.0

Martinique	0.0	0.0	11.2	29.5	0.0
Netherlands Antilles	0.0	0.0	160.5	624.6	0.0
St. Kitts-Nevis	0.0	0.0	0.5	8.3	0.0
St. Lucia	0.0	0.0	1.3	11.5	0.0
St. Vincent	0.0	0.1	0.6	6.2	16.7
Trinidad and Tobago	613.1	0.0	164.9	145.5	100.0
Virgin Islands	0.0	0.0	179.5	2679.0	0.0
Average	51.2	0.0	44.9	301.8	13.3

[a]Total energy includes domestic and imported fossil fuels, and hydroelectric power.

Table III.   Firewood/Charcoal Use and Sustainable Forest Yield for Selected Countries in the Caribbean Basin, 1977 and 1990 [4]

Country	1977					1990				
	Sustainable Forest Yield (10^15 J)	Demand for Forest Products				Sustainable Forest Yield (10^15 J)	Demand for Forest Products			
		Total (10^15 J)	Firewood/Charcoal		Nonfuel (10^15 J)		Total (10^15 J)	Firewood/Charcoal		Nonfuel (10^15 J)
			Total (10^15 J)	Per Capita (10^9 J)				Total (10^15 J)	Per Capita (10^9 J)	
Central America										
Costa Rica	194.1	39.4	24.5	11.8	14.9	161.3	76.3	33.4	11.9	42.9
El Salvador	20.3	37.0	35.6	8.4	1.4	22.1	39.3	35.4	5.9	3.9
Guatemala	492.1	63.6	57.2	8.6	6.1	428.9	83.2	76.8	8.1	6.4
Honduras	662.3	43.6	33.8	10.9	9.8	638.7	48.2	36.8	7.7	11.4
Nicaragua	536.7	34.0	24.1	9.9	9.9	537.6	43.3	25.3	6.8	18.0
Panama	346.2	16.9	15.6	8.9	1.3	337.8	18.0	17.1	7.4	0.9
Average	375.3	39.1	31.8	9.8	7.2	354.4	51.4	37.5	8.0	13.9
Greater Antilles										
Cuba	93.9	20.8	16.6	1.7	4.2	65.5	15.4	11.1	0.9	4.3
Dominican Republic	85.3	19.6	19.5	3.9	0.1	82.9	20.6	20.6	2.9	NA
Haiti	15.4	45.7	43.0	8.7	2.7	6.1	53.3	50.5	7.4	2.8
Jamaica	37.9	0.1	0.1	0.1	0.1	36.0	0.1	0.1	0.1	0.1
Average	58.1	21.6	19.8	3.6	1.8	47.6	19.9	20.6	2.8	1.8
Guyana	1680.6	9.7	0.2	0.2	9.5	1635.6	2.2	0.1	0.1	2.2
Lesser Antilles										
Trinidad & Tobago	17.3	1.0	0.1	0.1	0.9	16.4	0.3	0.1	0.1	0.3

of fuel wood is anticipated to decline by 1990, except in Costa Rica.

Variations in the use of fuel wood among the countries of the region are related to economic wealth, the accessibility of the forests to harvesting and the availability of forests from which to obtain fuel wood. Consumers tend to shift away from fuel wood and to fossil fuels as their income grows. Part of the reason for this shift is that fossil fuels are more convenient and have cleaner burning characteristics. Another and perhaps more important reason is that the use of fossil fuel imparts a degree of perceived status on those who can afford it. Forests located in very mountainous areas or remote from a majority of the population (e.g., Guyana) tend to be inaccessible for use as fuel wood. Firewood and charcoal use in such countries consequently tends to be relatively low. Significant forest reserves must exist in a country for fuel wood use to be an important part of the energy balance. Availability appears to be a limiting factor only in the Lesser Antilles, where the forests are small or almost nonexistent.

Firewood and charcoal consumption also varies between the urban and rural areas of a country. On a per-capita basis, urban dwellers consume less fuel wood than people in rural areas. Both the higher income and reduced accessibility to fuel wood to those in urban areas contribute to the lower rate of consumption. Of the fuel wood consumed by urban dwellers, charcoal is the primary type; whereas firewood predominates in rural sectors. Inhabitants of rural areas tend to be able to gather directly from forests the fuel wood needed for cooking. Thus, they have less concern about the low energy content per unit volume of firewood than those who transport, distribute and use fuel wood in urban areas. Conversion of firewood to charcoal improves the energy-content-to-unit-volume ratio. As a result, the quantity of fuel wood hauled to and used in cities is less with charcoal than would be the case with firewood.

Regardless of the variation in fuel wood composition among the countries listed in Table III, sustainable forest yield has been forecast to decline between 1977 and 1990, except in El Salvador and Nicaragua (Table III). Sustainable forest yield is the quantity of wood that can be harvested per year without diminishing. The decline in sustainable forest yield indicates that these countries are suffering from deforestation of forest reserves. Fuel wood consumption is only part of the demand for forest products. Nonfuel wood (e.g., construction, paper) are also an important part of the equation, especially in several of the Central American countries and Guyana. Another factor, which is excluded from Table III, is the conversion of forest lands to agricultural uses. Pressure for conversion is strongest in the more densely populated countries. Conversion of land to agricultural use has the greatest implications for sustained yield, since this represents the long-term loss of land from the forestry land base.

Aside from the biomass energy implications, deforestation indicates other potential problems for a country. After one area has been deforested, the ecosystem that retains the soil and allows for gradual release of rainfall is lost. During the rainy season, rapid runoff of rainfall occurs, which leads to downstream floods and soil erosion. Erosion is a serious concern in the Caribbean basin, since hydroelectric power is the only domestic commercial energy source for many of the countries. Eroded soil is deposited behind the hydroelectric dams and causes siltation, which over time reduces the generating capacity of a dam.

In 1977 deforestation was only a serious problem in two counties – El Salvador and Haiti (Table III). Although this problem will grow throughout the region, serious deforestation is expected to be a problem in these two countries through 1990. In Haiti, the problem has been projected to worsen, while in El Salvador, loss of sustainable forest yield is anticipated to improve slightly. This improvement will be due to an active reforestation program in the mountains. However, political strife in El Salvador probably will have an adverse impact on this program.

Various options exist to contain the decline in sustainable forest yield, or deforestation, including actions which affect both the supply and consumption of fuel wood. In terms of supply, government incentives or programs such as those in Cuba and El Salvador could be instituted to encourage replanting. Presently, deforested areas are rarely replanted. One problem with replanting is to prevent the cutting of the newly planted trees for fuel wood. On the demand side, current cooking devices for both firewood and charcoal are very inefficient throughout the region, especially in rural areas. Cooking in rural areas typically is done on a grate that is set on several stones. Such devices have a thermal efficiency of only about 7%. Charcoal stoves are more efficient and have a thermal efficiency of 15%. Improved stoves for both firewood and charcoal have been designed in both the Caribbean basin and in other parts of the world. Increased use of these stoves has been limited, partially due to cost and also to tradition. Another demand factor is the production of charcoal, which also is inefficient at present. Charcoal commonly is produced using an earthen kiln. Substantial combustion takes place during the conversion process because oxygen is able to enter an earthen kiln. Use of sealed kilns made of metal would improve the efficiency of the conversion process. However, sealed kilns are expensive for the small-scale charcoal producers in the region. A shift to organized charcoal production would improve the efficiency of fuel wood use, but it would also impact on those who currently produce charcoal.

Fuel Wood Plantations

The organized production of trees solely to be used for fuel is the concept of an energy plantation. The manner in which the wood would be converted to energy would depend on local conditions. Electrical generation and the production of firewood and charcoal are the two most common methods. Use of the fuel wood to generate electricity has attracted the most interest. This option would displace imported fossil fuels currently used to generate electricity. Organized production of firewood and charcoal would reduce the demand on natural forests for fuel wood.

No commercial-scale fuel wood plantations are in operation in the Caribbean basin, although test projects are being developed or under consideration in several countries. Even so, based on land area alone, the best prospects for fuel wood plantation development appear to be in Guyana and the countries of Central America. Limited potential appears to exist in the Greater Antilles. In the Lesser Antilles, this option holds virtually no promise of becoming a part of the energy supply.

In the development of a fuel wood plantation, the two major requirements are land and unskilled labor. The labor is required to plant, care for and harvest the trees when ready for use. To minimize the land area needed for a plantation, a fast-growing variety of trees would be substituted for the indigenous species. In tropical areas, such as the Caribbean basin, fast-growing species can produce up to about 20 metric ton/ha-yr of green wood. Natural tropical forests produce an annual average yield of about 4 metric ton/ha-yr.

Increased land use efficiency through fast-growing trees is important in the region. Land suitable for fuel wood plantations is limited. Very hilly or mountainous terrain, common in the region, should be avoided to prevent soil erosion. With the high population density and the dependence on export crops, competition for the available land from agriculture will be intense. This competition can only be expected to grow in the future as the population expands.

Agriculture

Agricultural biomass resources are divided into two broad categories: (1) the residue from crop and animal production and crop processing, and (2) ethanol. Included in the first category are bagasse, other milling residue, field crop residues and animal manures. Of these resources, only bagasse is currently used for energy on a large-scale in the Caribbean basin.

Bagasse

The fibrous residue of sugarcane milling is known as bagasse, and it accounts for 35% by weight of harvested cane. Bagasse is the traditional fuel used by the sugarcane industry to power the mills. In the smaller mills, bagasse is used to generate process steam only. Electricity and steam are cogenerated in the larger mills.

The energy content of the bagasse generated in the cane-producing countries of the region is shown in Table IV. These data, when compared with the commercial energy consumption data in Table II, indicate that bagasse is a significant energy resource in these countries. However, during the period 1970–1977, sugarcane (and thus bagasse production) experienced consistent growth only in the Central American countries. The economics of cane production favor large-scale farming operations, which has led to a decline of the industry in the smaller Caribbean islands. In fact, on two islands, Antigua and Martinique, cane production ceased between 1970 and 1977. Competition for land from food crops also has contributed to the decline of the industry on the islands. In addition, several of the island governments have sought to diversify the export crop base to include such products as bananas and citrus fruits. Even with the decline in sugarcane production in the island countries, it is still an important part of their economy. As long as cane is produced, bagasse will represent an important energy resource.

Theoretically, only part of the energy in bagasse should be needed for the refining of sugar. The remaining portion would, therefore, be available for other energy uses. In practice, however, the industry usually is able to meet only its own needs. Part of the reason for this is that many of the mills in the Caribbean have outdated, inefficient equipment. In addition, bagasse has been viewed as a waste product that requires disposal, rather than as an energy resource. Another factor which has prevented bagasse from becoming a more important energy resource is its moisture content. At the time of combustion, bagasse averages a moisture content of about 50%.

Modernization of the sugar mills would increase the efficiency with which bagasse is used. Increased cogeneration would enable the sugar mills to distribute their unused electricity through the utility grid. One approach to reducing the moisture content of bagasse which has received attention in the Caribbean basin is solar drying. Although experimental at this time, solar drying is anticipated to be a feasible process by 1990.

Table IV.   Production of Sugarcane, Sugar and Bagasse in the Caribbean Basin, 1977 [4]

Country	Sugarcane Production (10³ metric tons)	Annual Growth, (%) 1970–1977	Sugar Production[a] (10³ metric tons)	Bagasse[b] Production (10³ metric tons)	Energy Value (10¹⁵ J)
Central America					
Belize	784.0	5.0	98.0	274.4	2.3
Costa Rica	1,600.0	4.2	200.0	560.0	4.6
El Salvador	2,912.0	26.4	364.0	1,019.2	8.4
Guatemala	3,896.0	20.4	487.0	1,363.6	11.2
Honduras	800.0	11.1	100.0	280.0	2.3
Nicaragua	1,808.0	7.5	226.0	632.8	5.2
Panama	1,448.0	17.3	181.0	506.8	4.2
Greater Antilles					
Cuba	55,656.0	−1.0	6,957.0	19,479.6	160.0
Dominican Republic	10,024.0	3.0	1,253.0	3,508.4	28.8
Haiti	400.0	−3.0	50.0	140.0	1.1
Jamaica	2,376.0	−2.8	297.0	831.6	6.8
Puerto Rico	1,928.0	−5.1	241.0	674.8	5.5
Guyana	2,024.0	−3.0	253.0	708.4	5.8
Lesser Antilles					
Antigua	0.0	−100.0	0.0	0.0	0.0
Barbados	960.0	−3.1	120.0	336.0	2.8
Guadeloupe	728.0	−5.4	91.0	254.8	2.1
Martinique	0.0	−100.0	0.0	0.0	0.0
St. Kitts-Nevis	344.0	6.7	43.0	120.4	1.0
Trinidad and Tobago	1,424.0	−2.5	178.0	498.4	4.1

[a] Sugar production was estimated 12.5% of sugarcane production.
[b] Bagasse generation was estimated to be 35% of sugarcane production. Field residue was excluded from these figures.

Assuming the moisture content of bagasse were lowered to 30% with solar drying, the energy content per metric ton would be raised to 11.5 G.J. Compared with bagasse at a 50% moisture content (8.2 G.J. per metric ton), this is a 40% increase in the heating value.

Other Milling Residues

Various cereal crops, such as rice and corn, are grown in the Caribbean basin. Processing of these crops generates residue such as rice hulls. Unlike bagasse, these milling residues are commonly discarded or used as animal feed rather than converted to energy.

In several of the countries listed in Table V, the energy content of the cereal residues are substantial. These data, however, include both processing and field residues. On average, milling residue accounts for about 15–20% of the total cereal residue. Although milling residue is concen-

Table V.   Generation of Cereal and Livestock Residue in Selected Countries of the Caribbean Basin, 1977 and 1990 [5]

	Cereal Residue			Livestock Residue		
	Quantities (10^15 J)			Quantities (10^15 J)		
Country	1977	1990	Sources[a]	1977	1990	Sources[b]
Central America						
Costa Rica	6.6	9.4	R,C,S	31.2	42.2	C,H,P
El Salvador	18.9	37.4	C,S,R	20.6	24.9	C,P,H
Guatemala	29.5	43.2	C,S,R,W	39.8	48.2	C,P,H,S
Honduras	15.1	16.4	C,S,R	34.2	41.4	C,H,P,S
Nicaragua	10.1	14.8	C,S,R	46.0	55.6	C,P,H
Panama	7.4	8.3	C,R	23.6	28.6	C,H,P,F
Greater Antilles						
Cuba	1.2	1.8	R,C	102.3	124.0	C,H,P,S
Dominican Republic	9.0	15.7	R,C,S	39.0	47.3	C,H,P,S
Haiti	17.9	18.8	C,S,R	31.2	37.8	C,H,P,S
Jamaica	0.3	0.8	C,R	6.9	8.3	C,P,S,H,F
Guyana	8.1	8.9	R,C	6.2	7.5	C,F,P,S
Lesser Antilles						
Trinidad & Tobago	0.7	1.1	R,C	2.1	2.5	C,F,P,S,H

[a]Abbreviations: R = rice, C = corn, S = sorghum.
[b]Abbreviations: C = cattle, H = horses, donkeys and asses, P = pigs, F = chicken and fowl, and S = sheep and goats.

trated at the processing plant, little interest has been expressed in the expanded use of this material for energy.

Field Residues

All residues generated during the production and harvesting of crops were included in this category. No evidence was found of substantial use of this material for energy.

For cereal crops, about 80–85% of the energy content of total cereal residues is from field residue (Table V). The energy content of the field residue from sugarcane amounts to about 75% of the energy of bagasse on a per-unit-of-weight basis.

Although plentiful in the Caribbean basin, field residues appear to be unsuitable as an energy resource for two reasons. First, they are generated over a wide area in very low concentrations. Collection of the material and concentrating it for use as a fuel would pose a significant management problem. In addition, plant matter has a relatively low energy content per unit volume, which on the basis of energy balance prohibits the transportation of residue over more than a short distance. A second reason is that decaying residue returns nutrients to the soil. This is important, because little or no fertilizer is used on most farms in the region. If the present practice of leaving the residue on the fields were discontinued, the nutrients needed for biomass growth would be depleted.

Animal Residues

The raising of livestock is an important part of agriculture in the region, including the Lesser Antilles. Even so, animal manure is rarely used for fuel applications. In 1977 the energy content of the animal residues discarded in the countries listed in Table V was substantial.

The prospects for use of animal manures to generate methane appear to be limited to localized applications. The majority of the animals in the Caribbean basin are pastured. As a result, most animal manures are dispersed over wide areas. Only in situations where animals are confined would sufficient quantities of manure be generated. Under such conditions the manure could be used in a digester to produce methane, but digester costs might preclude wide-scale use.

Ethanol

Ethanol use as a motor fuel and as a chemical feedstock has recently received the most attention in the Caribbean region, especially since the advent of the Brazilian ethanol fuel program. Presently ethanol production in the basin is potable alcohol, except in Costa Rica. There, molasses, a sugarcane refining by-product, is used to produce fuel and potable ethanol. However, the region has a plentiful supply of biomass from which to produce ethanol. The three main types of biomass are: (1) sugar-bearing materials (e.g., sugarcane), (2) starches (e.g., corn, cassava, potatoes), and (3) celluloses (e.g., wood, agricultural residue). Given the dominance of the sugarcane industry in the Caribbean basin and the established technology for ethanol production, interest has focused on sugarcane. For sugarcane, the ethanol production yield is about 70 liter/metric ton of cane. Although this rate is low compared with other crops such as corn (370 liter/metric ton), the greater productivity of sugarcane and per hectare gives it the highest yield (3500 liter/ha-yr) [6]. Another factor favoring sugarcane is that bagasse can be used to fuel the process.

However, development of an ethanol-for-fuel industry in the countries of the region has been delayed. The cost of producing fuel–grade ethanol appears to be more costly at the present than the use of fossil fuels. Part of the reason is the cost of the biomass material used as a feedstock. When sugar prices are high, as in the later part of 1980, production costs are high. Unless sugarcane is grown expressly for ethanol production, cane will be very difficult to obtain during periods of high sugar prices. In addition, sugar is an important export commodity for most of the countries in the region. Use of sugar to produce ethanol would diminish the export earnings of a country. Of course, there would be some offset from the reduced demand for imported fuels.

Sugarcane could be used for ethanol with minimum disruption to the existing industry if cane were grown expressly for the purpose of ethanol manufacture, which is a concept identical to the fuel wood plantation. This is one approach being used in the Brazilian program. However, this approach implies a supply of surplus agricultural land. In the small and densely populated countries of the Caribbean basin, such land is less than plentiful. Land currently being used for another purpose could be converted to cane for ethanol production, but such a tradeoff could involve transferring land used for food crops to fuel production. The food-vs-fuel question must be weighed carefully.

Municipal Solid Waste

Households and commercial operations are the sources of MSW. Data on the quantity and composition of the refuse generated in the Caribbean basin are very limited. Consequently, it is difficult to accurately assess the potential for use of MSW as a fuel. Sufficient information does exist, however, to make a rough estimate of the potential.

The potential of converting MSW to energy is limited to the urban areas of the region as well as the islands of the Lesser Antilles. These areas generate sufficient quantities of refuse to justify waste-to-energy projects. In addition, use of MSW probably will be confined to those areas which also have a higher GNP per capita. On a nationwide basis, this level was estimated to be $2000 and above. Six countries in the Caribbean basin fit this category. Such areas will have higher percentages of complex organics (e.g., paper and plastics), which means a higher energy content for the MSW. In addition, the refuse will have a lower percentage of food waste, which means a lower moisture content, and thus a further improvement in the energy content.

In areas which meet the criteria listed above, refuse probably will be most suitable for direct combustion. Plans are currently being prepared for waste-to-energy plants for two of these countries, Puerto Rico and the U.S. Virgin Islands. Five potential projects have been identified in Puerto Rico. At the present time, one 1800 metric ton/day plant is under development in San Juan for electric power generation. On St. Thomas and St. Croix in the Virgin Islands, projects have been identified for desalination plants to be powered by refuse. As tourist islands, the Virgin Islands have large quantities of complex organics in the refuse. A problem in countries, or parts of countries, in which tourism is an important part of the economy is that tourism is seasonal. Therefore, both the quantity and composition of the refuse which might be used in a waste-to-energy system will vary during the year. Such variation can effect the viability of a project.

Urban communities in countries with a GNP per capita of less than $2000 probably will have refuse unsuitable for direct combustion. The MSW will have a high moisture content due to a high percentage of food waste. One option for such refuse is to convert the putrescible organics to compost and to use the separated residual combustibles as fuel. A project of this type is in operation in Rio de Janeiro, Brazil. The solid waste fuel is used in a cement kiln for preheating the clinker chamber.

CASE STUDY: THE DOMINICAN REPUBLIC

The Dominican Republic is a rapidly developing country (4.6%/yr real GNP growth from 1970 to 1977) with no domestic production of fossil fuels. To maintain the economic and social progress of the country, it was realized that efforts were needed to reduce dependence on imported energy. A result of this realization was the formation of the Comision

Nacional de Politica Energetica (CNPE) to develop an energy policy for the country. One goal of the CNPE is to promote the development of domestic energy resources such as biomass. Current biomass energy development projects by the CNPE include fuel wood plantations, solar drying of bagasse, biogas generation and ethanol. Other government agencies also are involved in biomass energy such as firewood, charcoal, bagasse and methane generation. These activities that are outlined below are specific to the Dominican Republic. Even so, the projects indicate the type of action that could be taken by other countries in the Caribbean basin.

Forest Products

Firewood and Charcoal

In 1967 the forestry law was modified to limit severely the commercial cutting of forests. This change was instituted to control deforestation. Only minor controls, however, were established on the consumption of fuel wood. Charcoal production is controlled only by limiting the number of trucks that can transport charcoal and the number of deliveries a truck can make each month. No controls exist on firewood consumption.

Since only minor controls exist for fuel wood, the consumption of firewood and charcoal has been forecast to grow from 2.3×10^{16} J in 1980 to 2.6×10^{16} J by 2000 [7]. Fuel wood consumption, however, is expected to peak around 1990. The shift in population from rural to urban areas, where less fuel wood is consumed per capita, is the reason for this peak. One result of the growth in fuel wood use will be a decline in forested land area of 41% or 565,400 ha between 1978 and 2000 [6]. Part of this loss will be the result of conversion of forested land to agricultural use.

Fuel Wood Plantations

Generation of electricity using the product from a fuel wood plantation has been a subject of much interest in the Dominican Republic. To test the concept in the country, the CNPE recently funded a demonstration project that is being implemented by a local university, Institute Superior de Agricultura. The fuel wood harvested from this project will be used for firewood and charcoal.

Should the demonstration prove the capability of fast-growing trees to meet expectations, there is interest in proceeding with commercial-scale operations for electricity generation. In a recent study, an estimated 58,200 ha, or about 1% of the land area in the country, would be needed to fuel a 50-MW generation plant [8]. This would be an important contribution to existing generating capacity. In 1980 this would amount to 7% of the generating capacity of the state power company, Corporacion Dominicana de Electricidad. Since the country is plagued with brownouts, a fuel wood–powered electric generator would be an important component of the utility system.

Agricultural

Bagasse

As mentioned in the overview section, one option for increasing the energy available from bagasse is to reduce the moisture content. The CNPE is planning to test the feasibility of lowering the moisture content of bagasse using a solar dryer. In a joint project with the state sugar company, Consejo Estatal del Azucar, a demonstration solar dryer is to be built at a sugar mill. The project will test the feasibility of this approach to drying and determine the parameters under which the system has the best operating characteristics.

This concept appears to be a workable approach to drying of bagasse. Assuming that it is workable and could be developed on a commercial-scale by 1990, the impact would be significant. With a 50% moisture content at the point of combustion, the available energy from bagasse has been projected to be 5.05×10^{16} J in 1990. A reduction in the moisture content to 30% by solar drying would increase the available energy about 9.4% to 5.52×10^{16} J.

Animal Manure

Presently, two or three biogas digesters are in operation in the Dominican Republic. CEAGANA, a part of the state sugar company, has the largest digester, which has a capacity of 90 m³.

Although the majority of the animals in the country are pastured, there are a number of livestock operations (e.g., chicken, dairy farms) that have numerous animals confined in small areas. A project has been initiated by CNPE to demonstrate biogas digesters at several farms. The purpose of this demonstration project will be to show farmers that this is a workable method of producing usable energy.

This project also was conceived with a view towards the future. As land competition increases, less space will be available for pasture. Consequently, to meet the demand for meat products, it is anticipated that there will be a partial shift to confinement of animals on feedlots prior to slaughter. Successful demonstration of this project will help to ensure that future animal feeding operations will consider anaerobic digestion as a source of fuel gas.

Ethanol

Considerable interest exists in the potential for producing ethanol as a partial substitute for gasoline. However, various studies have indicated that production would be uneconomical, so no plans have been announced to test this option. Given the level of gasoline consumption, an estimated 373.8 million liters by 1990, any plans to proceed in this area would be with a full-scale plant. The present approach is to continue to monitor the Brazilian ethanol program.

CONCLUSIONS

Firewood, charcoal and bagasse, which are presently the dominant biomass energy sources in the Caribbean basin, will continue to remain the major sources of biomass energy for the indefinite future. The importance of wood and bagasse as energy resources could increase if current concepts such as fuel wood plantations and solar drying of bagasse are found to be feasible on a commercial scale. If the results of both concepts, increased fuel wood production and higher energy content bagasse, are applied to electrical generation, it could be a significant development for the region. Currently, except for minor contributions from hydro projects, all electricity is generated using imported fuel.

In the other biomass areas, the prospects appear to be marginal, except in some instances. Large contributions of energy from other milling and field residues appear to be unlikely. Biogas from animal manure is a possibility on farms with sufficient numbers of confined animals. However, few farms have sufficient animals to justify the expense of digesters. Ethanol, although interesting, probably will remain uneconomical in the near term. Should the economics improve through new technology, the Caribbean basin has the raw materials with which to commercialize the technology. Finally, the prospects for municipal waste-to-energy projects appear to be good in about six countries. The success of the current project underway in San Juan, Puerto Rico, should promote interest in this resource in the other countries.

ACKNOWLEDGMENTS

I want to express my appreciation to the following people for their assistance on this effort: Franklin Viloria, Comision Nacional de Politica Energetica, Santo Domingo, Dominican Republic; and Andres Dorenberg and Airi Roulette, Energy/Development International, Port Jefferson, NY, and Washington, DC, respectively.

REFERENCES

1. "1979 World Bank Atlas," World Bank, Washington, DC (1980).
2. "World Energy Supplies, 1973–1978," United Nations, New York, Series J, No. 22 (1979).
3. Donaldson, G. et al. "Forestry: Sector Policy Paper," World Bank, Washington, DC (1978).
4. "Statistical Yearbook," United Nations, New York (1979).
5. Hughart, D. "Prospects for Traditional and Non-conventional Energy Sources in Developing Countries," World Bank, Washington, DC, Staff Working Paper No. 346 (1979).
6. "Alcohol Production from Biomass in the Developing Countries," World Bank, Washington, DC (1980).
7. Peterson, C. "Biomass Energy Resources, 1978–2000," in *Energy Strategies for the Dominican Republic*, Report of the National Energy Assessment, Comision Nacional de Politica Energetica, Santa Domingo, (1980).
8. Trehan, R. K., L. Newman, W. R. Park and M. K. Hallford. "Biomass Farming in the Dominican Republic: A Preliminary Analysis," U.S. Agency for International Development, Santa Domingo, (1980).

CHAPTER 30

MARKET POTENTIAL OF
WOOD FUEL IN THE SOUTHEAST

Tze I. Chiang and David S. Clifton, Jr.
Economic Development Laboratory
Engineering Experiment Station
Georgia Institute of Technology
Atlanta, Georgia

Since 1973 escalating fuel prices have forced more Americans to rediscover wood as a heating fuel. Not only is wood in plentiful supply and a renewable resource, it also is relatively cheap. In terms of heating capacity, it is estimated that a cord of hardwood burned in a sound stove will deliver as much heat as 160–170 gal of No. 2 fuel oil, or 26,000 ft^3 of natural gas, or 6300 kWh of electricity. With hardwood selling for less than $60/cord in much of the United States, this could constitute a substantial savings in winter fuel costs.

As a result of this recent back-to-wood movement, domestic demand for wood-burning stoves, for many years on the decline, is currently booming. Bureau of Census data show that in 1977 U.S. shipments of wood-burning stove-type residential heating devices totaled 234,000 units—about 2.7 times the number of units shipped in 1972 (Table I). Table I also shows that the expanded demand for wood burners has been at the expense of other types of burners (mostly oil and gas). Wood stoves as a percent of all domestic heating stoves increased from 6.5% in 1972 to 15.2% in 1977. The cost differentials between wood and other fuels are increasing and wood usage as a domestic fuel is expected to continue to increase, especially in rural locations.

Bureau of the Census data for 1970 show that there were 793,908 occupied housing units in the United States in which wood is used as the

Table I.   U.S. Shipments of Nonelectric Domestic
Heating Stoves [1] (thousands of units)

Year	Wood-Burning	All Fuels	Wood-Burning as % of All Fuels
1959	394	2648	14.9
1960	335	2191	15.3
1961	291	1977	14.7
1962	281	2112	13.3
1963	279	2218	12.6
1964	227	1935	11.7
1965	203	1618	12.6
1966	147	1629	9.0
1967	144	1518	9.5
1968	167	1587	10.5
1969	111	1573	7.1
1970	103	1454	7.1
1971	94	1393	6.7
1972	86	1317	6.5
1973	88	1284	6.8
1974	124	982	12.6
1975	129	1097	11.8
1976	118	1121	10.5
1977	234	1542	15.2

"principal" heating fuel. Of this number, almost one-third (254,618) were located in the South Atlantic states. Georgia led the region and the nation by burning wood in 66,604 housing units. The low cost of readily available firewood in Georgia has kept wood the "poor man's" fuel. In 1970 more than 24% of the occupied housing units in the 10 Georgia counties with the lowest median family incomes used wood as the principal heating fuel, as compared with less than 1% in the 10 most affluent counties [2]. It would appear that as heating costs rise and wood becomes increasingly attractive as an economical heating fuel, the market for wood-burning heaters will continue to expand. This should be especially true in low-income areas where fuel wood is plentiful and readily accessible.

A study was performed to examine the use of wood fuel in the Southeast in institutional and industrial markets [3]. Emphasis was given in this study to existing boiler systems in eight states (Alabama, Florida, Georgia, Kentucky, Mississippi, North Carolina, South Carolina and Tennessee). A summary of the highlights is presented in this chapter.

THE FOREST PRODUCTS INDUSTRY

The U.S. forest products industry is a major user of wood fuel because it has a large annual energy requirement, access to wood fuel, and the expertise and technology required to use wood fuel. It has been estimated that energy self sufficiency within the forest products industry is about 40–50% for pulp and paper mills, 20–40% for sawmills, and 50% for plywood and veneer mills. The industry uses about 3 quads of fuel annually and has been purchasing about 50% of this total in recent years. Nearly 75% of the self-generated fuel is produced from process wastes, principally pulping liquor. The remainder is wood and bark. Most of the fuel is burned to produce process heat and steam [4].

The U.S. forest products industry purchased 1.8 quads of energy in 1974. A total of 1.3 quads was in the form of oil and natural gas. This purchased energy can be replaced by wood fuel which is potentially available in the near term. The wood is available from several sources including noncommercial timber-growing on commercial forest land, forestry residues, surplus annual growth and mill residue [4].

The Southeast is the leading region in the production of pulp and paper products, and it is one of three major timber regions in the nation. The Southeast has 38.5% of the U.S. commercial forestland acreage, contributes 46.4% of the annual timber growth, and has 26.1% of the standing timber inventory in the nation [5]. The region is obviously a major candidate for energy self-sufficiency in the forest products sector.

INDUSTRIAL AND INSTITUTIONAL MARKETS

Fuels for industrial and institutional users are consumed mainly by boilers. Two series of data demonstrate the status of wood as boiler fuel by these users. The first data series was compiled from the U.S. Environmental Protection Agency (EPA) in eight southeastern states (Alabama, Florida, Georgia, Kentucky, Mississippi, North Carolina, South Carolina and Tennessee). Data were tabulated on the number of boilers, capacity and fuel type for industrial and institutional users. It should be noted that EPA data dealt only with boilers having capacities above 1,000,000 Btu/hr (1.05 GJ/hr). A summary of the tabulation is given in Table II for industrial boilers. Boiler capacities are given in different range groups along with the number of boilers and aggregated capacity. The purpose of the tabulation is to show wood-fired boilers as a percentage of all boilers. It is clear from the table that the smaller the range of boiler capacity, the higher the percentage of wood-fired boilers.

Table II.   Industrial Boilers in Eight Southeastern
States,[a] 1978 [6]

Capacity Range (MBtu/hr)[b]	Number			Aggregated Capacity (MBtu/hr)[b]		
	All Boilers	Wood-Fired	Wood-Fired (%)	All Boilers	Wood-Fired	Wood-Fired (%)
1–30	2,369	228	9.6	24,955	2,180	8.7
31–60	532	36	6.8	23,827	1,552	6.5
61–90	255	14	5.5	19,202	993	5.2
91–120	161	6	3.7	17,101	634	3.7
121–150	147	5	3.4	19,926	609	3.1
151–180	81	1	1.2	13,584	170	1.3
181 and over	301	7	2.3	112,020	2,164	1.9
Total	3,846	297	7.7	230,615	8,302	3.6

[a] Alabama, Florida, Georgia, Kentucky, Mississippi, North Carolina, South Carolina and Tennessee.
[b] 1 MBtu/hr = 1.05 GJ/hr.

Although wood-fired boilers constituted 7.7% of all boilers in the Southeast, wood boilers constituted 9.6% of the total within the 1- to 30-million-Btu/hr (1.05- to 31.6-GJ/hr) capacity range. In terms of aggregated capacity, wood boilers constitute only 3.6% of all boilers. As in the case of the number of boilers, the percentage of wood-fired boilers becomes progressively less when the capacity range increases.

The tabulation on institutional boilers in the Southeast is given in Table III. These boilers are operated by schools, hospitals, military services and governments. It is clear from the table that the percentage of wood-fired boilers is insignificant in this sector. Wood-fired boilers constitute only about 1% of the total and only 0.2% in terms of aggregated capacity.

A second series of boiler data was supplied by the American Boiler Manufacturers Association concerning new boilers ordered in the eight Southeastern states from 1968 to 1978. Those boilers with capacities over 100,000 lb/hr (45,400 kg/hr) are given in Table IV, and those under 100,000 lb/hr are given in Table V.

The tabulation reveals that the percentage of wood-fired boilers ordered has significantly increased since the energy crisis in 1973, both in number and in aggregated capacity. In terms of the total boilers, the percentage increased from 1.4% in 1972 to 9.9% in 1978 for units over 100,000 lb/hr capacity, and from 2.5 to 4.9% in the same period for units under 100,000 lb/hr capacity. In terms of aggregated capacity, the percentage increased from 0.5 to 2.4% for units over 100,000 lb/hr

Table III. Institutional Boilers in Eight Southeastern
States,[a] 1978 [6]

Capacity Range (MBtu/hr)[b]	Number			Aggregated Capacity (MBtu/hr)[b]		
	All Boilers	Wood-Fired	Wood-Fired (%)	All Boilers	Wood-Fired	Wood-Fired (%)
1–20	1,248	13	1.0	4,978	36	0.7
21–40	94			2,746		
41–60	31			1,575		
61–80	35			2,567		
81–100	8			695		
101 and over	36			6,048		
Total	1,452	13	0.9	18,609	36	0.2

[a] Alabama, Florida, Georgia, Kentucky, Mississippi, North Carolina, South Carolina and Tennessee.
[b] 1 MBtu/hr = 1.05 GJ/hr.

Table IV. New Boilers Ordered in Eight Southeastern States[a]
with Boiler Capacity over 100,000 lb/hr, 1968–1978 [7]

Year	Number			Aggregated Capacity (10^3 lb/hr)		
	All Boilers	Wood-Fired	Wood-Fired (%)	All Boilers	Wood-Fired	Wood-Fired (%)
1978	151	15	9.9	157,498	3,850	2.4
1977	171	10	5.8	125,762	1,769	1.4
1976	158	6	3.8	75,715	1,830	2.4
1975	204	5	2.5	127,616	2,235	1.7
1974	455	10	2.2	338,810	2,390	0.7
1973	458	15	3.3	310,318	2,815	0.9
1972	281	4	1.4	196,020	980	0.5
1971	369	3	0.8	176,172	412	0.2
1970	340	3	0.9	286,943	870	0.3
1969	338	7	2.1	154,414	2,020	1.3
1968	322	6	1.9	239,118	962	0.4

[a] Alabama, Florida, Georgia, Kentucky, Mississippi, North Carolina, South Carolina and Tennessee.

Table V.   New Boilers Ordered in Eight Southeastern States[a]
with Boiler Capacity Under 100,000 lb/hr, 1968-1978 [7]

Year	Number			Aggregated Capacity (10^3 lb/hr)		
	All Boilers	Wood-Fired	Wood-Fired (%)	All Boilers	Wood-Fired	Wood-Fired (%)
1978	350	17	4.9	15,508	807	5.2
1977	374	18	4.8	14,717	888	6.0
1976	298	7	2.3	11,559	340	2.9
1975	382	7	1.8	15,366	372	2.4
1974	538	28	5.2	24,292	1,287	5.3
1973	780	52	6.7	33,314	2,228	6.7
1972	645	16	2.5	27,323	807	3.0
1971	605	10	1.6	26,554	543	2.0
1970	666	4	0.6	28,643	155	0.5
1969	796	11	1.4	127,074	446	0.4
1968	714	3	0.4	22,960	144	0.6

[a] Alabama, Florida, Georgia, Kentucky, Mississippi, North Carolina, South Carolina and Tennessee.

capacity, and from 3 to 5.2% for units under 100,000 lb/hr capacity.

It is obvious from these data that wood as boiler fuel has been increasing, especially for small boilers and in industrial applications.

A SURVEY OF POTENTIAL MARKETS FOR WOOD BOILERS AND GASIFIERS

A survey was conducted in the eight Southeastern states during the two-month period February and March 1979 to assess the interest in purchasing boilers or gasifiers using wood as primary or secondary fuel. A list of boiler owners was compiled from EPA files which included only large-capacity systems (1,000,000 Btu/hr, 1.05 GJ/hr or more). The response rate to the survey was 11%.

Fuel Type

Fuels used for these boilers are oil, gas, coal, bark and wood, electricity, black liquor, hydrogen, and unspecified by-products. Oil of all grades and gas were the dominant fuels. Detailed fuel types, numbers of boilers, and the percentage of each fuel type are given in Table VI.

Table VI. Boiler Fuels Reported by Survey Respondents

Fuel Type	Number of Boilers	Percent
Oils	281	42
Gas	259	38
Coal	76	11
Bark and Wood	47	7
Electricity	4	a
Black Liquor	7	1
Hydrogen	1	a
By-products	1	a
Total	676	100

[a]Combined one percent.

Fuel Costs

Fuel costs (FOB plant) are presented in Table VII. They are given by fuel type, number of respondents and cost in dollars per million Btu. These costs were calculated on the basis of FOB plant, Btu content of each fuel and estimated boiler efficiency. Wood and bark included both self-generated and purchased material. "Others" denotes black liquor and unspecified by-products.

Table VII. Fuel Costs by Type on Per-Million-Btu Basis

Fuel Type	Number of Respondents	Fuel Costs ($/10^6 Btu)	
		Range	Average
Oils	136	1.41–4.68	3.21
Gas	107	1.58–4.18	2.58
Propane	9	3.20–5.66	4.57
Coal	25	1.16–2.86	1.97
Wood and Bark	11	0.24–3.85	1.79
Electricity	5	8.67–10.11	9.15
Others	4	1.67–3.06	2.38

Retrofitting Existing Boilers with Wood Gasifiers

After reviewing capital cost data for wood gasifiers provided in the questionnaire, 26 respondents indicated that they were interested in retrofitting their existing boilers with wood gasifiers. Negative answers were 145. Affirmative answers constituted 14% of the total respondents.

Installing New Wood Boilers

Of the total 186 respondents, 47 indicated that they were interested in installing wood boilers either as replacements for existing boilers or for further expansion, and 123 gave negative answers. The affirmative answers constituted 25% of the total responses.

Attractiveness of Equipment Prices

Capital costs for new wood boilers as well as wood gasifiers were shown in the questionnaire. Respondents were asked about price attractiveness to them concerning these two types of wood burners. Affirmative answers were 42 for wood burners or 23% of the total respondents, and 30 for wood gasifiers or 16%. Those giving negative answers were asked what percentage reduction on equipment prices would be necessary to interest them. Answers for wood boilers were in the range of 15–100% reduction with an average of 48%. For wood gasifiers, the reduction range was 10–100% with an average of 51%.

Price of Wood Fuel as Basic Consideration

Respondents were asked about their interest in a wood gasifier or a wood boiler if the price of the delivered wood fuel for the same amount of energy was above or below their current delivered fuel cost per unit. A majority of the respondents indicated that they would be interested in a wood gasifier or a wood boiler if the price of delivered wood fuel were substantially below their current fuel costs. To the majority, 30–40% below their current fuel costs would be necessary. To some, 20% below, 10% below or the same cost would be sufficient to interest them. Only a few were interested in wood burners even if the price of delivered wood fuel was above their current fuel cost per unit. The number of affirmative respondents for wood gasifiers and for wood boilers is very close. Details are shown in Table VIII. The respondents' perception of the price of delivered wood fuel as compared to other fuel prices seems to be based on insufficient information on fuel prices, since the price of delivered wood fuel was substantially below other fuel prices (Table VII).

Table VIII.   Level of Wood Fuel Price Below/Above
Current Fuel Cost for Interest in Wood-Fired Equipment

	Wood Gasifier	Wood Boiler
Below Current Fuel Cost (%)		
40	74	74
30	48	49
20	30	31
10	11	12
Same	4	7
Above Current Fuel Cost (%)		
10	2	2
20	1	1
30	1	1
40	1	1

Market Barriers

Major market barriers for installing wood boilers or gasifiers appear to be, in order of importance, (1) the difficulty of storing sufficient quantities of wood fuel; (2) the availability of wood fuel; (3) the lack of technology to allow the use of wood as a supplemental fuel in oil or gas boilers; (4) the availability of competitive fossil fuels; and (5) high initial capital cost.

Survey respondents considered that the most important incentives for installing new wood boilers or gasifiers are improved technology in equipment for storage, feeding and burning of wood, and an assurance of wood fuel supply. Tax write-offs and capital loans are surprisingly low incentive items.

SUMMARY

Wood as a source of domestic and commercial fuel has increased significantly in the Southeast since the energy crisis in 1973. An eight-state Southeastern survey concerning potential interests in installing new wood boilers and wood gasifiers among existing boiler users in 1979 revealed that 14% of the respondents were interested in retrofitting their existing boilers with wood gasifiers. About 25% of the respondents were interested in installing new wood boilers. It would require about 50% reduction of capital cost in a new wood boiler to be attractive to uninterested respondents. Major market barriers for installing wood boilers or gasifiers are equipment/fuel technology, storage of wood materials, high capital cost, wood fuel supplies and availability of competitive fossil fuels. The most important incentives for overcoming these barriers are improved technology in wood fuel and assurance of wood fuel supply.

ACKNOWLEDGMENTS

This study was sponsored by the Georgia Forestry Commission and funded by the Appalachian Regional Commission.

REFERENCES

1. "Current Industrial Reports, Selected Heating Equipment, MA-34N, 1959 to 1977," U.S. Department of Commerce, Bureau of the Census, Washington, DC.
2. "County and City Data Book, 1972. "A Statistical Abstract Supplement," U.S. Department of Commerce, Bureau of the Census (1974).
3. Chiang, T. I., and D. S. Clifton, Jr. "Market Potentials of Wood Fuel in the Southeast," Economic Development Division, Engineering Experiment Station, Georgia Institute of Technology, Atlanta, GA (1979).
4. Salo, D., L. Gisellman, D. Medville and G. Price. "Near Term Potential of Wood as a Fuel," Tech. Rep. MTR-7860, Materials Division, Mitre Corporation, McLean, VA (1978).
5. Burwell, C. C. "Solar Biomass Energy: An Overview of U.S. Potential," Science Vol. 199 (1978).
6. U.S. Environmental Protection Agency, Atlanta Regional Office. Unpublished data (1978).
7. Americal Boiler Manufacturers Association, Washington, DC. Unpublished data.

CHAPTER 31

DEVELOPMENT AND TESTING OF A SMALL, AUTOMATED WOOD COMBUSTION SYSTEM

Werner Martin and Daniel R. Koenigshofer
Integrated Energy Systems, Inc.
Chapel Hill, North Carolina

Much of the current interest in wood energy technologies centers around large, sophisticated systems for direct combustion. In fact, it is in the residential sector that wood use has skyrocketed in the past few years. Sales of residential heating units expanded from 0.25 million in 1972 to 1 million in 1979 [1]. A potential of 18–23 million households using wood energy in the coming years is claimed [2]. Further, there are uncounted small offices, shops, schools, etc. that could use or are using small wood combustion systems.

Considering that wood is a relatively disperse energy source, it may be that its most efficient use is precisely in such small installations. Unfortunately, such widespread, small-scale use is not without problems. Residential wood combustion has been cited as a major source of air pollution in many communities [1]. Likewise a recent study indicates that under a maximum wood use scenario for western North Carolina, "areas with a predominance of residential, commercial and institutional space heating users (in contrast to industrial) would suffer greatest from particulate emissions from wood fuel" [3]. Carbon monoxide and hydrocarbon emissions were estimated to increase by a factor of 100 in some areas.

While some of these small users are attracted to wood burning for esthetic reasons, many home owners and most business owners convert to wood to save money. If wood is to be used widely and seriously, combustion systems must be developed which offer at least the following features:

- easily retrofitted to liquid- or gas-only boilers and furnaces;
- fast, accurate heat modulation;
- reliable and convenient fuel and ash handling;
- low pollutant emissions;
- small scale [50–100 MJ/hr (47,000–95,000 Btu/hr)]; and
- low cost, long life.

For the past year and a half, we have been developing and testing a burner which, it is hoped, will have the above features.

The purpose of this chapter is to present the performance and emission data obtained from preliminary testing while burning manufactured wood pellets with the Integrated Energy Systems (IES) burner. The test methods and instrumentation are discussed, along with conclusions for design improvements and general observations regarding pellet combustion.

THE BURNER CONCEPT

The IES burner has a horizontal fuel flow bed with over- and under-fire, preheated combustion air. Combustion and ash removal take place in the burner itself; only the hot gases are introduced into the heat exchanger. Thus, the unit can be easily retrofitted to existing oil- and gas-fired boilers and furnaces without provision for ash handling within the firebox. Figure 1 is a view into the exhaust end of the burner showing the fuel bed and combustion chamber. Since a relatively small amount of fuel is burning at any one time, output modulation is easily achieved by control of fuel and/or air feed rate.

Wood pellets were chosen as a fuel due to their uniformity, availability, cleanliness and ease of handling. Another advantage of pellets is that a supply infrastructure is being established which could, in time, provide the convenience the heating oil supplier presently does. The system is not confined to pellet combustion and can, in fact, burn virtually anything that can be put in it such as leaves, nut shells, coal, green woodchips, etc. For the testing, the pellets were sorted in a metal hopper and fed by screw auger to the burner. Feed rate can be continuously adjusted automatically by a thermostat.

The prototype which was tested is residential scale [50 MJ/hr (47,000 Btu/hr)]. A second unit, with an output four times that has been built

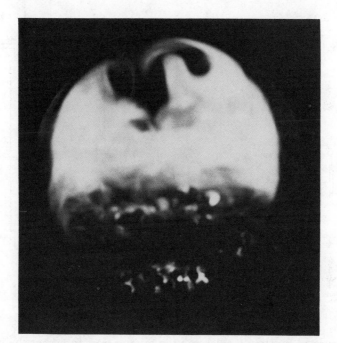

Figure 1. View into burner combustion chamber.

and is being tested. The upper size limit is unknown, but the current units would handle any residential application and be useful for many small commercial systems.

METHODS

The test unit and the system parameters chosen for measurement are schematically shown in Figure 2. A used Weil-McLean oil-fired boiler

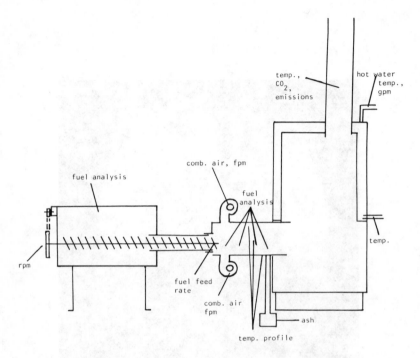

Figure 2. Test setup and measured system parameters.

was converted by simply enlarging the oil gun port to accept the wood burner. The only other modification was the installation of a glass observation port opposite the burner exhaust. This water wall unit was originally rated at 148 MJ/hr (140,000 Btu/hr) input. This unit was selected solely on the basis of availability and ease of testing.

The hopper and auger are sold commercially for grain handling. The hopper contained approximately 0.15 m³ (5 ft³), which fueled the burner at full rate for about 70 hr. Again, this size was based on convenience and availability. Actual storage volume would be determined by rate of consumption, fuel delivery schedule and space availability.

Water was drawn from a nearby pond, circulated on a single pass through the boiler, and dumped below the pond. Flowrate was constant at 11.4 liter/min (3 gal/min), inlet temperature at 26°C. Since the objective of the testing was not to determine the boiler efficiency, water flowrate and temperature were not varied.

The objective of the testing was to acquire sufficient data to prove the concept and provide a basis for design improvements. Measured parameters were temperature, fuel composition, and emissions.

Temperature

Temperatures were measured at the following points:

1. surface structure of the burner, outside* and inside the wall surface;
2. bottom, center and top of fuel bed;
3. air and gas streams: combustion air intake*, air at point of introduction into combustion chamber, and combustion gases over fuel
   bed, at exit of combustion chamber and in stack after boiler*; and
4. water stream; boiler inlet* and outlet*.

For all temperature measurements, a Syscon International Model 5200
Digital Temperature Indicator was used. All thermocouples were type K
(up to 1000°C) surface, air and lance probes. The parameters noted by
an asterisk were monitored continuusly while specific, though typical,
2- to 3-hr runs were made for the other measurements. The data were
hand-recorded. Intake air flow was measured with an Air Flow Corporation Model 54 A-1 hot-wire anemometer. Exhaust air flow was measured
with an Alnor Anemometer.

Fuel

To determine what changes occur in the wood pellets as they travel
through the tunnel, samples of the solid fuel stream were analyzed. After
running the combustion system for 2 hr to reach steady state, the system
was shut down and the samples were immediately withdrawn. Nine 50-g
samples from different points in the combustion chamber, along with a
sample of raw pellets from the auger-hopper and a sample of ashes
accumulated in the boiler base, were doublebagged in plastic and sent to
the North Carolina State University's Engineering Research Service Division. Moisture, ash and volatiles content, and calorific value were determined by standard proximate analysis methods. In this manner, a history
of the wood pellets during their 17-min travel through the combustion
chamber could be determined. An additional analysis in the bomb calori-
meter was performed to determine what portion of the total heating
value of the raw pellets reaches the actual combustion zone as wood
gases, and what portion as solid, charred wood residue.

Emissions

The largest portion of the combustion end products leaves the burner
with the exhaust gas stream. Therefore, emission tests are important not

only for potential environmental impact assessment, but also for combustion efficiency determination. The U.S. Environmental Protection Agency (EPA) Industrial Environmental Research Laboratory (IERL) at Research Triangle Park, NC, conducted emission tests on the prototype unit. The methods are described by Hall and DeAngelis [4]. The components analyzed in the continuous samples of the stack gases were:

- CO and CO_2 by nondispersive infrared methods;
- O_2 by pyromagnetic method;
- NO, NO_2, NO_x by chemiluminescence;
- total hydrocarbons (HC) by heated flame ionization method;
- SO_x by fluorescence;
- smoke by Bacharach scale; and
- total suspended particulates by EPA Method 5, modified with added resin column.

At various fuel feed rate/combustion air rate settings, CO, CO_2, O_2, NO/NO_x, HC and SO_2 were continuously monitored over 2 hr of operation. At least three smoke readings were taken randomly for each particular setting.

The total suspended particulate samples were withdrawn only at two specific settings. The combustion system was run at fuel feed and air intake settings which were believed to be the optimal operating points of the prototype system, considering both air pollution and heat output. A 2.5-hr sampling time was required to get a 1 m^3 sample (minimum). The collected particles in the cyclone, filter paper, and condensible components in the resin column and impingers were analyzed for total weight and chemical composition.

RESULTS AND DISCUSSION

Temperature Profiles

Figure 3 depicts the temperatures measured at various points of the tunnel burner, air and fuel stream. These temperature readings were taken at what is believed to be the rated capacity of the prototype tunnel (2.33 kg/hr) and at a level of 50–60% excess air.

As indicated, the fuel is 100–300°C near the inlet of the burner. It remains relatively constant, as moisture is driven off, until air is introduced at point e. Within a short space, combustion gases reach steady state at 871°C.

Combustion air was heated from 21 to 71°C. This was less than desired.

The burner surfaces reach temperatures of 500–550°C which, to assure

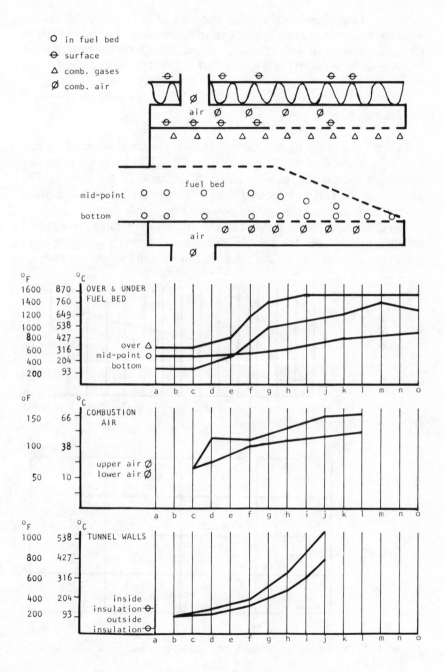

Figure 3. Temperature profiles through burner.

long life, may necessitate special materials. Further, the 2.5 cm of asbestos boiler insulation appears insufficient since the outer surface temperature reached 400°C and averaged 250°C. By the methods of Duffie and Beckman [5] this loss is 7.67 MJ or 15% of the input.

Fuel Analysis

Figure 4 shows the theoretical combustion phases of wood as a function of temperature [6]. It has been modified to correspond to the temperatures observed as a function of location in the burner. The analysis of the fuel at various stages of the pyrolysis and combustion process revealed the information presented by Figure 5.

The data show that at the end of the process the sample consisted of 94.7% ash and about 5% volatiles. This indicates nearly complete combustion. Thus, the margin for improvement of combustion is small.

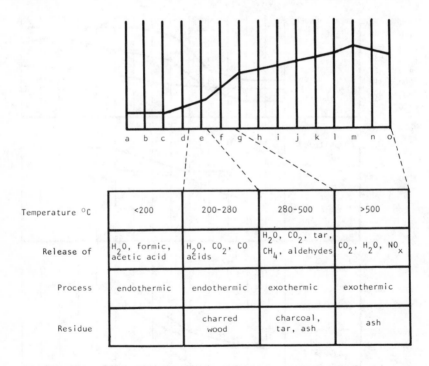

Temperature °C	<200	200-280	280-500	>500
Release of	H_2O, formic, acetic acid	H_2O, CO_2, CO acids	H_2O, CO_2, tar, CH_4, aldehydes	CO_2, H_2O, NO_x
Process	endothermic	endothermic	exothermic	exothermic
Residue		charred wood	charcoal, tar, ash	ash

Figure 4. Theoretical combustion processes at various points in the burner.

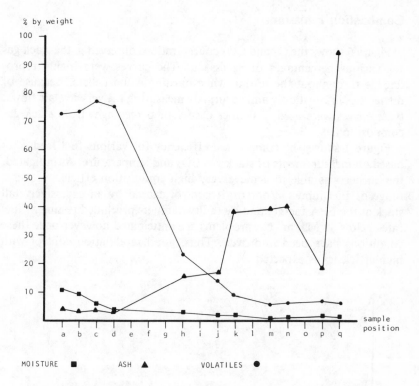

Figure 5.   Composition of wood pellets traveling through burner.

The other information in Figure 5 which should be highlighted is the rapid loss of volatiles from the solid wood pellets between points f and h. Translated into the residence time of the fuel in the tunnel, the data show that after the pellets have been in the tunnel for 6–9 min, they have lost 2/3 of the volatiles. The moisture and volatiles are almost completely evolved prior to introduction of combustion air. Thus, virtually all of the volatiles are burned as a free gas not within the fuel bed. In this respect, the burner performs as a gasifier. The moisture content of the pellets, which is 10.4% in the sample taken from the fuel hopper, has already dropped in the tunnel inlet (point b) to 9.1%, and in the first section of the tunnel itself to 3.6%. It is also interesting to note that after a low value of 0.4% at the hottest point of the combustion zone, the moisture increases again up to a level of about 1.8%. This may be caused by the high affinity for water of the ash silica particles. During cooling of the samples, they may have taken up water at varying rates.

Combustion Emissions

Figure 6 shows the O_2 and CO_2 concentrations observed in the stack gas for various percentages of excess air. The curves were highly reproducible throughout the testing when burning wood pellets. Because of difficulties in constantly and accurately measuring the very low air flows, these curves were used to deduce excess air percentages from stack gas composition.

Figure 7 shows the combustion efficiency for various fuel feed rates based on measurements of stack gas CO_2 and temperature. As indicated, the burner was able to achieve very high combustion efficiency over a range of fuel inputs. Poor draft control caused by an excessively tall stack at the EPA facility and by faulty dampers precluded testing at fuel rates below 31 MJ/hr. No problems are anticipated however once these conditions have been improved. Thus, good modulation control with high efficiency is expected.

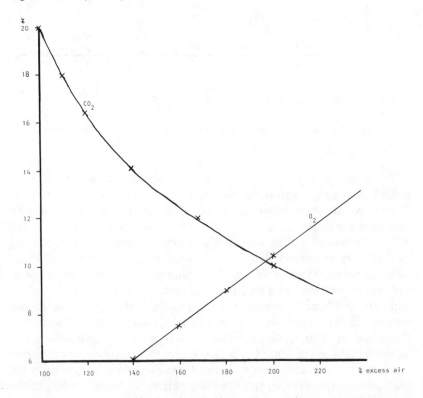

Figure 6. CO_2 and O_2 concentrations as a function of excess air.

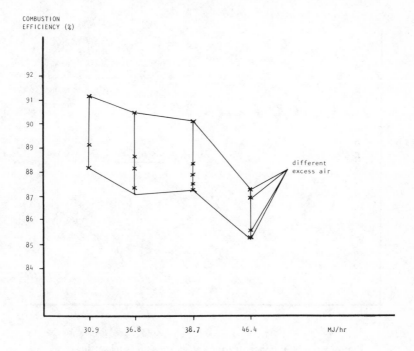

Figure 7.   Combustion efficiency for various fuel feed rates.

Two separate tests were run to determine system response rate. After steady-state operation was achieved, fuel feed was stopped with the combustion air blowers left on. The stack temperature dropped from 400 to 175°C in 8 min. At that point, fuel flow was restarted. Again, 8 min were required to reach steady state. Thus, the burner appears capable of 100% modulation over a 15–20 min span.

For cold starts, a propane ignitor is used. The burner requires about 1 hr to come from room temperature to steady state. The burner is capable of holding a fire for 3 hr, thereby minimizing cycling of the propane ignitor and reducing startup time.

The six gaseous components (O_2, CO_2, CO, HC, NO and NO_x) and smoke were monitored continuously as the fuel and air flowrates were varied. Figures 8 and 9 show the emission at various fuel feed rates with varying excess air as reflected by O_2 in the stack gas. Nitrogen dioxide ranged from 0 to 4 ppm and averaged 1 ppm; thus, only NO_x is shown. As expected, NO_x emissions increased at higher feed rates and higher O_2 (excess air). This will have to be optimized to balance heat output and emissions.

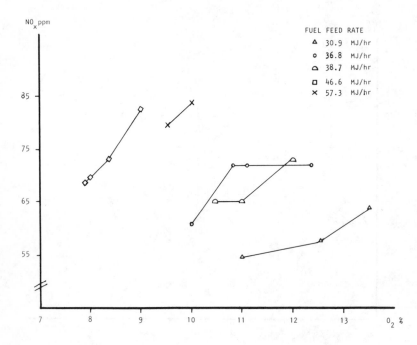

Figure 8.   NO$_x$ concentrations at various air/fuel settings.

Emissions for the tunnel burner are compared with those of other heating sources [4] in Table I.

The emissions are remarkably low for all pollutants except NO$_x$. Even NO$_x$ emissions are comparable to other wood-fired appliances, which are comparable to oil and gas heaters per MJ of heat input. The value for particulates is suspiciously low. Further testing is planned to verify our data. The range of pollutants is also quite low and relatively narrow even for CO. During the five runs, fuel feed rate varied as indicated in Figures 8 and 9 while the combustion dampers varied from full closed to full open. As indicated earlier, however, excessive draft and leakage made it very difficult to limit intake. Thus, starvation did not really occur. Measured stack gas flows were relatively constant varying by a factor of only 1.5.

No simulated full-cycle tests from cold start have been run. It should be noted, though, that the comparison data [4] were gathered at "good combustion" ranges of fuel and air after several charges had been burned. Thus, the data approximates best-case conditions.

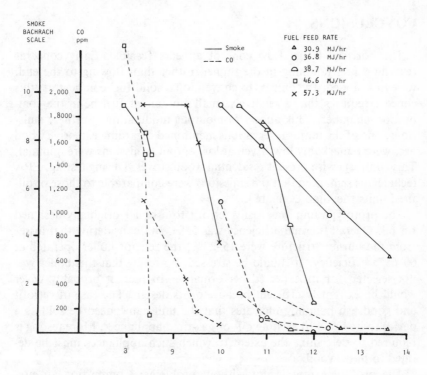

Figure 9. CO and smoke concentrations at various air/fuel settings.

Table I. Comparison of Emissions for Small Wood-Fired Heating Equipment at Steady State (g/kg)

	Particulate	HC	NO$_x$	CO
Fireplace	1.8–2.9	19	1.4–2.4	15–30
Baffled Stove	2.5–7.0	2.8	0.4–0.8	110–270
Unbaffled Stove	1.8–6.3	0.3–3.0	1.2–1.5	91–370
Tunnel Burner Range[a]	[b]	0.02–0.08	1.5–2.2	0.4–38
Tunnel Burner Optimal[c]	0.33	0.02	0.8	1.0

[a] Range of five 2.5-hr runs with continuous sampling at varying conditions.
[b] EPA method 5 was used only during the optimal run.
[c] Average of two 2.5-hr runs with continuous sampling at optimal conditions.

CONCLUSIONS

The horizontal fuel flow concept appears feasible. Early concerns regarding fuel jamming in the tunnel, rather than flowing to the end, were unfounded. In contrast to conventional solid fuel combustors, this concept requires that a relatively small amount of fuel be in the combustion chamber. This allows fast output modulation. Further, emissions, one of the major reservations mentioned regarding extensive wood use, were remarkably low. Even at low output, emissions were minimal. This contrasts with present residential wood stoves. During testing, EPA technicians remarked that the emissions were comparable to those of oil-fired units they had evaluated.

The prototype unit was easily retrofit to a boiler originally designed for oil. Overall thermal efficiency was 55%. But considering that losses from the burner structure were 15–20%, the retrofit boiler operated at 60–65% efficiency. It should be stressed, however, that this boiler was not selected for its expected performance. In fact, it probably never would be extremely efficient because of its design. The ease of retrofit and good ash handling indicates that the unit can indeed be sold as a device to be fitted to existing oil- or gas-fired appliances. More testing is required to determine the extent to which such appliances must be derated to burn wood.

The prototype unit is not without problems. Combustion air preheating was less than expected, and control of air flow was poor. Design improvements in the second generation burner should solve these problems. Heat loss from the tunnel structure itself at 15–20% was excessive. Improved insulation should remedy that. With 200 hr running time on the prototype, there are still questions about the durability of the materials and components. Clinkers have formed on occasion and caused severe problems for fuel flow in the tunnel. It is hoped that the cleaner pellets will eliminate this problem.

Finally, larger units must be constructed and tested. It is not known how well the concept will scale up. The upper limit appears to be the point at which particulate entrainment becomes excessive. In addition to building and lab testing larger units, field testing is needed to examine the performance of the entire system in actual use. Results to date are sufficiently encouraging to warrant further effort.

ACKNOWLEDGMENTS

The authors thank Mr. R. Hall, Mr. M. Osborne and Mr. G. Gillis at the Combustion Research Branch, IERL, EPA, for the excellent testing work they performed. Dr. L. Centofanti and Mr. C. Feltus of the U.S. Department of Energy Atlanta office provided funding under the Appropriate Technology Program. Dr. B. M. Gay of North Carolina State University provided valuable advice as well as fuel analysis.

REFERENCES

1. Cooper, J. A. "Environmental Impact of Residential Wood Combustion Emission and Its Implications," in *Wood Heating Seminar 6* (Washington, DC: Wood Energy Institute, 1980), pp. 26–55.
2. Neal, C. "Direct Response Advertising in the Wood Heat Industry," in *Wood Heating Seminar 6* (Washington, DC: Wood Energy Institute, 1980), pp. 258–269.
3. Integrated Energy Systems, Inc. "Increased Wood Combustion in Western North Carolina," in *Wood for Energy and Its Impact in Western North Carolina, Vol. III* (Raleigh, NC: North Carolina Dept. of Commerce, 1979), pp. 1–57.
4. Hall, R. E., and D. G. DeAngelis. "EPA's Research Program for Controlling Residential Wood Combustion Emissions." *J. Am. Poll. Control. Assoc.* 30(8):862–867 (1980).
5. Duffie, J. A., and W. A. Beckman. *Solar Energy Thermal Processes* (New York: John Wiley & Sons, Inc., 1974), pp. 61–83.
6. Browne, F. L. "The Course of Pyrolysis," in USDA Report: Theories of the Combustion of Wood and Its Control, No. 2136 (1963), p. 5.

AUTHOR INDEX

SUBJECT INDEX

585